Measurement *and* Evaluation *in* Human Performance

Third Edition

James R. Morrow, Jr., PhD
University of North Texas

Allen W. Jackson, EdD
University of North Texas

James G. Disch, PED
Rice University

Dale P. Mood, PhD
University of Colorado

Human Kinetics

Library of Congress Cataloging-in-Publication Data

Measurement and evaluation in human performance / James R. Morrow . . . [et al.].-- 3rd ed.
    p. cm.
  Includes bibliographical references and index.
  ISBN 0-7360-5540-1 (hard cover)
  1. Physical fitness--Measurement. 2. Physical fitness--Evaluation. I. Morrow, James
R., 1947-
    QP301.M3755 2005
    613.7'028'7--dc22

                  2004020218

ISBN-10: 0-7360-5540-1
ISBN-13: 978-0-7360-5540-6

Permission notices for material reprinted in this book from other sources can be found on page(s) xv-xvi

The Web addresses cited in this text were current as of September 27, 2004, unless otherwise noted.

Acquisitions Editor: Loarn D. Robertson, PhD
Developmental Editor: Judy Park
Assistant Editors: Lee Alexander and Bethany Bentley
Copyeditor: Julie Anderson
Proofreader: Kathy Bennett
Indexer: Bobbi Swanson
Permission Manager: Dalene Reeder
Graphic Designer: Fred Starbird
Graphic Artist: Tara Welsch
Photo Manager: Kareema McLendon
Cover Designer: Keith Blomberg
Art Manager: Kelly Hendren
Illustrators: Argosy and Sharon Smith
Printer: Edwards Brothers

Printed in the United States of America   10   9   8   7   6   5   4   3   2   1

Human Kinetics
Web site: www.HumanKinetics.com

United States: Human Kinetics
P.O. Box 5076
Champaign, IL 61825-5076
800-747-4457
e-mail: humank@hkusa.com

Canada: Human Kinetics
475 Devonshire Road Unit 100
Windsor, ON N8Y 2L5
800-465-7301 (in Canada only)
e-mail: orders@hkcanada.com

Europe: Human Kinetics
107 Bradford Road, Stanningley
Leeds LS28 6AT, United Kingdom
+44 (0) 113 255 5665
e-mail: hk@hkeurope.com

Australia: Human Kinetics
57A Price Avenue
Lower Mitcham, South Australia 5062
08 8277 1555
e-mail: liaw@hkaustralia.com

New Zealand: Human Kinetics
Division of Sports Distributors NZ Ltd.
P.O. Box 300 226 Albany
North Shore City, Auckland
0064 9 448 1207
e-mail: info@humankinetics.co.nz

Dedicated to the memory of Harold H. "Hal" Morris, 1938-2004.

He was an extraordinary person and a special friend who impacted all of us personally and professionally. JRM, AWJ, JGD, and DPM

Contents

Preface

In the third edition of *Measurement and Evaluation in Human Performance,* our primary goal remains to provide an interactive textbook focusing on the undergraduate student. We hope that you will take information, tools, and skills from this book that will help you in your daily activities as teachers, researchers, program developers and evaluators, physical therapists, health care providers, or whatever your career in human performance. We earnestly believe that the concepts learned in this course transcend all of the coursework that you will complete. The concepts contained in "measurement and evaluation" are important across all human performance topical areas in which you could be employed. We have updated and added new information and provided examples of the use and interpretation of measurement and evaluation concepts in the ever-changing environment of human performance. In addition to revisions within each chapter, the major changes include the following:

- ▶ Major rewriting of the first chapter to lay the groundwork for your learning through the remainder of the text
- ▶ Updating information about Statistical Package for the Social Sciences (SPSS)
- ▶ Inclusion of information about data set importing from Microsoft Excel® to SPSS
- ▶ Inclusion of specific data sets (downloadable from the World Wide Web) used for examples throughout many chapters
- ▶ Elimination of some advanced statistical concepts that are not typically used
- ▶ Inclusion of basic epidemiological statistics and their interpretation
- ▶ Reversed order of chapters 8 and 9
- ▶ Increased emphasis on international perspectives
- ▶ Enhancement of the text's World Wide Web site to facilitate learning
- ▶ A focus on decision making—decisions that you will have to make daily as a human performance professional

Testing and measurement are central to the field of human performance. In teaching a class, conducting a written or performance test, completing a fitness evaluation, or selecting team membership, each of us makes evaluative decisions daily. An important aspect of valid decision making is using evidence. Do the data support your decisions, and what have you done to confirm that your decisions are appropriate? Confirmation can come from research, testing, assessment, and interpretation of data you have collected. To make accurate decisions, you must have a firm foundation in basic measurement concepts. This book provides a wealth of measurement information that applies to all areas of human performance as well as to the health sciences. For example, epidemiological research has proven that a healthy lifestyle that includes regular physical activity and sufficient physical fitness will lower the risks of some chronic diseases and improve the quality of life. But without reliable and valid measurements of physical activity and fitness, such findings could not have been achieved, and the future health of millions of people would have been negatively affected.

To make appropriate decisions, you must be able to assemble information, sort it, evaluate it, and draw conclusions. Performing these processes requires an understanding of the concepts of basic statistics as they apply to reliability, validity, and objectivity. We provide you with the information and techniques you need to determine whether the measurement and evaluation information you collect and analyze is reliable, valid, and objective for decision making, whether you are a researcher, personal fitness trainer, school educator, coach, physical therapist, or simply someone interested in measurement associated with movement and human performance. We present this information in three main sections.

Part I, Basic Tools in Measurement and Evaluation, gives essential tools to organize, assimilate, and reduce information for analysis. Within part I, chapter 1 introduces measurement and evaluation and provides current examples of the types of problems that you will confront throughout the text and in your professional life. Chapter 2 introduces you to the Statistical Package for the Social Sciences (SPSS), which you will use to analyze data in your text. It also helps you become familiar with the World Wide Web site associated with this text. Chapter 3 presents descriptive statistics and the normal curve. Chapter 4 contains correlation and predictive statistics. Chapter 5 presents inferential statistical techniques. Each of these chapters prepares you to address questions related to measurement and evaluation, particularly those associated with reliability and validity. The important concepts that you learn in these three "statistics" chapters (chapters 3, 4, and 5) will be fundamental to the remainder of the text. We will continually return to the concepts learned in these chapters as we present information in subsequent chapters.

Part II, Reliability, Validity, and Grading, presents statistical information in a framework to help you judge the quality of data. Essentially, this information deals with two questions:

Are the data to be used in decision making being obtained accurately?

Are the decisions made based on truthful data?

These questions will be answered in a variety of situations. In chapters 6 and 7 we provide you with the understanding necessary to make decisions, and we illustrate the links between truthful decision making and the statistical knowledge and skills that you obtained in part I. For those of you who are employed or seeking employment in nonschool settings, such as YMCAs or YWCAs, health clubs, and corporate fitness programs, reliability and validity issues are central to data collection and decision making for individuals and groups of program participants. Chapter 8, Alternative Assessment, presents information about making reliable and valid decisions in performance-based assessment and addresses important issues associated with alternative methods of assessing achievement. For those of you who are employed or seeking employment in school-based settings, such as elementary, secondary, or higher education teachers and scholastic coaches, reliability and validity issues are central to making accurate decisions about student and student-athlete progress and achievement. Chapter 9 presents issues related to traditional grading practices. Each chapter in this unit ends with a real-life example of using the knowledge you have learned in combination with the statistical skills you learned in part I.

Part III, Applications of Measurement and Evaluation, applies basic statistical techniques, reliability, and validity to practical problems in the field of human performance and movement around the world. We include metric measurements in addition to English units, where appropriate, and we present evaluation examples that reflect the international importance of human performance assessment.

We are not mathematicians but are practical users of the information in your textbook. Thus, although we occasionally have to present information from a mathematical perspective, we try to keep the math to a minimum and focus on the concepts. Often students are apprehensive about the mathematical skills necessary for performing well in measurement and evaluation. Although having a strong mathematical background can be an advantage to a student, we present our material in such a way that minimal mathematical expertise is expected or required. Moreover, the use of computers makes the mathematics relatively simple. You should bear in mind, however, that you must understand the theory if you are to make effective decisions. We therefore present the theory and the skills that you will need to use when making decisions based on testing, measurement, and evaluation. We will show you how to take advantage of statistical software, SPSS, to set up a database and quickly and conveniently analyze data. More important, we will help you learn to interpret and evaluate the statistical results these programs generate. Our desire is that you will become competent with the software and have a good understanding of basic statistical concepts. Last, we will provide you with real-life examples of the use and interpretation of the concepts you learn in each chapter.

Each chapter is organized around key terms and definitions, as well as a measurement and evaluation "challenge" that begins each chapter with a problem you might encounter in your profession; the chapter then ends by returning to the challenge and suggesting methods of addressing it as a result of the content you have learned in the chapter. We provide you with measurement and evaluation "results" with which to make the decisions. Each chapter also includes clearly stated objectives at the outset of each chapter and an end-of-chapter review that ensures that you've met the objectives. Chapters also include summary tables, descriptive figures and graphs, key points, and mastery items (MI), the latter of which both emphasize key points made in the text and require you to apply the principles you have learned to solve problems. It is important that you complete all of the MIs presented to fully understand the material in the chapters. You should be able to "master" the item when you get to it. Selected MIs are located on the text's Internet site. MIs that involve the use of SPSS are identified with a computer icon. There computer-based MIs are used to tie chapter content to statistical and reliability and validity concepts learned in earlier chapters.

We use the World Wide Web as a tool to help you learn the concepts presented in this course. We have created a World Wide Web study page associated with your text (www.HumanKinetics.com/MeasurementAndEvaluationInHumanPerformance/SG/getstart_01.cfm). By accessing this Web site, you will be able to interact, obtain chapter outlines, practice (i.e., homework with answers), review, obtain examples, and develop your knowledge and skills about measurement and evaluation.

Our down-to-earth presentation will help you focus on the important concepts, and you will carry from this book important resources that you can use throughout your professional career. The measurement and evaluation process in kinesiology, physical education, athletics, human performance, and exercise and sport science is a complex one in which conventional knowledge and practice do not always mirror the findings and conclusions of scientific research. Our goal is for you to understand measurement and evaluation processes so you can make valid decisions.

The authors of your text have known each other for more than 25 years. Each of us has taught the course for which this book is intended many, many times, and we hope our insight and experiences will help you. We share common beliefs and understandings about the importance of the measurement and evaluation process. We know each other as coauthors and friends. We hope you enjoy and learn from our presentation. We use the skills, knowledge, and techniques presented in this book daily. We hope that you too will adopt the perspective that truthful measurement is important in decision making and continue to use these concepts throughout your professional career.

Acknowledgments

This text could not have been completed without much guidance, many suggestions, and a great deal of encouragement from the professionals at Human Kinetics. Our association with Human Kinetics has been most rewarding. Rainer and Julie Martens can be proud of their organization. We particularly acknowledge Loarn Robertson, Judy Park, and Lee Alexander for their assistance in bringing the text together. They were invaluable to us. We thank Julie Rhoda, who helped us greatly through the second edition and the associated web development phases. Additionally, we acknowledge the many members of the Human Kinetics team who worked specifically on our text. The Human Kinetics Web development team is a great resource to us and our readers. We greatly appreciate the many efforts of the various teams and individuals that help create the text and its ancillaries. We value the measurement and evaluation professionals from whom we have learned much. These (our mentors, friends, and students) include ASJ, ATS, BAM, CHS, DJH, GVG, HHM, HRB, JAS, JEF, JLW, JMP, KDH, LDH, LRO, LSF, MAL, MEC, MJL, MJS, MJS, MSB, MTM, RGF, RWS, SNB, SSS, TAB, TMW, VWS, and WBE. Finally, we acknowledge our families, who we love and appreciate. They tolerated the many hours we spent with our computers preparing this text and our other academic pursuits.

Credits

Figures 2.1, 12.3, 12.9, 12.10, and 12.11 and Table 12.10: Reprinted, by permission, from The Cooper Institute for Aerobics Research, 1999, *FITNESSGRAM test administration manual,* 2nd ed. (Champaign, IL: Human Kinetics).

Figure 12.2: Reprinted from Centers for Disease Control and Prevention, 1998, "Youth risk factor surveillance, 1997," *Morbidity and Mortality Weekly Report* 47: 1-89.

Figure 13.4: Reprinted, by permission, from J. Bartlett, L. Smith, K. Davis and J. Peel, "Development of a valid volleyball skills test battery," *Journal of Physical Education, Recreation and Dance* 62(2): 19-21.

Figures 13.5 and 13.6: Reprinted, by permission, from D.R. Hopkins, J. Schick, and J.J. Plack, 1984, *Basketball for boys and girls: Skills test manual.* (Reston, VA: AAHPERD).

Table 3.3: From E.F. Lindquist, 1942, *A first course in statistics.* (Boston: Houghton Mifflin). Copyright © 1942 by E.F. Lindquist. Reprinted by permission of Houghton Mifflin Company.

Table 6.1: Data based on an example from G. Sax, 1980, *Principles of educational and psychological measurement and evaluation* (Belmont, CA: Wadsworth).

Tables 6.8 and 6.9: Reprinted, by permission, from K.N. Green, W.B. East, and L.D. Hensley, 1987, "A golf skills test battery for college males and females," *Research Quarterly for Exercise and Sport* 58:72-76.

Table 7.3: Reprinted, by permission, from R.E. Rikli, C. Petray, and T.A. Baumgartner, 1992, "The reliability of distance run tests for children in grades K-4," *Research Quarterly for Exercise and Sport* 63: 270-276.

Table 8.5: Adapted from the Wichita Public Schools, Kansas.

Table 8.6: Adapted courtesy of Karen Nagle, Iowa City Schools, Iowa.

Table 8.7: From the Mid-continent Research for Education and Learning (McREL), Aurora, CO. Reprinted by permission of McREL.

Table 8.9: Adapted, by permission, from M.F. Kirk, 1997, "Using portfolios to enhance student learning and assessment," *Journal of Physical Education, Recreation and Dance* 68(7):29-33.

Table 9.9: Adapted, by permission, from M.F. Kirk, 1997, "Using portfolios to enhance student learning and assessment," *Journal of Physical Education, Recreation and Dance* 68(7): 29-33.

Tables 11.2, 11.3, and 11.5: Adapted from American College of Sports Medicine, 1995, *ACSM's guidelines for exercise testing and prescription.* (Philadelphia: Lea and Febiger).

Tables 11.6, 11.9, 11.13, 11.15, 11.20, 11.22, and 11.23: Adapted, by permission, from L. Golding, C. Myers, and W. Sinning, 1989, *Y's way to physical fitness,* 3rd ed. (Champaign, IL: Human Kinetics).

Table 11.7: Reprinted, by permission, from G. Borg, 1998, *Borg's perceived exertion and pain scales.* (Champaign, IL: Human Kinetics).

Table 11.8: Reprinted, by permission, from American Association of Health, Physical Education, Recreation, and Dance, 1985, *Norms for college students: Health related tests.* (Reston, VA: AAHPERD).

Table 11.14: Adapted, by permission, from American College of Sports Medicine, 1988, *Resource manual for guidelines for exercise testing and prescription.* (Philadelphia: Lea and Febiger).

Table 11.17: Adapted, by permission, from J. Graves, M. Pollock, D. Carpenter, S. Leggett, A. Jones, M. MacMillan, and M. Fulton, 1990, "Quantitative assessment of full range-of-motion isometric lumbar extension strength." *Spine* 15: 289-294.

Tables 11.18 and 11.19: Reprinted, by permission, from The Cooper Institute for Aerobics Research, 2002, *The physical fitness specialist certification manual* (Dallas, TX: The Cooper Institute).

Outline

Objectives

After studying this chapter, you will be able to

- define the terms *test, measurement,* and *evaluation;*
- differentiate norm- and criterion-referenced standards;
- differentiate formative and summative evaluation;
- discuss the importance of measurement and evaluation processes;
- identify the purposes of measurement and evaluation;
- identify the importance of objectives in the decision-making process; and
- differentiate among the cognitive, affective, and psychomotor domains as they relate to human performance.

Measurement and Evaluation Challenge

As you begin learning about measurement and evaluation processes in human performance, we present an overview of what you will study throughout this textbook. This first measurement and evaluation challenge presents a scenario that relates to most of the chapters and concepts you will study. We first describe a scenario, and then at the end of the chapter we explain how you might answer the questions that arise in the scenario.

Imagine that your father talks with you about his recent physical examination. It has been a number of years since he had a medical examination. His physician conducted a battery of tests and asked your father about his lifestyle. As a result, the physician told your father that he is at risk for developing cardiovascular disease. Your father was told that his weight, blood pressure, physical activity level, cholesterol, nutritional habits, and stress levels have increased his chances of developing cardiovascular disease. Your father tells you that he feels great, was physically active throughout high school and college, looks better than most people his age, and cannot imagine that he is truly at an elevated risk. Because he knows you are aware of cardiovascular disease risk factors, he asks you the following questions:

1. How does one know if the measures taken are accurate? (reliability)
2. What evidence proves that these characteristics are truly related to developing cardiovascular disease? (validity)
3. How likely is it that the physician is correct in his or her evaluation of the tests?
4. What aspect of the obtained values places one at increased risk? For example, how was a systolic blood pressure of 140 mmHg identified as the point at which one is at increased risk? Why not 130 mmHg or 150 mmHg? Similar questions could be asked about each of the measurements obtained.
5. What evidence exists that changing any of these factors will reduce risk?

Your father is concerned because he doesn't know what these numbers mean. Likewise, he and

(continued)

Measurement and Evaluation Challenge   (continued)

you are concerned about the accuracy of these measurements. You would like to explain to him how to interpret these results and encourage him to make the necessary lifestyle changes to reduce his cardiovascular risk.

Interpreting measurement results and determining the quality of the information one receives are what this course is all about. Information gained from this course will help you make informed decisions about the accuracy and truthfulness of obtained measures and decisions based on the measurements. In general, good measurement and subsequent evaluation should lead to good decisions, like changing one's lifestyle to improve health. We focus on measurements obtained in the cognitive, psychomotor, and affective domains.

Why is testing important? Is it really necessary to know many statistical concepts? What decisions are involved in the measurement process? How you answer these questions is important to your development as a competent professional in human performance.

Decision making is important in every phase of life whether it is related to professional or personal decisions. The manner with which one approaches decision making will affect the quality of a person's decisions. The statistical and measurement concepts presented in this text provide a framework for making accurate, truthful decisions.

We all gather data before making decisions, whether the decision-making process occurs in education or in other pursuits. For example, you might gather information for student grades, research projects, and fitness evaluation. Researchers gather data on fitness characteristics because of the relationships among fitness, physical activity, mortality, morbidity, and quality of life. The variables measured might include the amount of physical activity, blood pressure, and cholesterol levels. Weight loss and weight control are major health concerns, so you might be interested in measuring energy expenditure to estimate caloric balance. Likewise, you gather data about the weather before venturing out for a morning run, and you adjust your behavior based on the data you obtain (e.g., rain, warm, dark, cold). For example, if it is 40° F (4.5° C) and raining, you'll dress differently for your run than if it is 90° F (32.2° C) and sunny. Before purchasing a stock for investment, you gather data on the company's history, leadership, earnings, and goals. All of these are examples of testing and measuring. In each case, making the best possible decision is based on collecting relevant data and making an accurate decision.

The course you are embarking on has historically been called "tests and measurements." Although some students refer to it as "statistics," that does not accurately describe what the course is about. Some basic statistical concepts are presented in chapters 3, 4, and 5; however, the statistical and mathematical knowledge necessary for testing and measurement is not extensive. On the other hand, every chapter in this text focuses in some way on the important issues of reliability and validity. To make good decisions, you must measure and evaluate accurately. *Making effective decisions depends on first obtaining relevant information.* This is where testing and measurement enter the picture.

NATURE OF MEASUREMENT AND EVALUATION

The terms we use in measurement and evaluation have very specific meanings. *Measurement, test,* and *evaluation* refer to specific elements of the decision-making process. Although the three terms are related, each has a distinct meaning and should be used correctly. **Measurement** is the act of assessing. Usually this results in assigning a number to the character of whatever is assessed. A **test** is an instrument or tool used to make the particular measurement. This tool can be written, oral, physiological, or psychological, or it can be a mechanical device (such as a treadmill). **Evaluation** is a statement of quality, goodness, merit, value, or worthiness about what has been assessed. Evaluation implies decision making.

You can measure a person's maximal oxygen uptake ($\dot{V}O_2$max, a measure of aerobic capacity) in several ways. You might have someone perform a maximal run on a treadmill while you collect and analyze expired gases. You might collect expired gases from a maximal cycle ergometer protocol. You might have the subject perform either a submaximal treadmill exercise or a cycle exercise, and then you predict $\dot{V}O_2$max from heart rate or workload. You might measure the distance a person runs in 12 min or the time it takes to complete a 1.5 mi (2.4 km) run. Each of these tools results in a number, such as percent O_2 and CO_2, heart rate, minutes, or yards. Having assessed $\dot{V}O_2$max with one of these tools does not mean that you have evaluated it. *Obtaining and reporting data have little meaning unless you reference the data to something.* This is where evaluation enters the process.

Assume that you test someone's $\dot{V}O_2$max. Furthermore, assume that she has no knowledge of what the $\dot{V}O_2$max value means. Certainly, the subject might be aware that the treadmill test is used to measure fitness. However, the first question most people ask after completing some measurement is, How did I do? or How does it look? To simply report, "Your $\dot{V}O_2$max was 30 ml · kg^{-1} · min^{-1}" says little. You need to provide an evaluation. *An evaluative statement about the performance introduces the element of merit, or quality.*

Mastery Item 1.1

A physical education teacher records the number of sit-ups that a student completes in 1 min. Differentiate among the test, measurement, and evaluation characteristics reflected in this activity.

Norm- and Criterion-Referenced Standards

To make an evaluative decision, you must have a reference perspective. *You can make evaluative decisions from either norm-referenced (normative) or criterion-referenced standards.* An evaluative decision based on a norm-referenced standard means that you report how well a performance compares with that of others (perhaps of people of the same gender, age, or class). Thus, you might report that a $\dot{V}O_2$max of 30 is relatively poor for someone's particular age and gender. Conversely, you might simply report a person's performance relative to a criterion that you would like him or her to achieve. Assume that the $\dot{V}O_2$max of 30 was measured on someone who had had a heart attack. A physician may be interested in whether the patient achieved a $\dot{V}O_2$max of at least 25 ml · kg^{-1} · min^{-1}, which would indicate that the patient had achieved a functional level of cardiovascular fitness. This is a case of a criterion-referenced standard. You are not interested in how someone compares with others; the comparison is with the standard, or criterion. The criterion often is initially based on norm-referenced data and the best judgment of experts in the content area. You will learn much more about setting standards and the validity of standards in chapter 7.

Changes in youth fitness evaluation over the last decade provide a good comparison of norm-referenced versus criterion-referenced standards. Fitness scores used to be evaluated using a norm-referenced system, that is, relative to a child's classmates, by age and gender. Many youth fitness tests now are criterion referenced. Table 1.1 provides an example of the

Table 1.1 Criterion-Referenced and Norm-Referenced Standards for 12-Year-Old Boy Who Performed the 1 Mile Run

Measure	FITNESSGRAM criterion test score	AAHPERD Health-Related Fitness Test percentile
8:40	8:00	60th

A run time of 8:40 represents the boy's 50th percentile on the President's Challenge and is sufficient to receive the National Physical Fitness Award (a norm-referenced standard).

differential interpretation of norm-referenced and criterion-referenced standards for a 12-year-old boy who ran 1 mi (1.6 km) for time. The boy ran the mile in 8:40. His score (8:40) does not meet the minimum criterion for the FITNESSGRAM Healthy Fitness Zone (8:00).

Mastery Item 1.2

Are the following measures usually evaluated from a norm-referenced or a criterion-referenced perspective?

- Blood pressure
- Fitness level
- Blood cholesterol
- A written driver's license examination
- Performance in a college class

Mastery Item 1.3

List other examples of norm-referenced and criterion-referenced evaluative comparisons.

Formative and Summative Evaluation

Evaluations occur in two perspectives, formative and summative. **Formative evaluations** are initial or intermediate evaluations, such as the administration of a pretest and the subsequent evaluation of its results. Formative evaluation should occur throughout the instructional, training, or research process. Ongoing measurement, evaluation, and feedback are essential to the achievement of the goals in a program in human performance. For example, after shoulder surgery, your goal may be to regain range of motion (ROM) in the shoulder joint. Your physical therapist could assess your ROM and suggest alternative activities to improve it. These ongoing evaluations need not involve formal testing; simple observation and feedback sequences between the student or participant and the instructor or leader are often adequate. **Summative evaluations** are final evaluations that typically come at the end of an instructional or training unit. You, as a student in this course, are interested in the summative evaluation—the grade—that you will receive at the end of the semester.

The difference between formative and summative evaluations might seem to be merely the difference in timing of their data collection; however, the actual use of the data collected distinguishes the evaluation as formative or summative. Thus, in some situations the same data can be used for formative and summative evaluations.

A weight-loss or weight-control program provides a simple and useful example for applying formative and summative evaluations. Assume that you have measured a participant's body weight and percent body fat. Your formative evaluation indicates that he has a percent body fat of 30% and needs to lose 10 lb to achieve a desired percent fat of 25%. You establish a diet and exercise program designed to produce a weight loss of 1 lb per week for 10 weeks. Each week you weigh the participant, measure his percent body fat, and give him feedback on the formative evaluations you are conducting. The participant knows the amount of progress or lack of progress that is occurring each week. At the end of the 10-week program, you measure his body weight and percent body fat and conduct a simple summative evaluation. Were the weight-loss and percent-fat goals achieved at the end of the program?

The world-class athlete's summative evaluation might occur during the Olympic Games with the winning (or not winning) of a medal.

Mastery Item 1.4

Develop a scenario similar to the one in the previous section that is designed for a participant who wishes to increase his or her amount of daily physical activity and that involves formative and summative evaluation.

PURPOSES OF MEASUREMENT, TESTING, AND EVALUATION

Prospective professionals in kinesiology, human performance, physical activity, health promotion, and the fitness industry must understand measurement, testing, and evaluation because these professionals make evaluative decisions daily. Our students, athletes, clients, and colleagues ask us what tools are best and how to interpret and evaluate performance and measurements. Regardless of your specific area of interest, the best tools to use and how to interpret data may be the most important concepts that you will study. Related evaluation concepts are objectivity, reliability, relevance, and validity. These terms are discussed in greater detail in chapters 6 and 7.

There are many ways to use the evaluative process in human performance. For instance, consider the issue of accountability. Your employer might hold you accountable for a project; that is, you might be responsible for obtaining a particular outcome for an individual or program. Tests, measurement, and the evaluation process are used to show whether you are accountable. Obviously, you want the evaluation to accurately reflect the results of your work—assuming that you did a good job! Certainly, if you enter the teaching profession, you will hold your students accountable for learning and retaining the content of the courses you teach. Likewise, your students should hold you accountable for preparing the best possible tests for evaluating their class performance.

Mastery Item 1.5

Assume you are a weight-training instructor. How would you determine whether your program is effective?

As you will discover during your course of study, you need considerable knowledge and skill to conduct correct and effective measurement and evaluation. *As with any academic or professional effort, it is important to have a thorough understanding of the purposes of executing a measurement and evaluation process.* There are six general purposes of measurement and evaluation: placement, diagnosis, prediction, motivation, achievement, and program evaluation.

Placement

An initial test and evaluation allow a professional to group students into instructional or training groups according to their abilities. In some cases, instruction, training, and learning in human performance can be facilitated by grouping participants according to their abilities. All participants in a group can then have a similar starting point and can improve at a fairly consistent rate. Obviously, it is difficult to teach a swimming class if half the students are nonswimmers and the others are members of the swim team, but even less extreme differences can affect learning.

Diagnosis

Evaluation of test results is often used to determine weaknesses or deficiencies in students, medical patients, athletes, and fitness program participants. Cardiologists may administer treadmill stress tests to obtain exercise electrocardiograms of cardiac patients to diagnose the possible presence and magnitude of cardiovascular disease. Recall the measurement and evaluation challenge highlighted at the beginning of this chapter. The doctor made a diagnosis based on a number of physiological and behavioral measures. This was possible because of the known relationships between the measures and the incidence of heart disease.

Prediction

One of the goals of scientific research is to predict future events or results from present or past data. This is also one of the most difficult research goals to attain. You probably took the Scholastic Aptitude Test (SAT) or the American College Test (ACT) during your junior or senior year of high school. Your test scores can be viewed as predictors of your future success in college and perhaps were part of the admissions process used by your college or university. The exercise epidemiologist may use physical activity patterns, cardiovascular endurance measures, blood pressure, body fat, or other factors to predict your risk of developing cardiovascular disease.

Motivation

The measurement and evaluation process is necessary for motivating your students and program participants. People need the challenge and stimulation they get from an evaluation of their achievement. There would not be any athletes if there were only practices and no games or competitions. When are you going to be most motivated to learn the material for this course? What would motivate you to study and learn the material for this course or any other if you knew you would not be tested and evaluated?

Achievement

In a program of instruction or training, a set of objectives must be established by which participants' achievement levels can be evaluated. For instance, in this course, your final achievement level will be evaluated and a grade will be assigned on the basis of how well you met some objectives set forth by the instructor. Developing the knowledge and skills needed for proper grading is an important objective of this book; chapters 8 and 9 are devoted to the topics of assessment and grading. Improvement in human performance is an important goal in instruction and training programs, but it is very difficult to evaluate fairly and accurately. Is final achievement level going to be judged with criterion-referenced standards on a pass–fail basis or with norm-referenced standards and grades? Assessment of achievement is a summative evaluation task that requires measurement and evaluation.

Program Evaluation

You may have to conduct program evaluations in the future to justify your instruction and training programs. *The goal of program evaluation is to demonstrate (with sound evidence) the successful achievement of program objectives to your superiors.* If you are a physical education teacher, you may be asked to demonstrate that your students are receiving appropriate physical fitness training. You might compare your students' fitness test results with the test results of students in your school district or with national test norms. You might gather student and parent evaluations of your program. Professionals in YMCAs, YWCAs, and corporate or commercial fitness centers can evaluate their programs in terms of membership levels and participation, participant test results, participant evaluations, and physiological assessments. Your job and professional future could depend on your being able to conduct a comprehensive and effective program evaluation.

Mastery Item 1.6

You are starting a new competitive field hockey program at a high school or local sports club. What measurement and evaluation steps would you take to ensure that you have the best athletes available for participation in the program?

DOMAINS OF HUMAN PERFORMANCE

The purposes we've just discussed are related to the objectives of your program. Objectives are specific outcomes that you hope to achieve with your program. To be accurately measured and truthfully evaluated, these outcomes need to be measurable. Measurable outcomes are called behavioral objectives. Objectives in the area of human performance fall into three areas: the cognitive domain, the affective domain, and the psychomotor domain. The student of measurement and evaluation in education or psychology is concerned with objectives in the first two areas. *For students of human performance, the distinctive objectives are those in the psychomotor domain.* Bloom (1956) presented a **taxonomy** (classification system) of cognitive objectives (see table 1.2). A hierarchical list of Bloom's levels includes knowledge, comprehension, application, analysis, synthesis, and evaluation.

Table 1.2 **Taxonomy in Domains of Human Performance**

Taxonomy of the cognitive domain (Bloom, 1956)	Taxonomy of the affective domain (Krathwohl, Bloom, and Masia, 1964)	Taxonomy of the psychomotor domain (Harrow, 1972)
Knowledge • Of specifics • Of ways and means of dealing with specifics • Of the universals and abstractions in a field Comprehension • Translation • Interpretation • Extrapolation Application Analysis • Of elements • Of relationships • Of organizational principles Synthesis • Production of unique communications • Production of a plan for operations • Derivation of a set of abstract relations Evaluation • Judgments in terms of internal evidence • Judgments in terms of external evidence	Receiving • Awareness • Willingness to receive • Controlled or selected attention Responding • Acquiescence in responding • Willingness to respond • Satisfaction in response Valuing • Acceptance of a value • Preference for a value • Commitment Organization • Conceptualization of a value • Organization of a value system Characterization by a value complex • Generalized set • Characterization	Reflex movements • Segmental reflexes • Intersegmental reflexes • Suprasegmental reflexes Basic-fundamental movements • Locomotor movement • Nonlocomotor movement • Manipulative movement Perceptual abilities • Kinesthetic discrimination • Visual discrimination • Auditory discrimination • Tactile discrimination • Coordinated discrimination Physical abilities • Endurance • Strength • Flexibility • Agility Skilled movements • Simple adaptive skill • Compound adaptive skill • Complex adaptive skill Nondiscursive movements • Expressive movement • Interpretive movement

Objectives in the **cognitive domain** deal with knowledge-based information. Objectives in the **affective domain** concern psychological and emotional attributes. A taxonomy of these objectives, from Krathwohl, Bloom, and Masia (1964), is as follows: receiving, responding, valuing, organizing, and characterizing by a value complex. Affective objectives, which concern, for example, how people feel about their performance, are very important but often difficult to measure. Affective objectives are not normally measured for grading purposes. The third domain of objectives is the **psychomotor domain** (Harrow 1972); these are reflexive movements, basic locomotor movements, perceptual motor abilities, physical abilities, skilled movements, and nondiscursive movements. The measurement techniques and concepts associated with the psychomotor domain differentiate human performance students from students in other areas. There are other taxonomies for the cognitive, affective, and psychomotor domains; the three in table 1.2 are only examples.

When you are measuring and evaluating subjects with a specific test, you must take into account the level of the domain your participants have achieved. Each taxonomy is a hierarchy; each level is based on the earlier levels having been achieved. For example, it would be inappropriate for you to attempt to measure complex motor skills in 7-year-old children, because most of them have not achieved prior levels of the taxonomic structure. Likewise, it is difficult, if not impossible, for younger participants to achieve the higher-level cognitive objectives of a written test.

When measuring and evaluating individuals, consider the level of the domain that the individual has achieved.

Psychomotor Domain: Physical Activity and Physical Fitness

For many years, physical educators, exercise scientists, personal trainers, athletic coaches, and public health leaders have been concerned about the definition, the reliable and valid measurement, and the evaluation of physical fitness in people of all ages. This concern led to a growing number of fitness tests and protocols for both mass and individual testing. For example, The Cooper Institute's FITNESSGRAM, the President's Council on Physical Fitness and Sports, the President's Challenge, and the European test, Eurofit, are all youth fitness test batteries. Each of these test batteries consists of different test items, but all serve the purpose of assessing levels of physical fitness. You'll learn more about these test batteries in chapter 12. A large number of research studies have been conducted to demonstrate the feasibility, reliability, and validity of such fitness tests. Normative surveys have been conducted to establish the fitness levels of various populations. In this book we have devoted two chapters to physical fitness testing.

In the latter part of the 20th century, the health-related aspects of physical activity became a dominant concern of public health officials. The culmination of this concern was presented in the release of *Physical Activity and Health: A Report of the Surgeon General* (USDHHS 1996). This publication, led by senior scientific editor Steven N. Blair, presented a detailed case for the health benefits of a lifestyle that includes regular and consistent participation in moderate to vigorous levels of physical activity. Unfortunately, the report indicates that the majority of adults in the United States are not physically active enough for good health. As you will learn in chapter 12, there is considerable controversy over whether today's children are physically active and fit, or if children's fitness and activity levels have declined over the past few decades.

Physical activity is defined as bodily movement that is produced by the contraction of skeletal muscle and that substantially increases energy expenditure. Physical fitness, on the other hand, is a set of attributes that people have or achieve that relates to the ability to perform physical activity. Physical activity is something people do, whereas physical fitness is something people have or achieve. Heredity plays an important role in both factors but is probably more important in physical fitness. Physical activity is more difficult to measure reliably and validly than physical fitness. Measuring a behavior is generally more difficult than measuring an attribute.

Just as there are a variety of fitness tests and test protocols for measuring specific fitness attributes, there are a variety of techniques to measure physical activity. These techniques include motion sensors, written recalls, and heart rate biotelemetry. Techniques that work for adults are not always appropriate for children. An increasing body of scientific literature is appearing about the reliability and validity of physical activity measurement in a variety of situations and populations. In chapters 11 and 12 we explore the measurement and evaluation of physical activity and physical fitness, elements in the psychomotor domain.

Mastery Item 1.7

Select a position in your current field of study in which you might be employed. Develop two objectives for your job responsibilities from each of the cognitive, psychomotor, and affective domains.

Measurement and Evaluation Challenge

The issues just presented about standards, evaluation, purposes, and domains of human performance directly relate to the scenario about your father. For example, throughout the remainder of this course you will learn about the tools available to answer the questions your father raised. In chapter 2 you will learn about accessing the World Wide Web to obtain information and calculate health risks. You will also be introduced to powerful computer programs to help you analyze data and make decisions; you will use these computer programs throughout the remainder of the book. Chapters 3 through 5 will help you understand the statistical procedures necessary to make evidence-based decisions. Chapter 3 presents information about the distributions of measurements. Chapter 4 presents information about quantifying the relationship of one variable to another (e.g., relating decreased physical activity to increased cardiovascular risk). Chapter 5 presents an overview of research methods to help you decide if an intervention makes a significant difference in a specific outcome of interest (e.g., does moderate exercise reduce body fat?). Chapter 6 illustrates how to determine the accuracy (reliability) of measurement. You will learn how to determine the best measure to use, how to interpret the measurement, and what influences the errors of measurement that are always present when measures are taken. Chapter 7 illustrates how specific health standards are set and whether these standards affect the odds or risk of developing a specific disease. Chapter 8 provides examples of how to obtain alternative measures. For example, one could measure physical activity by asking for a self-report, by using a pedometer, or by directly observing a subject's daily behaviors. Chapter 9, although primarily concerned with grading, contains important information on how to properly add various measurements together to obtain a composite score. Chapter 10 provides tools with which to make an accurate assessment of knowledge. Knowledge is required but not sufficient for behavior change. For example, does your father know how to be physically active for a health benefit? A simple knowledge test might tell us this; however, knowledge tests must be accurate. You will learn how to determine if a knowledge test is accurate and truly reflects learning. Chapter 11 illustrates how to measure risk factors associated with cardiovascular disease. As an example, you will learn about some simple tests of aerobic capacity, body composition, physical fitness, and physical activity. Chapter 12 illustrates physical fitness and physical activity assessment in childhood, values that could be predictive of future health risk. Chapter 13 presents strategies for measuring and evaluating sport skills and physical abilities. Physical abilities, such as strength and flexibility, might be related to your father's overall health risks. Chapter 14 illustrates methods of measuring psychological stress levels, which might be pertinent in your father's case.

You and your father's interest in the measurements taken are right on target. How do you know they were accurate? How do you know they are truly predictive? How do you know if he is really at increased risk? How do you know if specific interventions help reduce risk? These and other questions are what your future learning in this book is all about.

SUMMARY

As a student, you are aware that nearly all educational decisions rely greatly on the processes of measurement and evaluation. Figure 1.1 illustrates the relationships among testing, measurement, and evaluation and indicates that as a human performance professional you will have to make a variety of decisions regarding the methods of collecting and interpreting data in the measurement process. A wide range of instruments (tests) are used to assess abilities in the cognitive, psychomotor, and affective domains. You will have to determine the domains in which you wish to have objectives and then develop specific objectives and select tests that produce objective, reliable, relevant, and valid measurements of your objectives. Once you collect the data, evaluative decisions can be either norm referenced or criterion referenced. In norm-referenced comparisons, the subject's performance is compared with that of others who were tested. Criterion-referenced standards are used to compare a subject's performance with a predetermined standard that is referenced to a particular behavior or trait. Evaluations can be formative (judged during the program or at intervals of the program) or summative (judged at the end of the program).

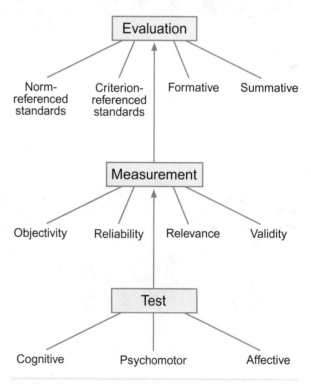

Figure 1.1 Relationships among test, measurement, and evaluation.

2

Using Technology in Measurement and Evaluation

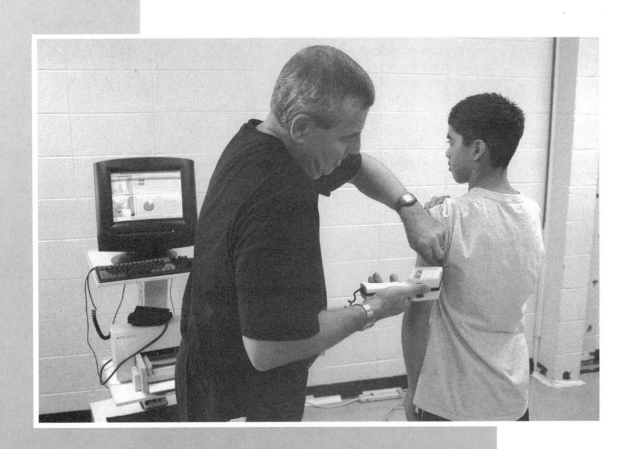

Outline

Measurement and Evaluation Challenge

Mike and Jessica are both teachers. Mike teaches biology and coaches golf. Jessica teaches physical education and coaches volleyball. They have been in their jobs for 2 years and are beginning to get overwhelmed by the amount of paperwork they have to do. Recording test grades, performance scores, and attendance takes considerable time and effort. They are interested in further developing their computer skills because they believe that doing this work on the computer will save them much time and effort. Additionally, both have recently been accepted to graduate school, where they will have to complete a thesis as part of their degree requirements. They are already thinking about thesis topics and how they might analyze data. They have heard about computer programs that serve as "electronic gradebooks," programs that record and analyze physical performance. They have home computers and computers at their schools. Can the computer help them become more efficient at recording and evaluating student cognitive and psychomotor performance? If so, in what ways?

Objectives

After studying this chapter, you will be able to

- identify the potential uses of microcomputers in your field;
- access the Web site associated with this book;
- identify sources of microcomputer software and hardware for use in exercise science and physical education;
- present examples of computer use in exercise science, physical education, and clinical health and fitness settings and describe how various testing procedures can be facilitated with computers;
- use SPSS to create and save data files; and
- use Excel to create a data file to be used in SPSS.

omputers have become commonplace in schools. Indeed, the vast majority of you probably have personal computers in your homes. Tasks that only a few years ago were tedious and time consuming are now completed in a matter of seconds by computers, and computers can now fit in a briefcase or even in your hand. Wireless communication is also expanding rapidly. As technology continues to expand, your ability to use a computer will affect you in your career and leisure activities.

Undoubtedly, computers are one of the most important technological advances of this century. Originally, mainframe computers were large machines that took up entire rooms (or even floors) in buildings. Users typically had to be connected via telephone or other electronic line to the mainframe, and mainframes were inaccessible to most people. However, the development of the microprocessor has resulted in small, powerful, relatively inexpensive **microcomputers.** Indeed, it has been suggested that if the change occurring in microcomputer technology over the past 50 years had also occurred in the automobile industry, cars would now get thousands of miles to a gallon of gas! However, even though microcomputers are now widely available, many students and professionals who measure and evaluate human performance have not taken full advantage of the power that computers provide.

Additionally, the development and worldwide use of the **Internet** and **World Wide Web** (WWW or Web) have had a significant impact on how individuals obtain information and communicate with one another. In 1999, the Pew Research Center reported that the percentage of Americans using the Internet went from 23% in 1996 to 41% in 1998. About one in five U.S. citizens use the Internet to get news and weather. The percentage of people using the Internet at least once a week for news increased from 6% in 1995 to 20% in 1998 to more than 60% in 2002. Approximately 90% of all college students and 60% of all adults use the World Wide Web. More than 80% of Internet users are trying to find the answer to a specific question or use a search engine.

There are numerous Web sites of specific interest and value to the human performance student. Some of these are professional organizations such as the American College of Sports Medicine (ACSM; www.acsm.org), the American Alliance for Health, Physical Education, Recreation and Dance (AAHPERD; www.aahperd.org), and the American Heart Association (AHA; www.americanheart.org). Others provide scientific and content-based information that may be related to your job responsibilities (e.g., courses you may teach, fitness and health information, athlete-training information).

Mastery Item 2.1

Speak with two or three faculty members in your major department. Ask them to provide examples of how they are using computers and the Internet. Consider how you might use computers in your daily activities.

Mastery Item 2.2

Obtain a copy of your local newspaper. Look through each section and see how many references you can locate that pertain to health reports where you can learn more about the story on the World Wide Web.

It is important to differentiate between hardware and software. Hardware consists of the physical machines that make up your microcomputer and its accessories. Software is the computer code, generated by a computer programmer, by which you interact with the computer as you enter data and conduct analyses, create text, and draw graphs. You don't have to be a software programmer to be a competent microcomputer user—the vast majority of expert microcomputer users do not do programming.

Use one of the WWW search engines to identify specific packages and resources that might be helpful to you as an exercise scientist.

USING MICROCOMPUTERS TO ANALYZE DATA

Microcomputer technology is now pervasive in schools and businesses. Many schools and businesses require students to be computer literate—able to interact with computers daily for work and pleasure. Computers have such a big influence in our daily lives (they're involved in everything from grocery shopping and banking to using the telephone) that we have to be able to use them. Computer literacy does not require one to be a computer programmer; one simply needs to be able to use computers in daily life, for example, to conduct daily tasks or for enjoyment (e.g., "surfing" the Web).

Go to a WWW search engine and enter someone's name and a particular topic. See how many sources you can identify related to this topic. Consider how you might use this information in this or another class or in your career.

Exercise scientists and physical educators must make many measurement and evaluation decisions that involve numbers, which computers are particularly adept at handling. Because the exercise and human performance professions require daily use of computers, you must familiarize yourself with their features and uses specific to your field so that you can understand and use the concepts presented in this text. Many of the decisions that you will make in your field require data analysis. Thus, we will introduce you to **SPSS,** a powerful data analysis program that will help you save, retrieve, and analyze much of the measurement and evaluation data that you will encounter daily.

SPSS makes number crunching fast, efficient, and almost painless. For example, the most important characteristics of any test are its reliability and validity. As you will learn in chapters 3, 4, and 5, computers can generate data related to reliability and validity in a matter of seconds. This will be illustrated in chapters 6 and 7 and used in MIs throughout the text book. Many statistics can help you make valid decisions. Chapters 3 through 7 provide you with many opportunities to practice using SPSS in scenarios similar to those you will use in your professions.

Additionally, we present information about how to create databases with **Microsoft Excel®**. These Excel databases can be easily read with SPSS. The benefit of creating your database with Excel is that it is readily available on computers. Thus, you could create your database while at work and then conduct the analysis with SPSS.

Measurement uses for a microcomputer in human performance, kinesiology, and physical education include the following:

▶ **Accessing the Web to obtain information relative to your specific job responsibilities.** Visit the student study Web site for some suggested sites to start with.

▶ **Determining test reliability and validity.** Statistics learned in chapters 3, 4, and 5 can be used to estimate the reliability (consistency) and validity (truthfulness) of test results in the cognitive, affective, and psychomotor domains. SPSS MI examples are provided in chapters 3 through 14.

▶ **Evaluating cognitive test results and physiological test performance.** Computers can help evaluate and report individual test results. Likewise, you can quickly retrieve, analyze, and return test results to subjects. SPSS computer examples are provided in chapters 11, 12, and 13. You can estimate your risk for development of diabetes from the American Diabetes

Association (www.diabetes.org) and your risk of cardiovascular disease from the American Heart Association (www.americanheart.org).

▶ **Conducting program evaluation.** Computers can calculate changes in overall student performance and learning across teaching units or track individual changes in a subject's performance. SPSS examples are provided in chapter 5.

▶ **Conducting research activities.** You can compare an experimental group of subjects with a control group to determine if your new experimental program has a significant effect on cognitive or physiological performance. SPSS examples are provided in chapter 5.

▶ **Developing presentations.** Specialized software can be used to create powerful presentations you can make before students, potential clients, patients, and professional peers. The presentations can include text, pictures, video, graphics, and sound to effectively present your message. Perhaps your instructor is using the presentation package that accompanies this textbook to illustrate specific points.

▶ **Assessing student performance.** Students and clients are always interested in how they perform on tests, whether the tests are cognitive, psychomotor, or physiological. Students, teachers, and clinicians are interested in what their individual score is, how it is interpreted, what it means, and what effect it has. Microcomputers make it easy to provide the answers to all these questions.

▶ **Storing test items.** Teachers always have to keep records of student grades. Programs that permit entry and manipulation of student data records are called spreadsheets. Spreadsheets are essentially computer versions of a data matrix with rows and columns of information. Students' names are often found in the first column, and data from course assignments fill the remaining columns. Thus, each row represents a different student and each column holds scores from tests and other assignments. If the instructor keeps a daily record of class grades, then average grades, final grades, printed reports, and so on can be generated with a few computer keystrokes. Likewise, health and fitness professionals can keep records of workouts and changes in weight, strength, aerobic capacity, and so forth.

▶ **Creating written tests.** Computers can serve as a bank for written test items. Rather than having to develop a new test each time you teach a unit, you can store test items on your microcomputer and generate a different test each time you teach the unit. Test banks can be built using word-processing or test-development software. Some test-development programs are very sophisticated and permit you to choose an item not only by content area but also by type of item, degree of difficulty, or date created. An SPSS computer example is provided in chapter 10.

▶ **Calculating numerous statistics.** Physiological measurements often involve equations for estimating values. For example, skinfolds are used to estimate percent body fat, and distance runs and heart rate measurements are used to estimate oxygen consumption. The microcomputer can greatly assist in calculating these values. Rather than substituting each number into an equation and going through the steps to complete the calculation, you can enter the formula into a microcomputer once and automatically calculate the desired value for each person. For example, go to the National Heart, Lung, and Blood Institute (NHLBI) of the National Institutes of Health (NIH) WWW site (http://www.nhlbisupport.com/bmi/) and calculate your body mass index. Mastery Items and illustrations in chapters 3, 4, and 5 will demonstrate how a microcomputer can greatly assist in calculating statistics. Mastery Items in chapters 6 through 14 provide you with opportunities to practice analysis and interpretation.

Mastery Item 2.5

Contact a local elementary, middle, or high school or local health club and determine the ways in which professionals use microcomputers. Find out how they use computers daily to help them to conduct their jobs more efficiently.

Mastery Item 2.6

Think of some time-consuming tasks that you need to do regularly. How would a micro-computer help you complete them more efficiently? What kinds of things can the kinesiology, exercise science, or physical education major do with a microcomputer?

Mastery Item 2.7

Talk with your classmates and see what types of test items they have access to. Together your class might consider developing a large bank of items, storing them on a disk, and sharing them with other preprofessionals so the items can be used once class members enter career employment.

Your Textbook and the World Wide Web

The development and growth of the Web has been phenomenal over the past few years. The WWW was conceptualized in the 1960s, but it was not until the 1990s that its use exploded in such a fashion that it is accessible nearly everywhere on earth. Today millions of people across the globe search the Web daily for information about their jobs and for resources related to their work and pleasure. Some courses and entire university degree programs are available over the Internet.

In writing this textbook, we have taken advantage of the Web to help you better learn, practice, and use the information presented in your textbook. A special Web study site has been developed for this textbook (www.HumanKinetics.com/MeasurementAndEvaluationInHumanPerformance/SG/getstart_01.cfm). This dynamic page will be periodically updated as new information is available. If your instructor has a personal Web site, he or she might even have a link to the class Web site from there. The key things that you can obtain at the Measurement and Evaluation section of the Web site for this textbook include

- outlines associated with each chapter in the textbook;
- answers to selected mastery items;
- practice problems with answers;
- quizzes;
- test review items for each chapter;
- mastery items data sets for chapters 2 through 14; and
- links to other relevant sites.

The Web site is organized so that you go to the site, choose the chapter you would like to obtain information about, and click on the link to that chapter. Once you reach the chapter itself, there are various links that get you to resources that help you prepare for class, review course material, take practice quizzes, complete practice problems, and prepare for examinations. Once your instructor provides you with the information necessary for accessing the site, you should go to the site, look around, and examine the resources there.

Physical fitness testing is an important component in most physical education, kinesiology, and exercise science programs. Not too many years ago, fitness test results were reported orally, on poorly prepared reports, or on purple mimeographed copies. Today, youth fitness programs use microcomputer software to analyze student results. The Cooper Institute's FITNESSGRAM is an example of an excellent software program. Figure 2.1 illustrates the type of microcomputer output that the teacher can give children (and their parents) to better inform them of their progress.

Teacher and student reviewing computerized FITNESSGRAM report.

USING SPSS

Many of the decisions that you will have to make about reliability and validity, whether you are a kinesiologist, physical therapist, clinician, coach, instructor, or educator, are based on statistical evidence. Now don't get scared! This is not a statistics text. It is a measurement and evaluation text. However, the statistics presented in chapters 3, 4, and 5 provide the framework for much of your decision making in measuring and evaluating human performance. Although we will use SPSS (a sophisticated statistical package that is widely available on many university campuses) throughout this text to assist you in calculating statistics used in reliability and validity decisions, your instructor may choose for you to conduct the analyses with another type of software. Regardless, the calculations will be nearly identical (within rounding error) and the interpretation of the results will be exactly the same, regardless of the particular statistical package you use. Nearly all of what we present in this text is available in both the full and student versions of SPSS. SPSS is continually updated, and new versions are available. You should check chapter 2 of this textbook's Web site to see if there is any specific information that you will need if an update of SPSS is released. Generally the updates do not have a major impact on the types of things that you will be doing in this class. You might need to be a little flexible when you access SPSS because SPSS is continually updated and the specific method that you use to access SPSS at your location may not be the same as other locations.

SPSS is software developed to analyze numbers (e.g., calculate the mean or draw a graph). However, SPSS must have a database on which to conduct the analysis. Thus, each analysis conducted by SPSS is run on a set of data created and saved through the SPSS data editor. The SPSS data editor lets you create a database (also called a data matrix) consisting of N rows of people with p columns of variables. Table 2.1 illustrates a data matrix consisting of 10 rows of people ("id" numbers 1 though 10) with six variables ("id," "gender," "age," "weightkg," "heightcm," "milesec"). Weight is measured in kilograms (kg) and height in centimeters (cm), and milesec is the total number of seconds (s) required to complete a 1 mi (1.6 km) walk-run.

FITNESSGRAM®

	Test Date	Height	Weight
Current	07/15/99	5'01"	105
Past	07/13/99	5'3"	122

 MESSAGES

AEROBIC CAPACITY

Healthy Fitness Zone

Needs Improvement | Good ———— Better | **My Scores**

Walk Test
Current
Past
42   52

VO_2max = **51** / **42**

VO_2max Indicates ability to use oxygen. Expressed as ml of oxygen per kg body weight per minute.

Time
Current   15:56
Past   16:34

Charlie, your scores on all test items were in or above the Healthy Fitness Zone. You are also doing strength and flexibility exercises. However, you need to play active games, sports or other activities at least 5 days each week.

Although your aerobic capacity score is in the Healthy Fitness Zone now, you are not doing enough physical activity. You should try to play very actively at least 60 minutes at least five days each week to look and feel good.

MUSCLE STRENGTH, ENDURANCE & FLEXIBILITY

(Abdominal) CurlUp   Number
Current   **45**
Past
15   28

Your abdominal strength was very good. To maintain your fitness level be sure that your strength activities include curl-ups 3 to 5 days each week. Remember to keep your knees bent. Avoid having someone hold your feet.

(Upper Body) Flexed Arm Hang   Seconds
Current   **49**
Past   **99**
6   13

Your upper body strength was very good, Charlie. To maintain your fitness level be sure that your strength activities include arm exercises such as push-ups, modified push-ups or climbing activities 2 to 3 days each week.

(Trunk Extension)
Current   **INC**
Past

Charlie, your flexibility is in the Healthy Fitness Zone. To maintain your fitness, stretch slowly 3 or 4 days each week, holding the stretch 20 - 30 seconds. Don't forget that you need to stretch all areas of the body.

If given, the flexibility test is performed on the right and left and is evaluated as 'Yes' or 'No' on both sides.

(Flexibility) Back Saver Sit and Reach   R,L (Inches)
Current   **Y,Y(9-10)**
Past   **Y,Y(8-10)**
N, Y   Y, Y
Y, N

BODY COMPOSITION

Percent Body Fat   Percent
Current   **16**
Past   **16**
25   10

Lower numbers are better scores on body composition measurement.

Charlie, your body composition is in the Healthy Fitness Zone. If you will be active most days each week, it may help to maintain your level of body composition.

ACTIVITY

	Number of Days
On how many of the past 7 days did you participate in physical activity for a total of 30-60 minutes, or more, over the course of a day?	4
On how many of the past 7 days did you do exercises to strengthen or tone your muscles?	3
On how many of the past 7 days did you do stretching exercises to loosen up or relax your muscles?	2

To be healthy and fit it is important to do some physical activity almost every day. Aerobic exercise is good for your heart and body composition. Strength and flexibilty exercises are good for your muscles and joints.

Good job, you are doing enough physical activity for your health. Additional vigorous activity would help to promote higher levels of fitness.

©The Cooper Institute for Aerobics Research

Figure 2.1 FITNESSGRAM reports such as this one can be given to parents and students to illustrate a student's improvement.

Table 2.1   **Sample Database (Data Matrix)**

id	gender	age	weightkg	heightcm	milesec
1	0	20	50	165	500
2	0	24	51	160	600
3	0	21	62	173	700
4	0	19	59	178	650
5	0	23	43	145	450
6	1	22	86	193	480
7	1	25	65	183	400
8	1	24	61	178	420
9	1	28	75	173	390
10	1	20	70	178	350

SPSS permits you to enter and manipulate data and conduct analyses that result in a variety of numbers, charts, and graphs. Each of the data tables used in your textbook is located on the Web site for each chapter. You can download them in SPSS or Microsoft Excel formats. You will learn more about this in the following paragraphs.

Getting Started

Locate and double-click on the icon for SPSS on your computer's desktop. Alternatively, you may have to go through the **Start** menu on the bottom left of your computer to locate SPSS. It will be different depending on your university or computer. Once you locate and begin SPSS, note that a blank data matrix appears. The upper left corner has the name "Untitled—SPSS Data Editor." The data editor lets you define and enter data (figure 2.2).

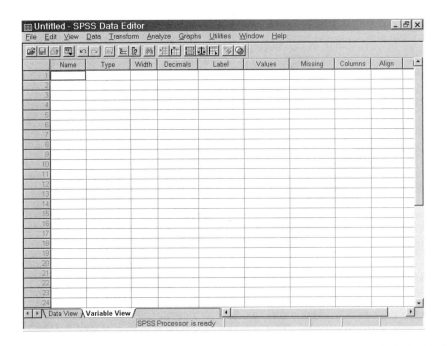

Figure 2.2   Screen capture of SPSS Data Editor.

Note there are two tabs near the bottom left of the SPSS window. One of them says **Data View** and the other says **Variable View.** The **Data View** window presents the data you have entered or provides a spreadsheet that permits you to enter data. The **Variable View** window permits you to define and name the variables themselves. It also permits you to identify variable labels, value labels, and missing values. These are illustrated in subsequent paragraphs.

Note also that there are several pull-down menus across the top of the data matrix. These generally provide you the following functions:

- ▶ **File**—Among other functions, permits you to create a new data matrix, open a data matrix that was previously saved, save the current data matrix, print the current data matrix or analysis results, and exit the program.

- ▶ **Edit**—Permits you to undo a previous command; cut, copy, or paste something from the window; or find a specific piece of data.

- ▶ **View**—Permits you to change the font in which your data appear and change the appearance of the data matrix window.

- ▶ **Data**—Among other functions, permits you to insert variables, sort the data, and select specific cases.

- ▶ **Transform**—Permits you to modify your variables in a number of ways. You will use the compute function often.

- ▶ **Analyze**—You will become familiar with the **Analyze** menu throughout this class. It lists the variety of statistical procedures that you will use. Don't worry—there are many listed here, but we will not be using all of them. Note that each of the options in this menu has an arrow next to it. The arrow indicates that additional submenus are available to you under this particular statistical procedure. You will become very familiar with these submenus as you work through the textbook.

- ▶ **Graphs**—Lists the various types of graphs you might use to present your data. We will use a limited number of these options with your textbook.

- ▶ **Utilities**—Permits you to modify your data matrix in a number of ways. We will not be using this menu in your course.

- ▶ **Window**—Lets you "hide" the data window when you are running several programs at a time and switch from the data matrix window to the SPSS output window once you have conducted an analysis.

- ▶ **Help**—Provides you with a variety of help sources when you are running SPSS and have a question. You might find the **Topics** (and then **Index**) submenus helpful.

Each of the pull-down menus has additional functions that you might enjoy investigating, but we provide you with sufficient information to conduct the SPSS processes you will use with your textbook. You are encouraged to investigate these various menus as you learn SPSS and use the **Help** windows provided within SPSS. The more you interact with SPSS, the better you will understand its capabilities and use it to make your work much easier. SPSS instructions are based on version 11.0. These instructions may change as SPSS updates its software. Changes will be updated on this text's Web site. We have provided an SPSS sample data set within at least one MI in each of the remaining chapters of your textbook.

Creating and Saving Data Files

Use the steps in mastery item 2.8 to create and save your first SPSS data file named table 2-1. Once you have the idea, you will be able to go through these steps quite rapidly. For the time being, trust us and follow the steps closely. Note that you will need to save the data matrix as table 2-1 (with a hyphen and not a period). This is because the computer will interpret the period as a "file extension" and cause you difficulty when you attempt to access the table at some later point. Thus, when you create and save your tables, name them with the following style: chapter number-table number. For example, in chapter 2, the second table would be table 2-2, the third table in chapter 2 would be 2-3, and so on.

SPSS variable names must begin with a letter and be no longer than eight characters in length. You cannot use spaces in the name, and you should avoid special characters when naming a variable. We have used a mnemonic name to identify our variables. Thus, "weightkg" is actually weight in kilograms (kg). The mnemonic helps you remember exactly what the variable is and the units in which it was measured (figure 2.3).

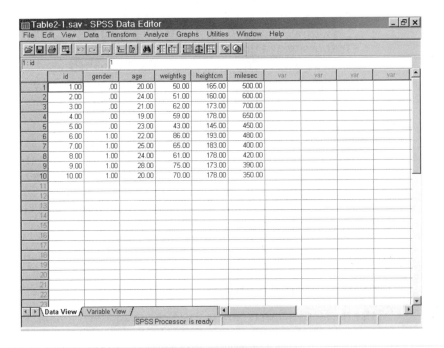

Figure 2.3   Screen capture of SPSS **Variable View** *(a)* and **Data View** *(b)* windows.

Mastery Item 2.8

Follow these step-by-step procedures to create an SPSS file named table 2-1 and save it to your floppy disk.

Creating and saving an SPSS data file

1. Be certain to have a floppy disk with you before you start this assignment. In some systems you may have to save your data to an electronic account.
2. Place the floppy disk in the machine and note the drive (A or B) location.
3. Locate the SPSS icon and click on it. (Alternatively, you might have to go to the Start button on the bottom left of your computer and locate SPSS among the programs listed in the Start menu.)
4. First you will name the variables, define the variables, and essentially build a "codebook" that helps you remember what the variables are.
5. Click on the "variable view" tab and note that the window now looks like that illustrated in figure 2.3a but without the information in it.
6. Name each of the variables in the first column. Note the variable name must start with a letter, contain no special characters, and be no longer than eight characters in length. Newer versions of SPSS do not limit variable names to eight characters.
7. For the time being, skip over the "Type," "Width," and "Decimals" columns.
8. You can expand on the variable names in the "Label" column.
9. Click on the right-hand edge of the "Values" column for the second variable (i.e., gender). Notice that you get a box that helps you define the values associated with numbers for gender. In our case, we have coded females as 0 and males as 1. Enter these values, click on the "Add" button each time, and then click on "OK." You have defined your variables and you are ready to begin entering data.
10. Click on the "data view" tab to get to the **Data View** window.
11. Enter the data from table 2.1 into SPSS. Your results should look like that in figure 2.3*b*.
12. You are now ready to save the data on the computer disk that you brought with you. Be certain the computer disk is inserted properly and not write protected. Go to the **File** menu and scroll down to **Save As** (see figure 2.4).

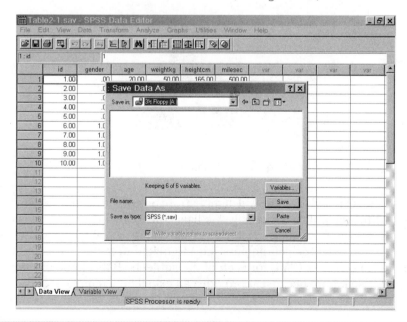

Figure 2.4   Screen capture of SPSS showing how to save a file.

13. In the File Name box, enter "table 2-1" (without the quotes).
14. Save the data to your disk (and not to the hard drive). Go to the **Save In** box at the top of the screen and click on the downward pointing arrow. Scroll down to the location where you just placed your disk and click.
15. Now click **Save.** Your table 2.1 data are now saved to your disk.
16. Go to the **File** menu, scroll down to **Exit SPSS,** and click. This will exit you from SPSS.

Mastery Item 2.9

Now that you have created and saved the data, let's recall it and conduct an analysis with it using the following procedures. Use table 2-1 that you have created with SPSS. First go to your floppy disk and locate table 2-1. Double click on it to begin SPSS.

1. Go to the **Analyze** menu.
2. Scroll down to **"Descriptive Statistics,"** over to **"Descriptives,"** and click.
3. When the **Descriptives** window appears, use the arrow to move "age," "weightkg," "heightcm," and "milesec" into the **Variables** box.
4. Click **OK** and then compare your results with those presented in figure 2.5.
5. If your results are different, go back to table 2.1 and compare the data in the table with what you have in the SPSS data editor.

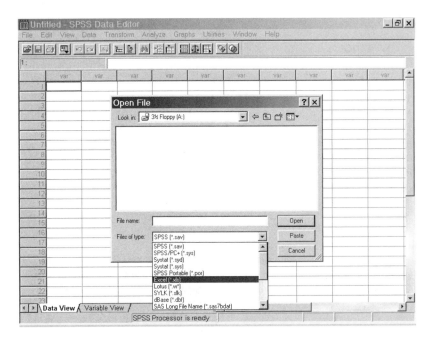

Figure 2.5   Descriptives for mastery item 2.9.

Our last example will demonstrate one of the most powerful functions of SPSS—the ability to manipulate data easily. We'll use the following instructions to create some new variables for the 10 subjects in table 2.1. We'll change weight in kilograms (kg) to weight in pounds (lb) and height in centimeters (cm) to height in inches (in.), and then we'll calculate the body mass index (BMI) for each of the subjects. BMI will be discussed further in chapter 11; we use it here because it provides an excellent example of SPSS data modification.

Using Compute Statements in SPSS

1. Follow the steps in mastery item 2.9 to access your data.
2. Go to the **Transform** menu, scroll down to **Compute,** and click—a new **Compute Variable** window will appear.
3. Type "weightlb" in the **Target Variable** box.
4. Put "weightkg" in the **Numeric Expression** box by using the arrow to move it.
5. Go to the keypad in the window and click on the "*" for multiplication.
6. Place the cursor next to the "*" in the **Numeric Expression** box and enter 2.2 (recall that to calculate weight in pounds from weight in kilograms, you simply multiply weight in kilograms by 2.2).
7. Click on **OK.**
8. Note that a new variable, "weightlb," has been created and added in a column to the right of "milesec."
9. Do the same thing to change height in centimeters to height in inches. Note that you divide height in centimeters by 2.54 to obtain height in inches.

Calculating BMI is a bit more involved. BMI is weight in kilograms divided by height in meters squared. There are several ways to get to this value. We'll take you step by step.

1. Use the **Compute** submenu (under **Transform**) to create a variable called "heightm" for height in meters by taking "heightcm" and dividing by 100. (Put heightm in as the "Target Variable.")
2. Use the **Compute** statement to create "BMI" from "weightkg/heightm ** 2" (the ** notation means square height in meters). (Put BMI in as the "Target Variable.")
3. Save the revised version of table 2-1 to your disk with the **Save** command under the **File** menu.

Mastery Item 2.10

Calculate the mean for the BMI you just created and confirm that the mean value is 20.6979. If you do not get this number, check the original numbers you entered and recheck how you created the variables at the various steps. If you find a variable that you created incorrectly, simply highlight the column for this variable and press the **Delete** key. The column will be removed from the data set and you can re-create it.

You may not always have SSPS available on the computer with which you are working. Thus, you can use Microsoft Excel to enter your data and then have SPSS read the Excel data file. We will give you an example of how to do this with the data from table 2.1. Follow the steps below to create an Excel database of table 2.1 and read the data into SPSS.

1. Open Excel on your computer. You will see a blank data sheet like that presented in figure 2.6.
2. Enter the variable names in the first row. Continue to use the SPSS restrictions on variable names. Each variable name must start with a letter, contain no special characters, and be no longer than eight characters in length.
3. Place your cursor in cell a2 and begin to enter the data from table 2.1. Once you have entered all of the data, your Excel data file should look like that presented in figure 2.7.
4. Go to the file menu, scroll down to **Save as,** and save the Excel version of table 2.1 to your disk just like you were instructed to do with the SPSS version of table 2.1.

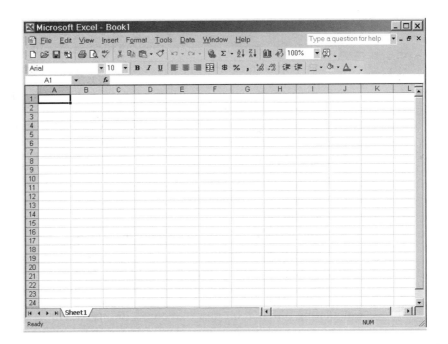

Figure 2.6   Blank Excel data sheet.

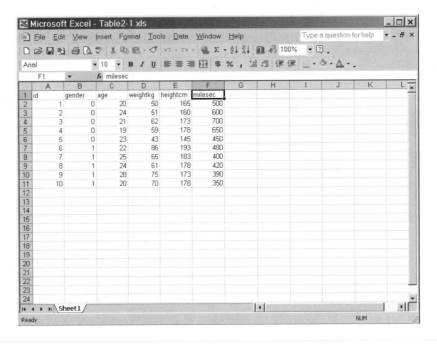

Figure 2.7   Excel data file with data from table 2.1 inserted.

You are now ready to access your Excel data with SPSS. SPSS is able to read data from Excel and place it into SPSS for data analysis. Do the following to read the Excel data file into SPSS.

1. Open SPSS as you have previously been instructed.

2. Go to the **File** menu and scroll down to **Open** and over to **Data** . . .; you will see the screen presented in figure 2.8.

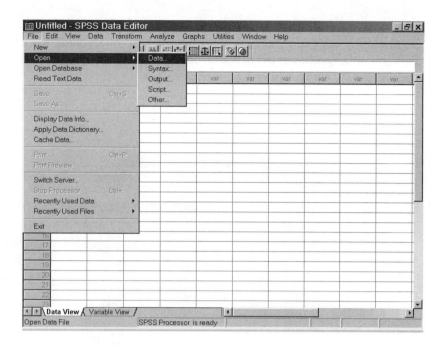

Figure 2.8   Opening an Excel data file in SPSS.

3. You will be presented with an **Open File** window. Go to the "Files of type" near the bottom of the window and click on the downward arrow so that you can then highlight the "Excel (*.xls)" indicator. This will indicate that you want to import an Excel data file into SPSS. This is illustrated in figure 2.9.

4. Locate the Excel file that you want to read into SPSS. Click on the file name and it will then appear in the "File name" box. Click on **Open.**

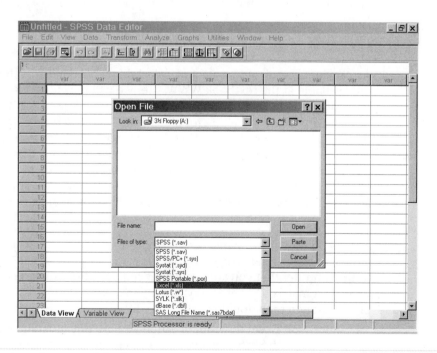

Figure 2.9   Accessing an Excel database to read into SPSS.

5. You will see an **Opening Excel Data Source** window. Click on the square that has "Read variable names from the first row of data." Recall that you placed the variable names in the first row of your Excel data file. Click **OK.** This screen is presented in figure 2.10.

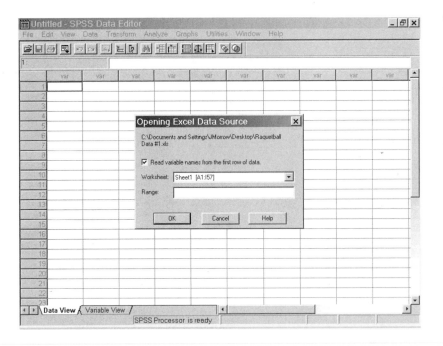

Figure 2.10 Opening an Excel file into SPSS.

Your data will automatically be placed into SPSS. Compare the results that you have just imported with those presented in figure 2.3*b,* where you entered the data directly into SPSS. Note that only the variable names and data have been imported into SPSS. You will need to go to the "variable view" tab and enter labels and values as you did originally with SPSS.

DOWNLOADING DATA MATRICES

As previously indicated, selected data tables from many of the chapters in this textbook are available at the textbook WWW site. We will now illustrate how you can download the data for use in class assignments, practice, and learning. Let's begin by having you log on to your textbook's WWW site for chapter 3. Once you reach the location where you find the data matrixes, you will notice that there are two columns with essentially the same names. The differences are in the file extensions (the left-hand column contains SPSS data files and the right-hand column contains Microsoft Excel data files). To download a file, simply click on the file name. Depending on the settings on your computer, either the file will be automatically opened for you in the specific file format (i.e., SPSS or Excel) or you will be able to save it to your disk.

▶ **SPSS download.** When downloading an SPSS file, you will get all of the data in the table as well as everything in the **Variable View** window that defines and describes the variables. If the computer that you are using does not have SPSS on it, you will not be able to view the SPSS data matrix. Do not worry. Simply take your data disk to a computer that does have SPSS on it and then double click on the SPSS data file to open the file in SPSS.

▶ **Microsoft Excel download.** When you are downloading Excel files, the process is the same as that with SPSS. However, recall that the Excel file has only the variable names and data and nothing that represents the codebook found in SPSS's **Variable View.**

Measurement and Evaluation Challenge

As you have learned, the computer can do much to assist Mike and Jessica with their tasks. Mike can record test grades, quiz scores, and attendance in a gradebook program or simply use a package such as SPSS to create a database of student performance. At the end of each grading period, he can easily have the program sum the scores, determine percentages, and assign grades. Jessica could purchase a statistical package that permits data storage and analysis of her volleyball team. Additionally, she might consider purchasing the FITNESSGRAM software package to record and report fitness test performance and improvement for her physical education classes. Regardless of how Mike and Jessica decide to use the computer, it can save them much time and effort in the completion of their daily activities. Last, both could use the World Wide Web to search for topics related to their positions (e.g., biology instruction, volleyball resources, health and fitness sites) and to search for literature for their graduate theses. They will be able to use SPSS or a similar statistical package to analyze their thesis data.

SUMMARY

Faster and more capable computers are continuing to change every aspect of our lives. Tasks that previously took hours now take only seconds. Whether used for research, testing, evaluating, or teaching, computers—in conjunction with statistics software—can greatly help the measurement and evaluation user to develop data on which to make decisions. Specialized software packages are available for developing written tests and for assessing adult and youth fitness tests.

Microcomputing skills, although perhaps difficult to learn at first, are some of the most valuable skills any professional can have. Development of the World Wide Web has had implications for the gathering and transmission of knowledge that affects every educator, allied health professional, and fitness instructor.

An excellent resource for you to learn more about the statistical methods that you will be studying in chapters 3, 4, and 5 is found on the WWW at the Rice Virtual Lab in Statistics (www.ruf.rice.edu/~lane/rvls.html).

3

Descriptive Statistics and the Normal Distribution

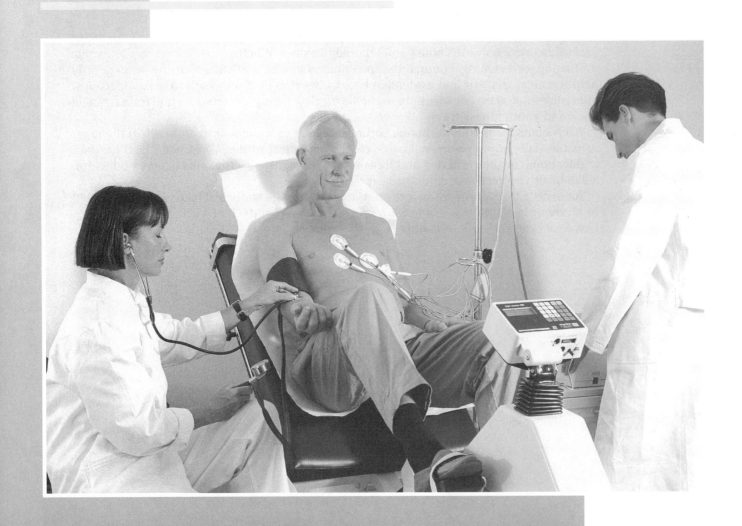

Outline

Measurement and Evaluation Challenge

James, a college student, recently had a complete health and fitness evaluation conducted at The Cooper Clinic in Dallas, Texas. Part of the evaluation required him to run to exhaustion on a treadmill. James ran for 24 min and 15 s. Using his treadmill time, the technician estimated James' $\dot{V}O_2$max to be 50 ml · kg^{-1} · min^{-1}. How does James interpret this value? Is this result high, average, or low? How does it compare with the results of others his age and gender? The concepts in this chapter will help James better interpret statistical results from any type of test he might complete.

Objectives

After studying this chapter, you will be able to

▶ illustrate types of data and associated measurement scales,

▶ calculate descriptive statistics on data,

▶ graph and depict data, and

▶ use SPSS computer software in data analysis.

Teachers and researchers often work with large amounts of data. Data can consist of ordinary alphabetical characters (such as student names), but data are typically numerical. In this chapter we cover the basics of data analysis to help you develop the skills necessary for measurement and evaluation. Understanding fundamental statistical analysis is required to accomplish this goal. If you can add, subtract, multiply, divide, and (with a calculator) take a square root, you have the mathematical skills necessary to complete much work in measurement theory. In fact, with the SPSS program presented in chapter 2, the computer does most of the work for you. However, you must understand the concepts of statistical analysis, when to use them, and how to interpret the results.

Descriptive statistics provide you with mathematical summaries of performance (e.g., the best score) and performance characteristics (e.g., central tendency, variability). They can also describe characteristics of the distributions, such as symmetry or amplitude.

SCALES OF MEASUREMENT

Taking a measurement often results in assigning a number to represent the measurement, such as the weight, height, distance, or time. However, not all numbers are the "same." Some types of numbers can be added and subtracted and the result has meaning. With other types of numbers, the result has little meaning. One method of classifying numbers is using scales of measurement, as presented here:

▶ **Nominal**—Naming or classifying, such as a football position (quarterback or tight end), gender (male or female), or type of car (sports car, truck, SUV). A nominal scale is categorical in nature, simply identifying differences among things on some characteristic. There is no notion of order, magnitude, or size.

▶ **Ordinal**—A ranking, such as the finishing place in a race. "Things" are ranked in order, but the differences between ranked positions are not comparable (e.g., the difference between rank number 1 and 2 may be quite small, but the difference between rank number 4 and 5 may be very large).

▶ **Continuous**—Numbers are said to be continuous in nature if they can be added, subtracted, multiplied, or divided, and the results have meaning. Continuous numbers can be either interval or ratio in form.

  ▶ Interval—Using an equal or common unit of measurement, such as temperature (°F or °C) or IQ. The zero point is arbitrarily chosen: That is, a value of zero simply represents a point on a number line. It does not mean that something doesn't exist. For example, in the centigrade temperature scale, 0° C does not indicate the absence of heat but rather the temperature at which water freezes. It is possible to have lower temperatures, referred to as "below zero."

  ▶ Ratio—Same as interval, except having an absolute (true) zero, such as weight or shot-put distance. If one person is 6 ft (1.82 m) tall and another is 3 ft (0.91 m) tall, the first person is twice as tall as the second.

An important concept to remember is that certain characteristics must exist before mathematical operations can be conducted. *Numbers can be perceived within a scale from nominal to ordinal to interval to ratio.* The scales of measurement are hierarchical in that each builds on the previous level or levels. That is, if a number is ordinal, it is also nominal; if the number is intervally scaled, it also conveys ordinal and nominal information; and if the number is a ratio, it conveys all three lower levels of information—nominal, ordinal, interval. *Only interval and ratio numbers can be subjected to mathematical operation (e.g., added, divided).* People sometimes use lower scales of measurement—ordinal and nominal—as if they were the higher-scale interval or ratio. For example, it is inappropriate to calculate an average from ordinal data. The average will always be the middle rank. This is another reason it is important to distinguish the level of measurement of the data before applying statistical tests.

Mastery Item 3.1

| Diving and gymnastic scores are on what scale of measurement?

SUMMATION NOTATION

To represent what they want to accomplish mathematically, mathematicians have developed a shorthand system called summation notation. Although summation notation can become quite complex, for present purposes you need learn only a few concepts. Three points are important for you to remember: n is the number of subjects, X is any observed variable that you might measure (e.g., height, weight, distance), and Σ (the capital Greek letter sigma) means the sum of. In summation notation $\Sigma X = X_1 + X_2 + \ldots Xn$, where n represents the nth (or last) observation. This reads, "The sum of all X values is equal to X_1, plus X_2, plus $\ldots X_n$." Note that when we work with the computer, the variables may appear in a form different from those in the text.

The only other major rules to remember concern parentheses and exponents. Recall that you do all operations within parentheses before moving outside them. If there are no parentheses, the precedence rules of mathematical operations hold: First, conduct any exponentiation; follow this with multiplication and division and then addition and subtraction. For example, ΣX^2 is read as "the sum of the squared X scores," whereas $(\Sigma X)^2$ is "the sum of X, the quantity squared." The distinction is important because the two terms represent different values.

Mastery Item 3.2

| Here is the number of points that John, Mary, Mike, and Karen, respectively, got correct on their last quiz: 4, 3, 2, and 5. Determine ΣX, $(\Sigma X)^2$, and ΣX^2.

Mastery Item 3.3

| Use the following scores to calculate ΣX, $(\Sigma X)^2$, and ΣX^2: 4, 5, 9, 3, 5, 15, 10, 8, 7, 8.

REPORTING DATA

After you measure your students or subjects on a variable, you may want to know how they performed. Don't you usually want to know how well you did on a test? If your teacher told you only your score, you would know little about how well you performed. You need some additional information. Often people want to compare themselves with others who have completed a similar test. Norm-referenced measurement allows just this. It tells you how well you performed relative to everyone else who completed the test. It is important to compare your performance with others in your group.

Mastery Item 3.4

| Provide some examples of other ways that you might classify yourself. That is, identify some nominal variables you might use to describe yourself.

One method for deciding how your performance compares with others is to develop a frequency distribution of the test results. A frequency distribution is a method of organizing data that involves noting the frequency with which various scores occur. The results from a written test for 65 students are presented in table 3.1. Look at them and assume you scored 46 on the test. How well did you do? This is difficult to determine when numbers are presented as they are in the table. But a frequency distribution can clarify how your score compares to others.

Table 3.1   65 Test Scores From a Written Examination

48	45	50	49	46	47	47	49	50	50
45	51	51	48	49	46	44	44	52	53
48	43	48	41	48	49	47	49	51	54
51	43	53	45	48	47	51	46	49	50
48	48	45	46	49	48	46	48	52	54
52	50	51	47	45	47	43	47	49	50
44	55	48	50	53					

Mastery Item 3.5

Use SPSS to obtain a frequency distribution and percentiles for the 65 scores shown in table 3.1. Confirm your analysis with the results presented in figure 3.1.
SPSS commands for obtaining a frequency distribution and percentiles follow:

1. Start SPSS.
2. Open table 3.1 data.
3. Click on the **Analyze** menu.
4. Scroll to **Descriptive Statistics** and across to **Frequencies** and click.
5. Highlight "score" and place it in the **Variables** box by clicking the arrow key.
6. Click **OK**.

Frequencies

Statistics
Test scores from a written examination

N	Valid	65
	Missing	0

Test scores from a written examination

		Frequency	Percent	Valid percent	Cumulative percent
Valid	41	1	1.5	1.5	1.5
	43	3	4.6	4.6	6.2
	44	3	4.6	4.6	10.8
	45	5	7.7	7.7	18.5
	46	5	7.7	7.7	26.2
	47	7	10.8	10.8	36.9
	48	11	16.9	16.9	53.8
	49	8	12.3	12.3	66.2
	50	7	10.8	10.8	76.9
	51	6	9.2	9.2	86.2
	52	3	4.6	4.6	90.8
	53	3	4.6	4.6	95.4
	54	2	3.1	3.1	98.5
	55	1	1.5	1.5	100.0
	Total	65	100.0	100.0	

Figure 3.1   Test scores from a written examination.

You may be asking yourself, Why do this? Recall that we were interested in determining how well you performed relative to the rest of the class. Your score of 46 appears in the lower half of the distribution. The fifth column, cumulative percentage, of the frequency distribution is a **percentile.** The percentile is obtained by summing the percentage ("percent," third column) of scores that fall at and below the percentile you are calculating. *A percentile represents the percent of observations at or below a given score.* This concept is extremely important because test results, such as standardized college admissions tests, are often reported in percentiles. If you achieve at the 90th percentile (P_{90}), this simply means that you have scored better than 90% of the people who took the test. Conversely, if you scored at the 10th percentile (P_{10}), 90% of the people taking the test scored better than you did. Your score of 46 is at the 26.2nd percentile. That is, 26.2% of the people scored lower than you did, and thus 73.8% scored higher.

The presentation in figure 3.1 assumes that a high score is better than a lower score. This is not always the case. If low scores are better—for example, in golf—you need to reflect this in your interpretation of the scores. For example, if the data presented in table 3.1 were golf scores for nine holes, your "lower" score of 46 represents the 73.8th percentile (100 − 26.2 = 73.8).

CENTRAL TENDENCY

Now you know where you fit into the distribution. Let's further consider how you can interpret your score. One way is to determine how you compare with the "typical" person who took the test. Basically, you are looking at where the scores tend to center—their **central tendency.** We will briefly describe three measures of central tendency:

▷ **Mean**—The arithmetic average; the sum of the scores divided by the number of scores. From summation notation, this definition may be represented as

$$M = \frac{\sum X}{n} \tag{3.1}$$

where M is the mean, X is the value of each observation, and n is the number of observations. Every score in the data is used to determine the mean—it is the most stable measure of central tendency. Using the four scores (4, 3, 2, 5), the mean is (4 + 3 + 2 + 5)/4 = 3.5.

▷ **Median**—The middle score; the 50th percentile. To obtain the median, order the scores from high to low and find the middle one. The median value for the data presented in table 3.1 is 48, which is the middle score. Note that the median is a specific percentile: P_{50}.

▷ **Mode**—The most frequently observed score. The mode is the most unstable measure of central tendency but the most easily estimated one. You should confirm from figure 3.1 that the mode is 48.

DISTRIBUTION SHAPES

Simply knowing a distribution's measures of central tendency does not tell you everything about the scores. Not all distributions have the same shape. Figure 3.2 illustrates several shapes that distributions may assume. The statistical term for the shape (or symmetry) of a distribution is **skewness.** Skewness values typically range from +1 to −1. A positively skewed distribution has a "tail" toward the positive end of the number line, and a negatively skewed distribution has a tail toward the negative end of the number line. Distributions with little skewness are characterized by values close to 0.

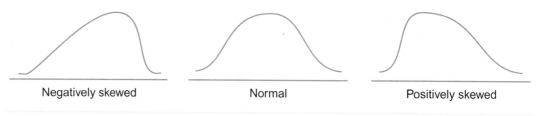

Figure 3.2   Distribution shapes.

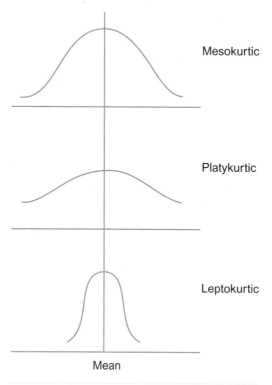

Figure 3.3   Three symmetrical curves.

There is another property of shape that is associated with a distribution. In figure 3.3, all three distributions are symmetrical curves with identical means, medians, and modes. However, there is an obvious difference in the amplitudes (or peakedness) of the distributions. The peakedness of a curve is referred to as its **kurtosis.** The middle curve (normal) is said to be mesokurtic (i.e., an average amount). The flatter curve is said to be platykurtic (flat) and the steep curve leptokurtic (peaked). A normal curve is a distribution with no skewness that is mesokurtic.

A good way to determine the shape of the distribution you are working with is to develop a histogram. A **histogram** is a graph that consists of columns to represent the frequencies with which various scores are observed in the data. It lists the score values on the horizontal axis and the frequency on the vertical axis.

Mastery Item 3.6 _____

What are some measures that are positively or negatively skewed?

Mastery Item 3.7 _____

Use the data in table 3.1 to create a histogram of the 65 scores (see figure 3.4).

SPSS commands for obtaining a histogram follow:

1. Start SPSS.
2. Open table 3.1 data.
3. Click on the **Analyze** menu.
4. Scroll to **Descriptive Statistics** and across to **Frequencies** and click.
5. Highlight "score" and place it in the **Variables** box by clicking the arrow key.
6. Click the **Charts** button.
7. Click the **Histogram** button.
8. Click **Continue**.
9. Click **OK**.

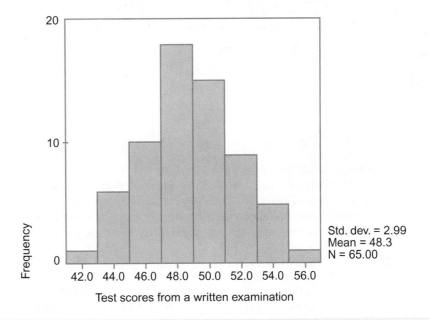

Figure 3.4   Histogram output for 65 scores.

Mastery Item 3.8

Use the data in table 3.1 to calculate the mean, median, and mode of the 65 scores (see figure 3.5).

SPSS commands for obtaining measures of central tendency follow:

1. Start SPSS.
2. Open table 3.1 data.
3. Click on the **Analyze** menu.
4. Scroll to **Descriptive Statistics** and across to **Frequencies** and click.
5. Highlight "score" and place it in the **Variables** box by clicking the arrow key.
6. Click on the **Statistics** button.
7. Click on "mean," "median," and "mode."
8. Click **Continue.**
9. Click **OK.**

Frequencies

Statistics

Test scores from a written examination

N	Valid	65
	Missing	0
Mean		48.31
Median		48.00
Mode		48

Figure 3.5   Mean, median, and mode for 65 scores.

Mastery Item 3.9

Obtain the height and age of 10 of your classmates. Use SPSS to calculate the mean, median, and mode for the 10 students. Delete the lowest score for each variable and recalculate the mean, median, and mode. Which measure of central tendency changes the most?

VARIABILITY

A curve's kurtosis leads directly into the next important descriptive measure of a set of scores. The platykurtic curve in figure 3.3 is said to contain more heterogeneous (dissimilar) data, whereas the leptokurtic curve is said to have observations that are more homogeneous (similar). The spread of a distribution of scores is reflected in various measures of variability. We present three measures of **variability** here: range, variance, and standard deviation.

Range

The **range** is the high score minus the low score. It is the least stable measure of variability because it depends on only two scores and does not show how the remaining scores are distributed.

Variance

The **variance** (s^2) is a measure of the spread of a set of scores based on the squared deviation of each score from the mean. It is used much more frequently than the range in reporting the heterogeneity of scores. The variance is the most stable measure of variability. Two sets of scores that have vastly different spreads will have vastly different variances. *Many types of variance (such as observed variance, true variance, error variance, sample variance, between-subject variance, within-subject variance, and systematic variance) will become important as measurement and evaluation issues are presented throughout this text.*

As an illustration of how to calculate the variance, consider the following scores: 1, 2, 3, 4, 5 (table 3.2).

The steps to calculate the variance by hand are as follows:

1. Calculate the mean.
2. Subtract the mean from each score.
3. Take each difference (deviation) and square it.
4. Add the results together and divide by the number of scores minus 1.

Table 3.2

X (observed score)	-	M (mean)	=	x (deviation score)	x^2 (squared deviation score)
1	-	3	=	-2	4
2	-	3	=	-1	1
3	-	3	=	0	0
4	-	3	=	1	1
5	-	3	=	2	4
Total				0	10

Equation 3.2 (the didactic formula) is used to illustrate the variance.

$$s^2 = \frac{\Sigma(X - M)^2}{n - 1} = \frac{\Sigma x^2}{n - 1} = \frac{10}{4} = 2.5 \tag{3.2}$$

That is, the variance is the average of the squared deviations from the mean (hence, the term *mean square* is sometimes used for a variance). Note that you divide by $n - 1$ and not n, so it isn't exactly the mean. You should use the following calculation formula to obtain the variance from a set of scores because it is generally easier to use:

$$s^2 = \frac{\Sigma X^2 - \frac{(\Sigma X)^2}{n}}{n - 1} \tag{3.3}$$

You should confirm that using the calculation formula will result in the same value for the variance as the didactic formula. (Hint: $\Sigma X^2 = 55$; $\Sigma X = 15$.) *Note that when all of the scores are identical, the variance is zero. Typically, this is not something you desire in measurement.* Variation among scores is preferable. The reasons for this will be illustrated throughout this book.

Variance in general can be illustrated with a square, as in figure 3.6. The square (total variance) is divided into two types of variance: *true variance* and *error variance.* You might think of the scores on a recent test as having the three types of variance. Not everyone scored the same, so there is **observed score** variance. Not everyone has the same "true" knowledge about the content of the test, so there is **true score** variance. Last, there is some "error" in the test, but not everyone has the same amount of error reflected in their scores, so there is **error score** variance (error could result from several sources, such as the student guessing or the teacher misscoring the test). These important concepts will be further discussed in chapter 6, when you study reliability theory.

The types of variance will become increasingly important throughout the remainder of this book, but it is important that you understand variance at this point. Although it is a calculated *number,* we will often speak of variance in conceptual terms (i.e., variability). It may be helpful for you to remember the square in figure 3.6 and the fact that not all scores are identical.

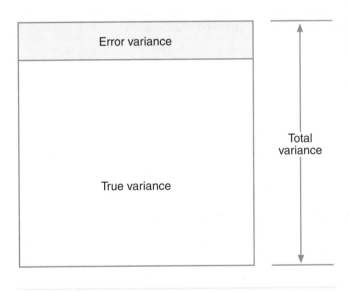

Figure 3.6 Three types of variance.

Standard Deviation

Whereas the variance is important, a related number—the **standard deviation**—is often used in descriptive statistics to illustrate the variability of a set of scores. *The standard deviation is the square root of the variance.* It is helpful to think of the standard deviation as a linear measure of variability. Consider the square (figure 3.6) used to illustrate the variance. When you take the square root of a square, you get a linear measure. The same concept holds true for the standard deviation. The standard deviation is important because it is used as a measure of linear variability for a set of scores. Knowing the standard deviation of a set of scores can tell us a great deal about the heterogeneity or homogeneity of the scores.

You are encouraged to use the calculation formula (equation 3.3) to calculate the standard deviation. Simply calculate s^2 and take its square root.

Why is the standard deviation so important, and what exactly does it mean? For example, in a **normal distribution,** a bell-shaped, symmetric probability distribution (figure 3.7), knowing the standard deviation tells you a great deal about the distribution. In any normal distribution if you add to and subtract from the mean the value of 1 standard deviation, you will obtain a range that encompasses approximately 68% of the observations. If you add and subtract the value of 2 standard deviations, you will obtain a range that encompasses approximately 95% of the observations, and adding and subtracting the value of 3 standard deviations will give a range encompassing 99.74% of the observations. This is true regardless of what the mean and standard deviation are, as long as the distribution is normal. To summarize:

$$M \pm 1s \rightarrow 68.26\% \text{ of observations}$$

$$M \pm 2s \rightarrow 95.44\% \text{ of observations}$$

$$M \pm 3s \rightarrow 99.74\% \text{ of observations}$$

Using the information obtained about the mean and standard deviation for a set of observations, you can approximate the percentile for any observation. Looking again at figure 3.7, assume that it illustrates the observations on a recent maximal oxygen uptake ($\dot{V}O_2$max) test that had a mean of 60 ml · kg⁻¹ · min⁻¹ and a standard deviation of 5 ml · kg⁻¹ · min⁻¹ and that the values are normally distributed. Use the figure to approximate percentiles for $\dot{V}O_2$max values of 50, 55, 60, 65, and 70. How likely would it be to obtain a $\dot{V}O_2$max in excess of 70 ml · kg⁻¹ · min⁻¹ on similar subjects?

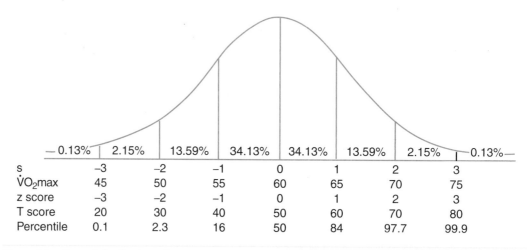

	−3	−2	−1	0	1	2	3
	0.13%	2.15%	13.59%	34.13%	34.13%	13.59%	2.15%	0.13%
s	−3	−2	−1	0	1	2	3
$\dot{V}O_2$max	45	50	55	60	65	70	75
z score	−3	−2	−1	0	1	2	3
T score	20	30	40	50	60	70	80
Percentile	0.1	2.3	16	50	84	97.7	99.9

Figure 3.7 Normal distribution of scores.

Mastery Item 3.10

Use the data in table 3.1 to calculate descriptive statistics for the 65 scores (see figure 3.8).

1. Start SPSS.
2. Open table 3.1 data.
3. Click on the **Analyze** menu.
4. Scroll to **Descriptive Statistics** and across to **Descriptives** and click.
5. Highlight "score" and place it in the **Variables** box by clicking the arrow key.
6. Click on the **Options** button.

7. Click on all of the descriptive statistics that you have learned so far.
8. Click **Continue.**
9. Click **OK.**

Descriptives

Descriptive statistics

	N	Range	Minimum	Maximum	Sum	Mean
Test scores from a written examination	65	14	41	55	3140	48.31
Valid N (listwise)	65					

Descriptive statistics

	Std. deviation	Variance
Test scores from a written examination	2.99	8.935
Valid N (listwise)		

Figure 3.8 Descriptives output for 65 students.

STANDARD SCORES

Knowing the mean and standard deviation makes it easy to calculate a **standard score.** *A standard score is a set of observations that have been standardized around a given M and standard deviation.* A commonly used standard score is the z score. It is calculated as follows:

$$z = \frac{X - M}{s} \tag{3.4}$$

In other words, you obtain the z score for any observation by subtracting the mean, *M,* from the observed score, *X,* and dividing this difference by the standard deviation, *s.* If you do this for all observations in a set of data, you get a set of scores that themselves have a mean of 0 and a standard deviation of 1 (i.e., they have been standardized). Note that we have included z scores in figure 3.7 and that z scores always have a mean of 0 and a standard deviation of 1.

Another commonly used standard score is the *T* score, which is calculated as follows:

$$T = 50 + \frac{10(X - M)}{s} \quad \text{or}$$
$$T = 50 + 10z \tag{3.5}$$

The mean of the *T* scores for all the observations in a set of data is always 50, and the standard deviation is always 10. Note that in a normal distribution, 99.74% of the scores will fall between a z score of −3 and +3 and between *T* scores of 20 and 80. Can you see why this is the case? (Hint: See figure 3.7.)

You are probably asking yourself, What is the point of standard scores? To answer this, let's assume you are a physical education teacher teaching a basketball unit in which you are going to grade the students on only two skill items (we're not recommending this!). Assume

Using a standard form of scoring, such as *z* scores or *T* scores, can help you appropriately weigh and compare scores measured in different units (i.e., number of shots made in 1 min and the length of time it takes to complete a dribbling obstacle course).

that you think shooting and dribbling are two skills important to basketball and that you can validly measure them. For shooting, you measure the number of baskets a student makes in 1 min; for dribbling, you measure the amount of time a student takes to dribble through an obstacle course. You assume dribbling and shooting are equally important in basketball (we probably don't agree with you!), so you wish to weight them equally. A quick look at the scores illustrates that you cannot simply add these two scores together for each student to determine who is the best basketball player. High scores are better on the shooting test, but low scores are better on the dribbling test. In addition, each score is measured in different units (i.e., number of baskets made and number of seconds in which dribbling was completed).

This is exactly where standard scores can help you. To weight these two skills equally on this test, you must first convert each person's scores to a standard form, such as *z* scores or *T* scores. Note that the dribble test is a timed test, so you must correct the obtained *z* score so that a faster time results in a higher *z* score. You do this by changing the sign of the *z* score (e.g., 2 becomes -2; -1.5 becomes 1.5). You can then add each student's two *z* scores together and obtain a total *z* score that is based on the fact that each test now has the same weight. You could not do this using the original raw scores, nor could you do it with any sets of scores having unequal variances. Generally, the set of scores with the larger variance would be weighted more heavily in the total if you simply added two raw scores together for each student. But if you convert them to a standard score, they have the same weight because they have the same standard deviation (remember that *z* scores have the same standard deviation—i.e., 1.0).

You can take this example one step further and decide to give the shooting test twice the weight of the dribbling test. All you need to do is multiply the *z* score for the shooting test by 2 and add it to the *z* score for the dribbling test. The key concept is that whichever test is most variable will carry the greater weight. In fact, if there is no variability on an examination (i.e., if everyone scores the same), the examination contributes absolutely nothing toward differentiating student performance. This concept is further illustrated for you in chapter 10 on grading.

NORMAL-CURVE AREAS (*z* TABLE)

If you examine equation 3.4, you'll see that converting a score to a *z* score actually expresses the distance a score is from its own mean in standard deviation units. That is, a *z* score indicates the number of standard deviations that a score lies below or above the mean. *In addition to weighting performance for grades, z scores can be used for a number of other purposes, mainly determination of (a) percentiles and (b) the percentage of observations that fall within a particular area under the normal distribution.* Consider again the normal distribution presented in figure 3.7. The area under the distribution is represented as a total of 100%. A student who scores at the mean has achieved the 50th percentile (i.e., has scored better than 50% of the group) and has a *z* score of 0 (and a *T* score of 50).

Observed scores alone may tell you very little about performance (and, in fact, tell you nothing about relative performance). However, if the teacher reports your score in *z* score form and you have a positive *z* score, you know immediately that you have achieved better than average; a negative *z* score indicates you scored below the mean. Using the *z* score and the table of normal-curve areas (table 3.3), you can determine the percentile associated with any *z* score for data that are normally distributed.

Table 3.3   Normal-Curve Areas

z	0.00	0.01	0.02	0.03	0.04	0.05	0.06	0.07	0.08	0.09
0.0	00.00	00.40	00.80	01.20	01.60	01.99	02.39	02.79	03.19	03.59
0.1	03.98	04.38	04.78	05.17	05.57	05.96	06.36	06.75	07.14	07.53
0.2	07.93	08.32	08.71	09.10	09.48	09.87	10.26	10.64	11.03	11.41
0.3	11.79	12.17	12.55	12.95	13.31	13.68	14.06	14.43	14.80	15.17
0.4	15.54	15.91	16.28	16.64	17.00	17.36	17.72	18.08	18.44	18.79
0.5	19.15	19.50	19.85	20.19	20.54	20.88	21.23	21.57	21.90	22.24
0.6	22.57	22.91	23.24	23.57	23.89	24.22	24.54	24.86	25.17	25.49
0.7	25.80	26.11	26.42	26.73	27.04	27.34	27.64	27.94	28.23	28.52
0.8	28.81	29.10	29.39	29.67	29.95	30.23	30.51	30.78	31.06	31.33
0.9	31.59	31.86	32.12	32.38	32.64	32.90	33.15	33.40	33.65	33.89
1.0	34.13	34.38	34.61	34.85	35.08	35.31	35.54	35.77	35.99	36.21
1.1	36.43	36.65	36.86	37.08	37.29	37.49	37.70	37.90	38.10	38.30
1.2	38.49	38.69	38.88	39.07	39.25	39.44	39.62	39.80	39.97	40.15
1.3	40.32	40.49	40.60	40.82	40.99	41.15	41.31	41.47	41.62	41.77
1.4	41.92	42.07	42.22	42.36	42.51	42.65	42.79	42.92	43.06	43.19
1.5	43.32	43.45	43.57	43.70	43.83	43.94	44.06	44.18	44.29	44.41
1.6	44.52	44.63	44.74	44.84	44.95	45.05	45.15	45.25	45.35	45.45
1.7	45.54	45.64	45.73	45.82	45.91	45.99	46.08	46.16	46.25	46.33
1.8	46.41	46.49	46.56	46.64	46.71	46.78	46.86	46.93	46.99	47.06
1.9	47.13	47.19	47.26	47.32	47.38	47.44	47.50	47.56	47.61	47.67
2.0	47.72	47.78	47.83	47.88	47.93	47.98	48.03	48.08	48.12	48.17
2.1	48.21	48.26	48.30	48.34	48.38	48.42	48.46	48.50	48.54	48.57
2.2	48.61	48.64	48.68	48.71	48.75	48.78	48.81	48.84	48.87	48.90
2.3	48.93	48.96	48.98	49.01	49.04	49.06	49.09	49.11	49.13	49.16
2.4	49.18	49.20	49.22	49.25	49.27	49.29	49.31	49.32	49.34	49.36
2.5	49.38	49.40	49.41	49.43	49.45	49.46	49.48	49.49	49.51	49.52
2.6	49.53	49.55	49.56	49.57	49.59	49.60	49.61	49.62	49.63	49.64
2.7	49.65	49.66	49.67	49.68	49.69	49.70	49.71	49.72	49.73	49.74
2.8	49.74	49.75	49.76	49.77	49.77	49.78	49.49	49.79	49.80	49.81
2.9	49.81	49.82	49.82	49.83	49.84	49.84	49.85	49.85	49.86	49.86
3.0	49.87									
3.5	49.98									
4.0	49.997									
5.0	49.99997									

Reprinted from Lindquist (1942).

The values down the left side of table 3.3 are *z* scores in whole numbers and tenths, and the numbers across the top represent *z* scores to the hundredth's place. (For example, for a *z* score of 1.53, you find 1.5 on the left and read across to the 0.03 column.) The numbers in the body of the table are percentages of observations that lie between the mean and any given standard deviation distance from the mean. Verify that 34.13% of the scores lie between the mean and 1 standard deviation above the mean (i.e., a *z* score of 1.00). Note that no negative *z* scores are shown in the table: They are not necessary because the normal distribution is symmetrical. Therefore, 34.13% of the observations also fall between the mean and 1 standard deviation below the mean (i.e., a *z* score of −1.00). See if you can determine the percentage of scores that fall between 1 and 1.5 standard deviations above the mean. Figure 3.9 provides an illustration of this area. Use table 3.3 and figure 3.9 to help you determine the answer (43.32 − 34.13 = 9.19%).

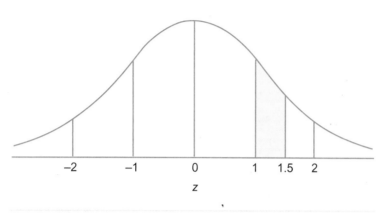

Figure 3.9   Area of normal curve between 1 and 1.5 standard deviations above the mean.

In summary, the important points to remember when using the table of normal-curve areas are that

▸ the reference point is the mean,

▸ *z* scores are presented to the nearest one-hundredth, and

▸ the numbers in the body of the table are percentages.

Use table 3.3 to confirm that a *z* score of 1.23 is (approximately) the 89th percentile. Remember that the reference point is the mean and that a percentile represents all people who score at or below a given point. (Hint: This includes those scoring below the mean if your observed score is above the mean or the percentile is greater than 50 or the *z* score is greater than 0.00.)

Use table 3.3 to confirm that a *z* score of -1.23 represents (approximately) the 11th percentile. You can do this using the following method:

1. Find 1.2 in the left column of the table and 0.03 at the top. They intersect at the percent value of 39.07.

2. Recall what this number means: 39.07% of the scores lie between the mean and a *z* score of either +1.23 or −1.23.

3. Recall that 50% of the scores fall below the mean.

4. Thus, 50 − 39.07 gives the "leftover" area below a *z* score of −1.23; the answer is 10.93 (or approximately the 11th percentile).

Use table 3.3 to confirm that 14.98% of a set of normally distributed observations fall between a *z* score of 0.50 and 1.00. (Draw a picture to help you see this.)

Mastery Item 3.11

Use table 3.3 to determine the following:

▸ Whether a *z* score of 1.95 represents the 97.44th percentile

▸ The *z* score for the 45th percentile

▸ The percentage of scores that fall between *z* scores of 1.30 and 2.00

Mastery Item 3.12

Use table 3.3 to confirm that if the cutoff to achieve an A in this course is a z score of 1.35, 8.85% of the people would be expected to earn a grade of A.

If a score is reported in T score form, simply transform it to z score form and then estimate the percentile. To change from a T score to a z score, simply substitute the T score mean (50) and standard deviation (10) into the z score formula. If your T score is 30, your z score would be $(30 - 50)/10 = -2.00$.

Mastery Item 3.13

Verify that the percentile associated with a T score of 68 is 96.41.

Mastery Item 3.14

Verify that the percentile associated with a T score of 47.5 is 40.13.

Although you can use table 3.3 to convert from z scores to percentiles, you can also use the table to convert from percentiles to z scores for normally distributed data. This might be helpful if you wanted to create a summation of z scores to determine the best overall performance, as in our earlier basketball example. Assume you were told that you performed at the 69th percentile on a test.

You can determine your z score on the test as follows:

1. Calculate 69% − 50% = 19% to get the area between your percentile and the mean.
2. Find 19 in the body of the table—the closest value is 19.15.
3. Find what z score that 19.15 corresponds to by going over to the left margin and up to the top row. The z score is 0.50. (It is positive, because you were above the mean.)

If you wanted to confirm that this is a T score of 55, you would simply substitute 0.50 into the T score formula (3.5).

Mastery Item 3.15

Use SPSS to compare the data in table 3.1 to the normal distribution (see figure 3.10).

1. Start SPSS.
2. Open table 3.1 data.
3. Click on the **Analyze** menu.
4. Scroll to **Descriptive Statistics** and across to **Frequencies** and click.
5. Highlight "score" and place it in the **Variables** box by clicking the arrow key.
6. Click on the **Charts** button.
7. Click on **Histograms**.
8. Click on **With normal curve.**
9. Click **Continue.**
10. Click **OK.**

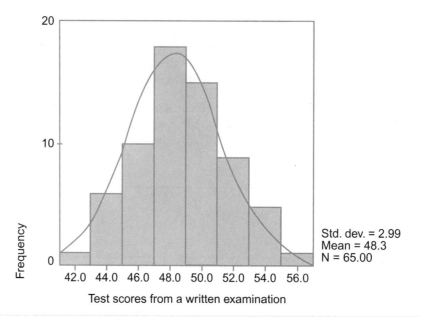

Figure 3.10   Histogram output for mastery item 3.15.

Mastery Item 3.16

Use SPSS to calculate *z* scores for the data in table 3.1.

1. Start SPSS.
2. Open table 3.1 data.
3. Click on the **Analyze** menu.
4. Scroll to **Descriptive Statistics** and across to **Descriptives** and click.
5. Highlight "score" and place it in the **Variables** box by clicking the arrow key.
6. Click on the **Save standardized values as variables** box.
7. Click **OK.**

The scores are converted to *z* scores, and a new column of *z* scores is added to your data file. Be certain to save the modified data file because you will need the *z* scores in the next mastery item.

Mastery Item 3.17

Use SPSS to calculate *T* scores.

1. Start SPSS.
2. Open table 3.1 data.
3. Click on the **Transform** menu.
4. Click on **Compute**.
5. In the **Target Variable** box, type "tscore."
6. In the **Numeric Expression** box, type "50 + 10*zscore."
7. Click **OK.**
8. Now the *T* scores have been added to your data file. Save the data file.

Use the data that you saved from mastery items 3.16 and 3.17 and SPSS to verify that the mean and standard deviation for the z scores and T scores are 0 and 1 and 50 and 10, respectively.

Measurement and Evaluation Challenge

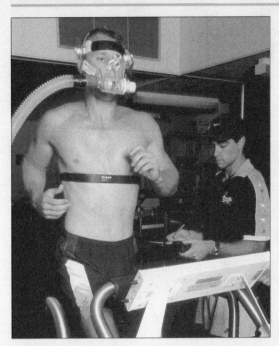

What has James learned in this chapter that will help him interpret the results from his treadmill test? James can compare his $\dot{V}O_2$max to that of the average individual who is his age and gender. He can determine his percentile based on his score, the mean score, and the variability in $\dot{V}O_2$max. James has learned how to develop and interpret scores based on score distributions and norms. Assuming the average $\dot{V}O_2$max for a male of his age is 45 ml · kg^{-1} · min^{-1} with a standard deviation of 5 ml · kg^{-1} · min^{-1}, James can easily determine that his performance is at the 84th percentile. Only 16% of people his age and gender have $\dot{V}O_2$max values that exceed 50 ml · kg^{-1} · min^{-1}.

SUMMARY

The descriptive statistics presented in this chapter are foundational to the remainder of your text. Of particular importance is your ability to understand and use the concepts of central tendency and variability and to use and interpret the normal curve and areas under sections of the normal distribution. If you are more interested in statistical methods, see Glass and Hopkins (1996). Thomas and Nelson (2001) provide excellent examples of research and statistical applications in human performance.

At this point you should be able to accomplish the following tasks:

1. Differentiate between the four levels of measurement and provide examples of each.
2. Calculate and interpret descriptive statistics.
3. Calculate and interpret standard scores.
4. Use a table of normal-curve areas (z table) to estimate percentiles.
5. Use SPSS to enter data and generate and interpret
    a. frequency distributions and the percentiles associated with observed scores,
    b. histograms for observed scores,
    c. descriptive statistics (mean, standard deviation) on variables, and
    d. z and T scores.

Correlation and Prediction

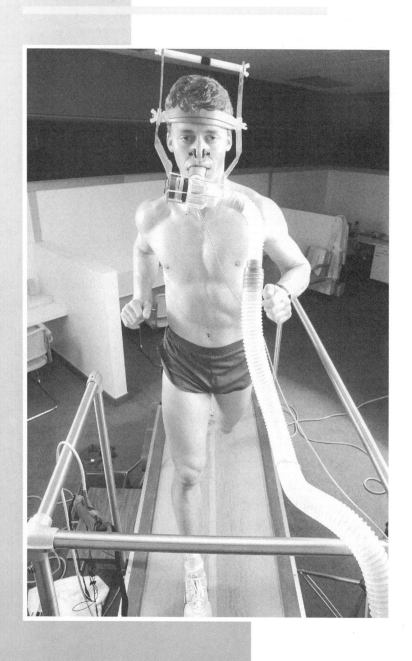

Outline

Measurement and Evaluation Challenge

Now that you know how to calculate descriptive statistics, you are ready to learn other statistical procedures that are important. In chapter 3, a technician reported James' $\dot{V}O_2max$ as 50 ml · kg^{-1} · min^{-1}. However, James recently read that $\dot{V}O_2max$ should actually be measured by collecting expired air and analyzing it for oxygen and carbon dioxide content. Yet James did not have any of this conducted during his treadmill run. He recalls the technician saying, "James, your treadmill time is associated with a $\dot{V}O_2max$ of 50 ml · kg^{-1} · min^{-1}." James now realizes that his treadmill time was related to his actual $\dot{V}O_2max$. Correlational procedures are used to determine the relationship between and among variables. Indeed, predictions and estimations of one variable from one or more other variables are common in sport and exercise science. For example, in James' case, his treadmill time was used to estimate his $\dot{V}O_2max$ without having to actually collect expired oxygen and carbon dioxide.

Objectives

After studying this chapter, you will be able to

▶ calculate statistics to determine the relationships between variables,

▶ calculate and interpret the Pearson product-moment correlation coefficient,

▶ calculate and interpret the standard error of estimate (SEE),

▶ use scatterplots to interpret the relationship between variables,

▶ differentiate between simple correlation and multiple correlation, and

▶ use SPSS computer software in data analysis for correlational and regression analysis.

H ow are two tests related? As performance on one test increases, does performance on the other increase? The relationship between variables is examined statistically by **correlation coefficients. Correlations** help you describe relations and in some cases predict outcomes; they are very useful in measurement theory.

CORRELATION COEFFICIENT

In chapter 3 you learned to describe data using measures of central tendency and variability, and we discussed one variable or test at a time. However, teachers, clinicians, and researchers often measure more than one variable. They are then interested in describing or measuring the **relationship**—the statistical association between these variables. This is a fundamental statistical task that data analysts, kinesiologists, teachers, clinicians, and researchers should be able to perform. An example would be, *What is the relationship between the bench press and the leg press strength tests?* That is, do these strength measures have anything in common? To measure this relationship, we would calculate the correlation coefficient, specifically the **Pearson product-moment correlation coefficient (PPM),** symbolized by *r*.

The correlation coefficient is an index of the **linear relationship** (an association that can best be depicted by a straight line) between two tests. It indicates the magnitude, or amount of relationship, and the direction of the relationship. Figure 4.1 illustrates these aspects of the correlation coefficient. *As you can see, the correlation coefficient can have either a positive or negative direction and a magnitude of –1.00 ≤ r ≤ 1.00.*

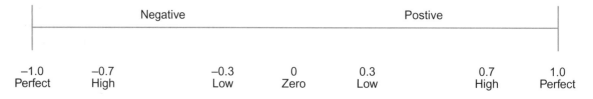

	Negative				Postive		
−1.0 Perfect	−0.7 High	−0.3 Low	0 Zero	0.3 Low	0.7 High	1.0 Perfect

Figure 4.1 Attributes of *r*.

The terms *high* and *low* are subjective and are affected by how the correlation was obtained, by the subjects tested, by the variability in the data, and by how the correlation will be used. There is nothing bad about a negative *r*. The sign of *r* simply indicates how the two variables covary (i.e., go together). A positive *r* indicates that people scoring above the mean on one variable, *X,* will usually be above the mean on the second variable, *Y.* A negative *r* indicates that those scoring above the mean on *X* will generally be below the mean on *Y.* Thus, a correlation of –.5 is not less than one of +.5. In fact, they are equal in strength but opposite in direction.

Mastery Item 4.1

If you were calculating a correlation coefficient and computed it to be +1.5, what must have happened?

Let's examine the factors of direction and magnitude of the correlation coefficient. Table 4.1 provides the scores for 10 students on three measures: body weight, chin-ups, and pull-ups. As you can see by looking at the data, low values for chin-ups are generally paired with low values for pull-ups. If you were to calculate the correlation between these values (described later), you would find *r* to be indicating a **direct relationship** (positive relationship). In contrast, if you examine body weight versus pull-ups, you can see that high body weights are paired with low pull-up scores. These measures have an **indirect relationship** (or negative or inverse relationship). Figures 4.2 and 4.3 provide illustrations in the form of scatterplots of these relationships. A **scatterplot** is a graphic representation of the correlation between two variables. (You can create a scatterplot by labeling the axes with names and units of measurement. Then plot each pair of scores for each subject.)

Table 4.1　Sample Correlation Data

Subject	Body weight	Chin-ups	Pull-ups
1	130	10	8
2	130	9	7
3	140	15	12
4	150	9	10
5	150	7	6
6	160	5	3
7	160	3	4
8	160	8	7
9	170	4	5
10	170	6	3

Note: During a pull-up, palms face away from you; during a chin-up, palms face toward you.

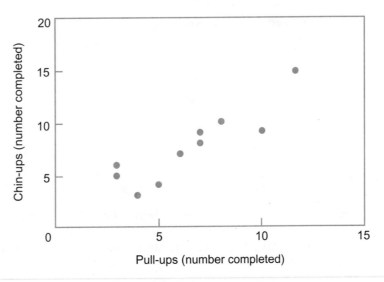

Figure 4.2　Scatterplot of correlation between pull-ups and chin-ups.

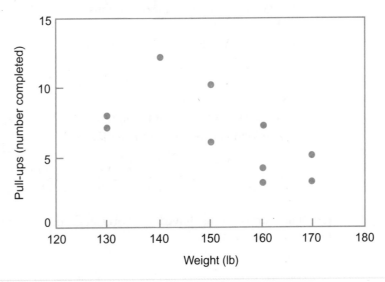

Figure 4.3　Scatterplot of correlation between body weight and pull-ups.

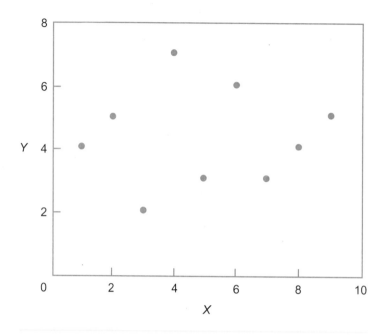

Figure 4.4 Scatterplot of zero correlation.

Keep in mind that the correlation coefficient is an index of linear relationship. All paired points must be on a straight line for a correlation to be perfect, –1 or +1. If two tests have a correlation coefficient of 0 (or **zero correlation)**, their scatterplot would demonstrate nothing even resembling a straight line. Figure 4.4 illustrates a sample scatterplot for a zero correlation.

Mastery Item 4.2

Draw a scatterplot of the relationship between body weight and chin-ups presented in table 4.1. Describe the correlation between these two measures.

CALCULATING *r*

Having developed a fundamental understanding of the correlation coefficient and the ability to illustrate correlations with simple scatterplots, let us turn to the calculation of *r*. Using the data in table 4.2, follow the steps here to calculate the correlation coefficient between chin-ups and pull-ups.

Steps in Calculating r

1. Arrange your data in paired columns (here, *X* is number of chin-ups, and *Y* is the number of pull-ups).

2. Square each *X* value and each *Y* value and place the results in two additional columns, X^2 and Y^2.

3. Multiply each *X* by its corresponding *Y* and place the results in a new column (called cross product, or *XY*).

4. Determine the sum for each column (*X, Y, X^2, Y^2, XY*).

5. Use the following formula to calculate *r*:

$$r = \frac{n(\Sigma XY) - (\Sigma X)(\Sigma Y)}{\sqrt{[n(\Sigma X^2) - (\Sigma X)^2][n(\Sigma Y^2) - (\Sigma Y)^2]}} \qquad (4.1)$$

Note that everyone must have two scores. If a student does only one test, his or her score cannot be used to calculate *r*.

Table 4.2   **Calculating the Correlation Coefficient**

	Chin-ups		Pull-ups		
Subject	X	X^2	Y	Y^2	XY
1	10	100	8	64	80
2	9	81	7	49	63
3	15	225	12	144	180
4	9	81	10	100	90
5	7	49	6	36	42
6	5	25	3	9	15
7	3	9	4	16	12
8	8	64	7	49	56
9	4	16	5	25	20
10	6	36	3	9	18
Σ	76	686	65	501	576

$$r = \frac{10(576) - (76)(65)}{\sqrt{[(10 \times 686 - (76)^2)(10 \times 501 - (65)^2)]}} = .89$$

Mastery Item 4.3

How would you describe the magnitude of the correlation computed between chin-ups and pull-ups? (Hint: Examine figure 4.1 and 4.2.)

Mastery Item 4.4

Hand calculate the correlation between the 10 paired observations of weight and pull-ups in table 4.1.

Mastery Item 4.5

Use the data in table 4.1 and SPSS to calculate the correlation between chin-ups and pull-ups to verify the correlation calculated in table 4.2.

1. Start SPSS
2. Open table 4.1.
3. Click on the **Analyze** menu.
4. Scroll down to **Correlate** and across to **Bivariate** and click.
5. Highlight "chinups" and "pullups" and place them in the **Variables** box by clicking the arrow key.
6. Click **OK.**

The Pearson product-moment correlation coefficient *(r)* can be used to calculate the relationship between two variables, such as between an individual's scores for chin-ups and pull-ups.

Mastery Item 4.6

 Use SPSS to create a scatterplot of the relationship between body weight and chin-ups.

1. Start SPSS.
2. Open table 4.1.
3. Click on the **Graphs** menu.
4. Scroll down to **Scatter** and click.
5. Click on **Simple.**
6. Click on **Define.**
7. Put "bodywt" in the **X-axis** box.
8. Put "chinups" in the **Y-axis** box.
9. Click **OK.**

Coefficient of Determination

An additional statistic that provides further information about the relationship between two measures is r^2. The square of the correlation is called the **coefficient of determination.** This value represents the proportion of shared variance between the two measures in question. To define shared variance, let us examine a specific example. If the correlation between a distance-run test and $\dot{V}O_2max$ is $r = .9$, then r^2 would be .81; the percentage of shared variance would be $0.81 \times 100 = 81\%$. This would mean that performance in the distance run accounts for 81% of the variation in $\dot{V}O_2max$ values. Nineteen percent of the variance (100% $- 81\%$) is the unique variance in $\dot{V}O_2max$ that is unexplained by performance in the distance run. Thus, 19% is error or residual variance. That is, it is variance remaining after you have used a predictor *(X)* to account for variation in the criterion *(Y)* variable. The coefficient of determination is important in statistics and measurement because it reflects the amount of variation found in one variable that can be predicted from another variable.

Mastery Item 4.7

If *r* = .5, calculate the percentage of shared variance. Draw a box representing the variance and depict the percentage of shared variance.

Negative Correlations

Two measures can have a negative correlation coefficient for one of two reasons. First, a negative correlation coefficient can result from two measures having opposite scoring scales. For example, the distance covered in a 12 min run and the time required to run 1.5 mi (2.4 km) would be negatively correlated. Runners with more endurance will cover more distance and have greater distance scores on the 12 min run; the same runners will run the 1.5 mi faster and have lower time scores. Runners with less endurance will have opposite results.

A second reason for a negative correlation coefficient is that two measures can have a true negative relationship. A good example is provided by the measures of body weight and pull-ups in table 4.1. Heavier people have a more difficult time lifting or moving their body weight than do lighter people.

Mastery Item 4.8

What would be the reason for a negative correlation between marathon run times and $\dot{V}O_2$max?

Limitations of *r*

The correlation coefficient is an index of the linear relationship between two variables. If two variables happen to have a **curvilinear relationship,** such as the one depicted in figure 4.5 between anxiety level and performance, the PPM correlation coefficient would be near 0, indicating no linear relationship between the two variables. However, to say that there is no relationship at all between the variables would obviously be incorrect. Both low- and high-anxiety scores are related to lower performance, and midrange anxiety scores are associated with higher performance. This limitation of *r* is one reason for graphing the relationships between variables using the scatterplot technique.

A second limitation of—or disclaimer about—*r* is that correlation is not necessarily an indication of a cause-and-effect relationship. Even if the correlation coefficient for two variables is +1 or −1,

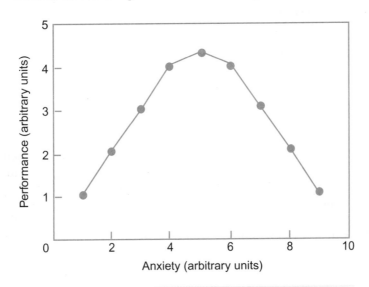

Figure 4.5 Graph of curvilinear relationship.

a conclusion, on the basis of r alone, that one variable is a cause of a measurable effect in another is incorrect. Some third variable may be the cause of the relationship detected by a high r value. For example, if we found the correlation coefficient between body weight and **power** to be +1 and assumed a cause-and-effect relationship, we might be tempted to use the following logic: (a) Higher body weights cause higher power; therefore (b) all athletes who need power should gain weight. Athletes might then become so heavy that they would suffer decreases in power.

Mastery Item 4.9

An epidemiologist has found a strong **positive correlation** between the amount of time that children watch television and their 1.5 mi (2.4 km) run times. He argues for decreased television watching to improve aerobic capacity in children. What is wrong with this logic?

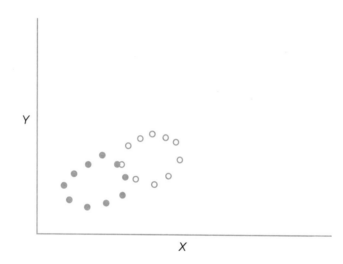

Another limitation of r is the effect of the variance or range of the data on the magnitude of r. Variables with larger variances or ranges tend to have higher r values than variables with restricted variances or ranges. Figure 4.6 demonstrates this phenomenon. The value of r in either the solid or open dots in the figure is much smaller than the value of r for the combined set of data. This is because the variance is greater in the complete set of data than it is in either subset.

Figure 4.6 Correlation example of range restriction.

Mastery Item 4.10

Generally, will the magnitude of r between $\dot{V}O_2$max and times for a 1 mi run be higher in a group of seventh-grade children or in a group of children in grades seven through nine? Why?

PREDICTION

In science, one of the most meaningful research results is the successful forecast. The most intriguing use of correlations is in **prediction**—that is, estimating the value of one variable from one or more other variables. From a mathematical standpoint, if there is a relationship between X and Y, then X can predict Y and vice versa. This does not mean that X and Y are causally related, however. To establish a cause-and-effect relationship between X and Y, another type of study and analysis would be necessary (i.e., establishing and testing a hypothesis with an experimental study).

Straight Line

As you may recall from high school geometry, any point on a plane marked with an x-axis and y-axis can be identified by its coordinates (X, Y) on the plane, and a straight line can be defined by the equation $Y = bX + c$, where b is the slope of the line and c is where the line intercepts the y-axis. The slope indicates how much Y changes for a unit change in X. The Y-intercept represents the value of Y when $X = 0$. In figure 4.7 we have marked five coordinate points: (0,1), (1,2), (2,3), (3,4), and (4,5); these points fall on a straight line. In figure 4.7, the Y-intercept is 1 and the slope of the line is 1. When all of the paired points do not fall in a straight line, a line of best fit will be plotted through the set of points—this is called the regression line.

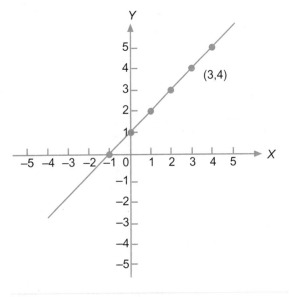

Figure 4.7   Plot of a straight line.

Mastery Item 4.11

What is the correlation between the X and Y coordinate points in figure 4.7?

Simple Linear Prediction

Simple linear prediction, also called *regression,* is a statistical method used to predict the criterion, outcome, or **dependent variable,** Y, from a single predictor or **independent variable,** X. If the two variables are correlated, which indicates that they have some amount of linear relationship, we can compute a prediction equation. The prediction equation has the same form as the equation for a straight line in plane geometry:

$$Y' = bX + c \tag{4.2}$$

We must think in terms of Y' because, unless the correlation between X and Y is –1 or +1, Y' is only an estimate of Y. Generally, Y' will not equal Y. The following formulas are used for calculating b and c:

$$b = \frac{n(\Sigma XY) - (\Sigma X)(\Sigma Y)}{n(\Sigma X^2) - (\Sigma X)^2} \tag{4.3}$$

$$c = \bar{Y} - b\bar{M} \tag{4.4}$$

For example, using the data in table 4.1, we can calculate b and c and present the prediction equation for predicting pull-ups from body weight. Note the prediction line presented in figure 4.8 represents the prediction equation and illustrates the error of prediction.

Mastery Item 4.12

Using the equation provided in figure 4.8, find the value of Y if X is 160.

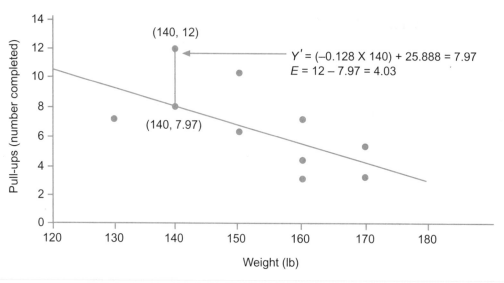

Figure 4.8   Error (residual) scores.

Getchell, Kirkendall, and Robbins (1977) provided an excellent example of the use of simple linear prediction in human performance. They found the correlation coefficient between $\dot{V}O_2$max and 1.5 mi run time for young women joggers to be .915. They calculated a linear prediction equation to predict $\dot{V}O_2$max from run time for the young women joggers. Their equation is

$$\text{Predicted } \dot{V}O_2\text{max} = -4.182X + 98.3$$

where –4.182 is the slope, 98.3 is the Y-intercept, and X is the jogger's run time for the 1.5 mi run. The units of $\dot{V}O_2$max and run time are ml · kg⁻¹ · min⁻¹ and min, respectively.

Simple linear prediction has many applications in the field of human performance. For example, $\dot{V}O_2$max can be predicted based on 1.5 mi run time.

Mastery Item 4.13

Use the data in table 4.1 to hand calculate *b* and *c* and provide the equation for predicting the number of chin-ups from body weight.

Mastery Item 4.14

Use the data in table 4.1 to verify the linear prediction equation to predict pull-ups from body weight (see figure 4.9).

1. Start SPSS.
2. Open table 4.1.
3. Click on the **Analyze** menu.
4. Scroll down to **Regression** and across to **Linear** and click.
5. Place "pullups" in the **Dependent** box.
6. Place "bodywt" in the **Independent(s)** box.
7. Click **OK**.

The SPSS output indicates that the slope *(b)* is –0.128 and the Y-intercept *(c)* is 25.888. Therefore, the equation to predict the number of pull-ups one can do from knowing body weight would be

$$\text{Predicted pull-ups } (Y') = -0.128 \times \text{body weight in lb} + 25.888$$

Regression

Coefficients[a]

Model		Unstandardized coefficients		Standardized coefficients	*t*	Sig.
		b	Std. error	Beta		
1	(Constant)	25.888	8.323		3.111	.014
	Body weight in pounds	—.128	.055	—.637	—2.339	.047

a. Dependent variable: PULLUPS

Figure 4.9   Regression output for mastery item 4.14.

Errors in Prediction

Unless the correlation coefficient is –1 or +1, Y' will not necessarily equal Y. The following equation summarizes this:

$$E = Y - Y' \tag{4.5}$$

E represents the inaccuracy of our predictions of Y based on the prediction equation. Figure 4.8 provides a demonstration of the error, or residual score. Residual scores are important for several reasons. First, they represent pure error of estimation or prediction. If these can be minimized, prediction can be improved. Second, residual scores can represent lack of fit, which means that the dependent variables do not predict a portion of the crite-rion. The predictors should then to be examined to reduce this problem. Finally, residual scores could represent a pure measure of a trait with the predictor statistically removed. In our earlier example of the correlation between pull-ups and body weight, pull-ups could be predicted from body weight. The resulting residual score is interpreted as a person's ability to do pull-ups with body weight statistically controlled. Or, in other words, if everyone was the same weight, how many pull-ups could you do? Positive residual scores would indicate that you did more pull-ups per unit weight than predicted. An important point to remember is that E has a zero correlation with X. This means that the prediction equation is equally accurate (or inaccurate) at any place along the X score scale.

Mastery Item 4.15

Can you think of a potential application of residual scores?

The **standard error of estimate (SEE),** also called the *standard error of prediction (SEP),* is a statistic that reflects the average amount of error in the process of predicting Y from X. Technically, it is the standard deviation of the error, or residual scores. The following formula is used for its calculation:

$$s_e = s_y \sqrt{1 - r^2}$$ (4.6)

If we use the data in table 4.1, the *SEE* for predicting pull-ups (Y) from body weight (X) would be as follows:

$$s_e = 2.953\sqrt{1 - (.637)^2}$$

$$s_e = 2.953\sqrt{1 - (.4058)}$$

$$s_e = 2.276$$

Because this is a standard deviation of the error or residual scores, it could be used as follows: If we predicted 7.97 pull-ups for someone weighing 140 lb (63.6 kg), about 95% of the people who weigh 140 lb will have pull-up scores between approximately 12 and 3. Remember that ~95% of the time a score is located ±2 standard deviations from the mean of a normal distribution and error scores are assumed to be normally distributed. The stan-dard deviation here is actually the SEE because SEE is the standard deviation of the errors of prediction. Note also that the 7.97 predicted from 140 lb is the mean value of pull-ups for those who weigh 140 lb.

Mastery Item 4.16

Use SPSS and the data in table 4.1 to calculate the standard error of estimate (SEE) when predicting pull-ups from body weight (see figure 4.10).

1. Start SPSS.
2. Open table 4.1.
3. Click on the **Analyze** menu.
4. Scroll down to **Regression** and across to **Linear** and click.
5. Place "pullups" in the **Dependent** box.
6. Place "bodywt" in the **Independent(s)** box.
7. Click **OK.**

Model Summary

Model	R	R square	Adjusted R square	Std. error of the estimate
1	.637[a]	.406	.332	2.4138

a. Predictors: (constant), body weight in pounds

Figure 4.10 Output for mastery item 4.16.

Note that SPSS calculates the SEE in a slightly different method. Thus, the 2.276 cal-culated in the text differs from the SPSS value (i.e., 2.41).

MULTIPLE CORRELATION

Correlation and prediction are two interrelated topics that are based on the assumption that two variables have a linear relationship. We have examined the notion of simple linear prediction that has one predictor, X, of the criterion, Y. A more complex prediction of Y can be developed based on more than one predictor: X_1, X_2, and so on. This is called **multiple correlation,** multiple prediction, or multiple regression. The mathematics of this approach are much more complicated. If X and Y have a curvilinear relationship, then nonlinear regression can be used to predict Y from X. Although the mathematics involved in these techniques are beyond the scope of this book and will not be discussed here, we will provide one brief example of a multiple regression equation. Jackson and Pollock (1978) published prediction equations that combined both multiple and nonlinear prediction. Their equation predicted hydrostatically measured body density (BD) (dependent variable) of men from age (A), sum of skinfolds (SK), and sum of skinfolds squared (SK²). The three predictors (independent variables) are the multiple predictors, and the sum of skinfolds squared is the nonlinear component in the prediction. The equation follows:

$$BD = 1.10938 - 0.0008267(\Sigma SK) + 0.0000016(\Sigma SK^2) - 0.0002574(A)$$

Measurement and Evaluation Challenge

James now understands the statistical methods used to estimate his $\dot{V}O_2$max from treadmill time. The type of treadmill test that James completed is called a Balke protocol. Treadmill protocols are further described in chapter 11. Research has shown the correlation between treadmill time with the Balke protocol and measured $\dot{V}O_2$max to be >.90. The prediction equation is $\dot{V}O_2$max' = 14.99 + 1.444 X (= treadmill minutes in decimal form). Note that $\dot{V}O_2$max' is actually Y' (i.e., the predicted value of Y, based on X). Thus, James, who ran for 24 min and 15 s (i.e., 24.25 min), has a predicted $\dot{V}O_2$max of 14.99 + 1.444 X (24.25) = 50 ml · kg⁻¹ · min⁻¹. However, James also realizes that there is some error in the prediction equation because the correlation is not perfect (i.e., ±1.00). The standard error of estimate (SEE) reflects the amount of error in the prediction equation. With this equation, the SEE is about 3 ml · kg⁻¹ · min⁻¹. Thus, James can be 68% confident that his actual $\dot{V}O_2$max is between 47 and 53 ml · kg⁻¹ · min⁻¹ (his predicted score ±1.00 SEE).

SUMMARY

The correlation and prediction statistics presented in this chapter lay the foundation and provide the many necessary skills for the measurement and evaluation process. As you will see, these skills are necessary for generating measurement theory, as well as applying reliability and validity concepts to practical problems in exercise and human performance.

At this point you should be able to accomplish the following tasks:

1. Calculate and interpret measures of correlation.
2. Calculate and interpret a prediction equation.
3. Calculate the SEE.
4. Use SPSS to enter data and to generate and interpret
    a. correlation coefficients,
    b. scatterplots of variables, and
    c. simple linear prediction equations.

5

Inferential Statistics

Outline

Measurement and Evaluation Challenge

Not only is James taking a course in measurement and evaluation in human performance, but he is also enrolled in a course titled "Physiological Basis of Human Performance." His instructor has assigned a research article for the class to read. In the article the researcher hypothesized that drinking a carbohydrate solution in water would improve cycling endurance performance beyond that which would result from drinking water only. The research study compared two groups of endurance cyclists. One of the groups drank water only. The other group drank water that contained a 4% solution of carbohydrate. The cyclists were then tested to see how long they could cycle at a given workload.

A t test indicated that the carbohydrate-drinking group cycled "significantly ($p < .05$)" longer than the water-only group. James wants to understand what a t test is and what "significance" means in this context. He wonders what $p < .05$ means. Does it mean that the researchers proved that the carbohydrate drink was better than drinking only water if you want to increase your cycling endurance? You will find out how to interpret these and other results and will learn about other statistical methods in this chapter.

Objectives

After studying this chapter, you will be able to

▶ understand the scientific method and the hypotheses associated with it,

▶ perform an inferential statistical analysis to test a hypothesis, and

▶ use selected programs from the SPSS computer software in data analysis.

he descriptive statistical techniques that we have presented thus far are those that you will most commonly use for measurement problems. There are, however, a number of other statistics that you may need to use in various measurement situations. The most common of these examine group differences. When these tests are used to relate the characteristics of a small group **(sample)** to those of a larger group (population), they are referred to as inferential statistics. Much human performance research is conducted using **inferential statistics.**

HYPOTHESIS TESTING

The scientific method uses inferential statistics for obtaining knowledge. The **scientific method** requires both the development of a scientific hypothesis and an inferential statistical test of that hypothesis versus another competing hypothesis. A **hypothesis** is a statement of a presumed association between at least two variables in a population. A **population** is the entire group of individuals or observations in question (e.g., college seniors). A measure of interest in the population is called a **parameter.** Inevitably, because entire populations are so large and unwieldy (imagine surveying all U.S. college seniors on a certain matter), you study hypotheses about a population by using a subgroup of the population, called a **sample.** The measure of the variable of interest in the sample is called a **statistic.** Using various techniques, you can make an inference—but not an absolute statement—about the whole population from your work with a sample. (Table 5.1 contains the symbols that are commonly used to distinguish sample statistics from population parameters.)

Consider the following examples. A teacher was interested in the jumping ability (parameter) of fifth graders in the school. There were 200 fifth graders (population). The teacher randomly selected 50 students (sample) and administered a standing long jump test. The scores on this test were analyzed, and the sample values (statistic, e.g., the mean) were considered to be representative of the population parameter. Surveys taken before presidential elections use samples to estimate the percentage of people preferring a particular candidate. Note, however, that there is error in this technique, which you will learn more about later in this chapter.

Table 5.1   **Statistical Symbols**

Measure	Population parameter	Sample statistics
Mean	μ	M
Standard deviation	σ	s
Correlation	ρ	r

Mastery Item 5.1

Create a research problem related to something of interest to you. Identify the following: (a) population, (b) sample, (c) parameter, and (d) statistic.

Hypotheses are the tools that allow research questions to be explored. A hypothesis may be one of several types:

▸ **Research hypothesis**—What the researcher actually believes will occur. For example, assume you believe that training method is related to oxygen uptake. Your research hypothesis is, *oxygen uptake is related to the type of aerobic training one uses.* You can investigate this hypothesis with a *t* test or ANOVA.

> ▶ **Null hypothesis** *(H₀)*—A statement that there is no relation (association, relationship, or difference) between variables. In this example, your null hypothesis is that the mean oxygen uptake is not different for training groups who use different training methods. This is the hypothesis that you will actually test (and hope to discredit) using the techniques of inferential statistics.

> ▶ **Alternative hypothesis** *(Hᵢ)*—A statement that there is a relation (association, relationship, or difference); typically the converse of H_0. Here, your alternative hypothesis is $\mu_1 \neq \mu_2$ where μ_1 is the population mean for group 1 and μ_2 is the population mean for group 2. Remember that you actually obtain data on samples only and then infer your results to the population. In this example, the research hypothesis is H_1.

Before you perform the appropriate statistical test, a probability level is selected beyond which the results are considered to be "statistically significant." This probability value is called the **significance,** or **alpha level (α),** and allows you to test the probability of the actual occurrence of your result. The alpha level is conventionally set at .05 or .01 (i.e., 5% or 1%). For example, if the investigator sets the alpha level at .05, he or she is saying that the probability of obtaining the statistic just by chance must be less than 5 times out of 100 before he or she will decide the null hypothesis is not tenable. In effect, you assume no relationship between variables until you have evidence to the contrary. The statistical data may provide the contrary evidence.

However, the researcher might reach an incorrect conclusion (i.e., say there is a relationship, a correlation, or difference when in fact there is not). The probability of making such an error is the alpha level. This error is referred to as a **type I error.** The alpha level is set at .05 or .01 to make the probability of a type I error extremely small. You can also make a second type of error, a **type II error,** by concluding that there is no relationship between the variables in the population when in fact there truly is. This brief introduction to significance testing will be expanded in subsequent sections of this chapter.

The SPSS computer program (or other statistical software) will calculate the actual alpha level for you. If the probability is less than the preset alpha level of .05 or .01, you conclude that there is a significant relationship between the variables. Thus, H_0 is rejected and H_1 is accepted. Figure 5.1 illustrates the type of decisions and errors you might make. One can never know the "true state of the null hypothesis" in the population, so one always runs the risk of making a type I or type II error. You cannot make both a type I and type II error in the same research study. Can you look at figure 5.1 and see why this is the case?

True state in population

	H_0 is true H_1 is false	H_0 is false H_1 is true
Reject H_0, accept H_1	Type I error (alpha)	Correct decision
Accept H_0, reject H_1	Correct decision	Type II error (beta)

Your decision

Figure 5.1   Type I and type II errors.

Mastery Item 5.2

A coach wishes to study the relationship between leg strength and jumping ability. Write the appropriate null and alternative hypotheses for this problem.

Selection of the appropriate statistical technique is based on your research question and the level of measurement of the variables. The number of groups and the level of measurement of the data determine the appropriate statistics to use.

Some of the most common are as follows:

▶ χ^2 **(chi-square test)**—Used to examine associations in nominal data

▶ *t* **test**—Used to examine a difference in a continuous (interval or ratio) variable among two and only two groups

▶ **ANOVA (analysis of variance)**—Used to examine differences in continuous (interval or ratio) variables among more than two groups

INDEPENDENT AND DEPENDENT VARIABLES

The differences between independent and dependent variables are important. The dependent variable is the "criterion" variable; its existence is the reason you are conducting the research study. The independent variable exists solely to determine if it is related to (or influences) the dependent variable. There are a number of ways that independent and dependent variables can be characterized; these are presented in table 5.2.

Table 5.2 **Variable Classification**

Independent	Dependent
Presumed cause	Presumed effect
The antecedent	The consequence
Manipulated or measured by researcher	Outcome (measured)
Predicted from	Predicted to
Predictor	Criterion
X	*Y*

If the dependent variable is nominally scaled, the differences between groups (or cells) are measured by frequencies or proportions. If you are dealing with continuous (interval or ratio) data, then differences in mean values are often examined. For example, suppose you want to examine the difference between the training effect produced by circuit weight training and that produced by aerobic dance. The variable you select to measure the training effect is $\dot{V}O_2$max. This is the dependent variable. The independent variable is method of training and has two levels: circuit weight training and aerobic dance.

Training effect variables. The independent variable is method of training and has two levels: circuit weight training and aerobic dance.

A track coach is interested in examining two methods of cross country training: (a) over-distance training and (b) distance intervals. The coach decides to evaluate the methods based on the runners' times at the district meet. Identify the dependent and independent variables in this study.

SELECTED STATISTICAL TESTS

The following are statistical tests that examine associations or differences between groups. The techniques selected represent common basic inferential tests.

Chi-Square (χ^2)

Purpose: To determine if there is an association between levels (cells) of one or more nominally scaled variables.

Example: An aerobics instructor is teaching two classes, aerobic dance and circuit weight training. The instructor wants to know if the proportion of males and females is the same in each class. The null hypothesis is that there is no association between gender and type of class in which one is enrolled. The alternative hypothesis is that there is an association. One can only reject the null hypothesis and believe that the alternative hypothesis is the true state of circumstances in the population when the probability of the null hypothesis being true is very small (i.e., <.05) based on the sample data. The data are found in table 5.3. Use them in conjunction with the following SPSS commands to calculate the chi-square test of association and verify the results with those presented in figure 5.2.

1. Start SPSS.
2. Open table 5.3.
3. Click on the **Analyze** menu.
4. Scroll down to **Descriptive Statistics** and across to **Crosstabs** and click.
5. Put "class" in the rows and "gender" in the columns by clicking the arrow keys.
6. Click **Statistics.**
7. Check the **Chi-Square** box.
8. Click **Continue.**
9. Click **OK.**

The resulting SPSS printout is presented as figure 5.2. Although a number of statistics are calculated for us, the test statistic in which we are interested is the Pearson chi-square. The observed chi-square value is 22.5. The probability associated with this test is reported to be .000. In actuality, however, you can never have a probability of 0. It is simply the case that the computer program calculates the probability (i.e., significance) to three decimal places. In any case, you should interpret this as .0001. Thus, it is exceedingly rare to find the cell frequencies being distributed as they are when, in fact, there is no association between type of class and gender (the H_0). Because of this exceedingly small probability, the teacher can conclude that there is an association between gender and type of class. The null hypothesis (H_0) of no association is rejected, and it is concluded that there is an association between gender and type of class in which one enrolls. Figure 5.2 shows that 10 of the 12 males registered for circuit weight training, whereas all 18 of the 18 females registered for aerobic dance. This association might help the instructor in planning the types of activities for the classes.

Table 5.3 Data Entry for χ^2 Example

Id	Gender	Class
1	1	1
2	1	1
3	1	1
4	1	1
5	1	1
6	1	1
7	1	1
8	1	1
9	1	1
10	1	1
11	1	2
12	1	2
13	2	2
14	2	2
15	2	2
16	2	2
17	2	2
18	2	2
19	2	2
20	2	2
21	2	2
22	2	2
23	2	2
24	2	2
25	2	2
26	2	2
27	2	2
28	2	2
29	2	2
30	2	2

Note: Gender code: 1 = male, 2 = female. Class code: 1 = circuit, 2 = dance.

Case processing summary

	Cases					
	Valid		Missing		Total	
	N	Percent	N	Percent	N	Percent
Class enrollment * Gender of the subject	30	100.0%	0	.0%	30	100.0%

Class enrollment * Gender of the subject crosstabulation

Count

		Gender of the subject		Total
		Male	Female	
Class enrollment	Circuit weight training	10		10
	Aerobic dance	2	18	20
Total		12	18	30

Chi-square tests

	Value	df	Asymp. sig. (2-sided)	Exact sig. (2-sided)	Exact sig. (1-sided)
Pearson chi-square	22.500[b]	1	.000		
Continuity Correction[a]	18.906	1	.000		
Likelihood ratio	27.377	1	.000		
Fisher's exact test				.000	.000
Linear-by-linear association	21.750	1	.000		
N of valid cases	30				

a. Computed only for a 2 x 2 table.
b. 1 cells (25.0%) have expected count less than 5. The minimum expected count is 4.00.

Figure 5.2 SPSS crosstabs output.

t Test for Two Independent Groups

Purpose: To examine the difference in one continuous dependent variable between two (and only two) independent groups. Independent groups are groups that are not in any way related.

Example: A high school volleyball coach is selecting players for the varsity team and is using serving accuracy as a selection factor. After the team is selected, the coach wants to quantify the differences in serving accuracy between varsity and subvarsity players. The serving scores are presented in table 5.4.

Table 5.4   Serving Scores

Varsity	20, 18, 17, 19, 20, 16, 18, 19
Subvarsity	16, 15, 17, 14, 15, 13, 14, 12

The research hypothesis for this study is, *There is a difference in serving accuracy between varsity (v) and subvarsity (sv) players*. The null hypothesis to be tested is that the mean volleyball serving score for the varsity players is equal to the mean for the subvarsity players (i.e., there is no difference between the two):

$$H_0 : \mu_v = \mu_{sv} \qquad (5.1)$$

The alternative hypothesis is that the mean for the varsity players is not equal to the mean of the subvarsity players:

$$H_1 : \mu_v \neq \mu_{sv} \tag{5.2}$$

For the coach's purpose, the alpha level is set at .05. The SPSS procedure for the *t* test may be used to analyze the data. Use the data in table 5.4 to calculate an independent groups *t* test and confirm your results with those presented in figure 5.3.

1. Start SPSS.
2. Open table 5.4.
3. Click on the **Analyze** menu.
4. Scroll down to **Compare Means** and across to **Independent-Samples T Test** and click.
5. Put "score" in the **Test Variable(s)** box by clicking the arrow key.
6. Put "group" in the **Grouping Variable** box by clicking the arrow key.
7. Click the **Define Groups** button.
8. Put "1" in the **Group 1** box and "2" in the **Group 2** box.
9. Click **Continue.**
10. Click **OK.**

SPSS output is displayed in figure 5.3. Inspection of the means indicates that the varsity players (group 1; mean = 18.38) served significantly (Sig. [2-tailed] = .000) better than the subvarsity players (group 2; mean = 14.50).

A number of statistics in the output from the *t* test are beyond the scope of this class. For our purposes, you can ignore the results under "Levene's Test for Equality of Variances." Focus your attention on the areas below "t-test for Equality of Means." Notice the *t* presented with a value of 5.136 (actually presented twice). Think of this *t* as if it is a *z* score that you

Group statistics

	Team level	N	Mean	Std. deviation	Std. error mean
Serving score	Varsity	8	18.38	1.408	.498
	Subvarsity	8	14.50	1.604	.567

Independent samples test

		Levene's Test for Equality of Variances		t-test for Equality of Means					95% confidence interval of the difference	
		F	Sig.	t	df	Sig. (2–tailed)	Mean difference	Std. error difference	Lower	Upper
Serving score	Equal variances assumed	.095	.763	5.136	14	.000	3.88	.754	2.257	5.493
	Equal variances not assumed			5.136	13.769	.000	3.88	.754	2.254	5.496

Figure 5.3 Sample *t* test output: group statistics and independent samples test.

learned about in chapter 3. It is not a z score but it is much like a z score. If the z score was large (e.g., greater than 3 in absolute value), you know that the probability of finding a value this large is very small. The same can be said of a "t value." Thus, you can see that the t is relatively far into the tail of the distribution. This t is generally a rare occurrence. Of most importance to you is the box titled "Sig. (2-tailed)." This is the probability that the null hypothesis is true given the data from the sample. Because the probability is less than .05, the coach would reject the null hypothesis and retain the alternative. This is an example of a t test with independent groups. Suppose the coach wanted to examine the serving accuracy of the varsity during preseason as opposed to the end of the season. A paired t test would be used because the same group is being measured at two points in time.

Dependent t Test for Paired Groups

Purpose: To compare two related (paired) groups on one dependent variable. Groups can be paired by matching them on some external characteristic (e.g., siblings) or by measuring the same group twice (i.e., pre- and postperformance).

Example: A basketball coach is concerned with the jump training program that he has used. To check the effectiveness of the program, he tests the players at the beginning and the end of the season. The research hypothesis is that there is a difference in preseason and postseason jumping ability. The null hypothesis is that there is no difference in the players' jumping abilities over time. To test the null hypothesis, the SPSS t test procedure is used again. However, the data are entered differently from the previous example because each person is tested twice (compare tables 5.4 and 5.5). This allows SPSS to properly pair the data so that the correct result is calculated.

Use the data in table 5.5 to calculate a paired (dependent) t test, and confirm your results with those presented in figure 5.4.

1. Start SPSS.
2. Open table 5.5.
3. Click on the **Analyze** menu.
4. Scroll down to **Compare Means** and across to **Paired-Samples T Test** and click.
5. Put "presea" and "postsea" in the **Paired Variables** box by clicking the arrow key.
6. Click **OK.**

Table 5.5   **Data Format for Paired t Test**

Presea	Postsea
18	20
20	24
17	20
16	19
15	20
18	22
19	21
17	21

The mean difference between the posttest and the pretest was found to be 3.38. The observed t value was found to be 9.000, with an associated probability (alpha level) that approaches 0 (Sig. [2-tailed]). Thus, the null hypothesis is rejected and the alternative hypothesis is retained. The coach can conclude that the differences in jumping from the

Paired samples statistics

		Mean	N	Std. deviation	Std. error mean
Pair 1	Post-season performance	20.88	8	1.553	.549
	Pre-season performance	17.50	8	1.604	.567

Paired samples correlations

		N	Correlation	Sig.
Pair 1	Post-season performance & Pre-season performance	8	.775	.024

Paired samples test

		Paired differences							
					95% confidence interval of the difference				
		Mean	Std. deviation	Std. error mean	Lower	Upper	t	df	Sig. (2–tailed)
Pair 1	Post-season performance - Pre-season performance	3.38	1.061	.375	2.49	4.26	9.000	7	.000

Figure 5.4 Sample *t* test output: paired samples statistics, paired samples correlations, and paired samples test.

beginning of the season to the end can be attributed to something other than chance and that the jump training appears to be effective.

Note that the differences in jumping performance could have been caused by some factor other than the training program. Because a time lag occurred between the first and second administrations of the test, the differences could have been attributable to growth, maturation, or some other factor not under the control of the researcher. In an actual experiment, these factors would have to be controlled.

Mastery Item 5.4

Develop your own data set to be analyzed with a dependent *t* test. Run SPSS and interpret your results.

One-Way ANOVA

Purpose: To examine group differences between one continuous (interval-scaled or ratio-scaled) dependent variable and one nominal-scaled independent variable. Unlike the *t* test, ANOVA can handle independent variables with more than two levels (groups) of data.

Example: The data for this example were collected at a youth baseball camp. The players were tested for distance throwing and classified according to their defensive ability. In this example, the independent variable is defensive ability (coded 1, 2, and 3), and the dependent variable is distance throwing ability measured in feet. The problem to be examined is whether there are differences in distance throwing ability across the three defensive-ability groups.

   The null hypothesis is that means for the distance throw for the three defensive-ability groups are equivalent:

$$H_0 : \mu_1 = \mu_2 = \mu_3 \tag{5.3}$$

The alternative hypothesis is that the means are not equivalent:

$$H_1 : \mu_1 \neq \mu_2 \neq \mu_3 \tag{5.4}$$

   The alpha level was set at .01 rather than .05, indicating that the researcher wanted to reduce the probability of making a type I error and increase confidence that if differences were found among the means they were not attributable to chance. The data for this problem are presented in table 5.6.

Table 5.6   Input Format for One-Way ANOVA

Id	Group	Score
1	1	93
2	1	90
3	1	95
4	1	75
5	1	88
6	2	48
7	2	70
8	2	72
9	2	68
10	2	65
11	3	70
12	3	57
13	3	40
14	3	48
15	3	50

Mastery Item 5.5

Explain why testing at the .01 level rather than the .05 level indicates that the researcher wants to be more certain that the observed differences are true differences. Now use the data in table 5.6 to calculate a one-way ANOVA and confirm your results with those presented in figure 5.5.

1. Start SPSS.
2. Open table 5.6.
3. Click on the **Analyze** menu.
4. Scroll down to **Compare Means** and across to **One-Way ANOVA** and click.
5. Put "score" in the **Dependent List** box by clicking the arrow key.
6. Put "group" in the **Factor** box by clicking the arrow key.
7. Click **Options.**
8. Click on the **Descriptives** box.
9. Click **Continue.**
10. Click **OK.**

Using one-way ANOVA, a coach can determine if a player's throwing distance (dependent variable) is related to his or her defensive ability (independent variable with more than two levels).

The key information presented in figure 5.5 is the Sig. (significance or probability). The other information is used to obtain the significance. For ANOVA, the significance test is an F ratio. Again, think of this F value like the z score you learned about in chapter 3. It is not a z score but it is somewhat like a z score. High values are rare, and the probability of obtaining a high value is reduced when the groups do not differ much from each other. Because the probability level for the observed event is less than .01 (the computer gives it as .000), the null hypothesis is rejected and the alternative hypothesis is retained.

Examination of the group means in figure 5.5 indicates that the players in the highest ability group (group 1) threw the farthest (M = 88.2 ft [26.9 m]); players in the lowest ability group (group 3) threw the shortest distance (M = 53 ft [16.2 m]); and players in the middle group (group 2) threw somewhere in between (M = 64.6 ft [19.7 m]). Statistical tests called multiple-comparison tests exist to compare the specific groups to one another; however, they are beyond the scope of this text.

Descriptives

Distance thrown

	N	Mean	Std. deviation	Std. error mean	95% confidence interval for mean		Minimum	Maximum
					Lower bound	Upper bound		
High	5	88.2000	7.8549	3.5128	78.4468	97.9532	75.00	95.00
Middle	5	64.6000	9.6333	4.3081	52.6387	76.5613	48.00	72.00
Low	5	53.0000	11.2694	5.0398	39.0072	66.9928	40.00	70.00
Total	15	68.6000	17.6141	4.5479	58.8456	78.3544	40.00	95.00

ANOVA

Distance thrown

	Sum of squares	df	Mean square	F	Sig.
Between groups	3217.600	2	1608.800	17.145	.000
Within groups	1126.000	12	93.833		
Total	4343.600	14			

Figure 5.5   One-way ANOVA results.

Measurement and Evaluation Challenge

James has learned that a *t* test was conducted in the research article because there was a single independent variable, consisting of a control group who drank water only and an experimental group who drank the carbohydrate solution. He also learned that the researcher hypothesized that cycling duration (the dependent variable) was a function of which drink was consumed. He now knows that $p < .05$ means that a null hypothesis was rejected and an alternative hypothesis was accepted. He realizes that the researcher may have made a type I error, but the probability of such an error is less than 5 times in 100.

Thus, it is highly likely that carbohydrate drinks will improve cycling endurance for most cyclists, but it is not a certainty because a type I error could have been made (although the probability of doing so is <5/100). Because the null hypothesis was rejected, it is impossible for the researcher to have made a type II error. Additionally, James has learned that the researcher's hypothesis about the influence of carbohydrate drinks can never be proven.

SUMMARY

This chapter provided a brief overview of the tests used in inferential statistics; however, many assumptions regarding these techniques were not discussed. Statistical tests of significance often obscure practical differences. There is no substitute for blending statistical findings with intuitive logic. In-depth treatment of statistical methods can be found in Glass and Hopkins (1996). Thomas and Nelson (2001) provide excellent examples of research in human performance.

At this point you should be able to accomplish the following tasks:

1. Understand and interpret the scientific method.
2. Write and interpret research, null, and alternative hypotheses.
3. Use SPSS to enter data and to generate and interpret
    a. chi-square tests of association,
    b. independent and dependent *t* tests, and
    c. one-way ANOVA.

part II

Reliability, Validity, and Grading

In this part, what you learned about basic statistics and computer applications in part I will be extended and applied to issues related to valid decision making. We all make decisions in life, and each of us attempts to make the best decisions possible. A decision you make in the field of human performance might be about a person's aerobic capacity, muscular strength, or amount of daily physical activity. You may also need to make valid decisions about cognitive knowledge and report grades or levels of achievement to students, clients, or program participants. Alternatively, you may need to evaluate a program that you direct. Good decisions are based on sound data, which in turn reflect the characteristics of reliability, validity, and objectivity. You will use the SPSS skills that you gained in part I to accomplish specific tasks related to these characteristics. MIs in each chapter provide you the opportunity to use SPSS procedures to illustrate and analyze measurement problems.

Chapter 6 provides the important steps you will need to judge the quality of norm-referenced data. To make accurate decisions about individuals or groups, you must use data that are sufficiently reliable, valid, and objective. For example, when reporting someone's aerobic capacity, you will need to be certain that the value is truthful. Invalid data can result in inappropriate decisions. Chapters 6 and 7 help you analyze data so that you can report them in such a way as to make valid interpretations and decisions. No measurement technique is perfectly reliable or valid, but you need to know how to interpret the amount of reliability and validity reflected in your measurement protocol so that you can make appropriate decisions. Chapter 7 addresses these issues from a criterion-referenced perspective.

Chapter 8 focuses on alternative, or authentic, assessment, which includes a number of techniques used to assess performance in schools and other settings. These assessment techniques are based more on "real-life," application-based evaluations than on traditional statistical evaluations of a test's worthiness. And yet, many of the instruments used in authentic assessment will be discussed from the perspective of reliability and validity. Of particular interest is objectivity in the methods often used with authentic assessment. For example, authentic assessment may use portfolios developed by students, clients, or participants. The ability to evaluate individual portfolios without having one's biases affect the interpretation of the results is important, as is the concern for bias in any evaluation. Although it is never possible to eliminate bias totally, it is possible to recognize bias and attempt to minimize it.

Chapter 9, directed to those of you who will be teaching and evaluating student performance in instructional settings, provides guidelines for valid assignment of grades and factors important to valid assessment and reporting of achievement.

6

Norm-Referenced Measurement

Outline

Measurement and Evaluation Challenge

Kelly, a YMCA fitness director, wants to assess the cardiovascular fitness of young adult members. Kelly is interested in determining the $\dot{V}O_2$max of her members. However, she has heard that the best measure of cardiovascular fitness is to have an individual run on a treadmill until he or she is exhausted. This requires collecting the air that these individuals expire while on the treadmill and, therefore, requires considerable and expensive equipment. Because of these difficulties, Kelly is interested in using a surrogate (field test) measure such as the YMCA 3-Minute Step Test (see chapter 11, p. 233) in place of treadmill performance. However, Kelly is concerned that a field test might lack the accuracy that a treadmill test provides. Kelly's concern is a very real one. How can she determine if the measure that is being obtained with the field test is reliable (i.e., consistent) and valid (i.e., truthful)?

Objectives

After studying this chapter, you will be able to

▶ discuss the concepts of reliability and validity,

▶ differentiate among the types of reliability and how to calculate them,

▶ identify the types of validity evidence that can be used to provide information about a test's truthfulness and calculate the appropriate statistics,

▶ describe the relationship between reliability and validity and comment on why these concepts are so important to measurement,

▶ evaluate the evidence for reliability and validity typically presented in the measurement of human performance, and

▶ use SPSS to calculate reliability and validity statistics.

egardless of the area of human performance in which you work, you will need to make decisions based on the data you collect. Often, these decisions require you to make comparisons among different people or report test results to an individual. For example, Kelly may need to report the results of her testing to the programming director or the YMCA's board of directors to keep a particular fitness program funded. It is important that your decisions and reports be accurate. The accuracy of your decisions relates to the norm-referenced characteristics of your variables. As you learned in chapter 1, the most important measurement characteristics are reliability, objectivity, and validity. (Recall from chapter 1 that a norm-referenced standard is a level of achievement relative to a clearly defined subgroup.)

Reliability and **validity** are the most important concepts presented in this book. The many computational, theoretical, and practical examples presented throughout the book can be traced to these concepts. *Reliability relates to the consistency or repeatability of an observation; it is the degree to which repeated measurements of the same trait are reproducible under the same conditions.* Reliability can also be described as accuracy, consistency, dependability, stability, and precision. A test is said to be reliable if one obtains the same (or nearly the same) score each time the test is administered to the same individual. As you can see from this terminology, test reliability will be extremely important as Kelly determines which field test to administer.

Validity is the degree of truthfulness of a test score. That is, does the test score, once found to be reliable, accurately measure what it reports to measure? *Validity is dependent on two characteristics: reliability and relevance.* **Relevance** is the degree to which a test pertains to its objectives. Thus, for a measure to be valid, it must measure the particular trait, characteristic, or ability consistently, and it must be relevant. That is, the instrument or test must be related to the characteristic reportedly being measured.

Thus, you can see that reliability and validity are important concerns for Kelly. She must be certain that the field test produces results that are consistent from trial to trial yet also accurately estimates what the $\dot{V}O_2max$ would be if it were measured on the treadmill.

Mastery Item 6.1

Can you think of some field or surrogate measures that might be used to estimate $\dot{V}O_2max$?

A test may be valid in one set of circumstances but not in another. There are many tests that have sufficient reliability but poor validity. For example, assessment of total body weight is typically a very reliable measure. It changes little from day to day, and two different evaluators would likely report the same or nearly the same value when measuring it. However, total body weight is not a valid measure of total body fatness, because total body weight is made up of fat, bone, and lean tissue. Thus, one's weight depends on the relative proportions of these body components.

Objectivity is a special kind of reliability. *Objectivity is interrater reliability.* You have probably taken "objective" tests (such as multiple-choice items) and "subjective" tests (such as essay items). These tests are classified as such because of the type of scoring system used when grading the examination. Multiple-choice, true–false, and matching test items are said to be objective because they have a high amount of interrater reliability. That is, regardless of who the grader is, the scores on these types of items are very consistent from one grader to another because there is a well-defined scoring system for the correct (or best) answer. However, a test can be objective in nature yet not accurate or reliable. If the questions are poorly written, a multiple-choice examination may be an unreliable and invalid measure of student knowledge. The scoring of essay tests tends to be more subjective—different readers will give different scores to an answer—but there are ways to make the scoring of essay tests more objective (see chapter 10).

Mastery Item 6.2

Consider Olympic gymnastics or diving events and the judging system used for them. Why are the high and low scores eliminated?

RELIABILITY

Many of the basic statistical concepts presented in chapters 3, 4, and 5 help us determine whether a test is reliable and valid. Teachers and researchers generally need specific evidence about a test's reliability and validity—and not mere general statements suggesting that it is reliable and valid. Numerous statistics are used to provide evidence of reliability and validity. *The variance (presented in chapter 3) and the Pearson product-moment (PPM) correlation coefficient (presented in chapter 4) are used to provide evidence of a test's reliability and validity and thus need to be understood fully.* Before getting into the number crunching associated with reliability and validity, however, we need to consider these concepts from theoretical perspectives so you are clear about exactly what these constructs are. With a deeper understanding, you will be better able to determine which statistical procedure you should use and how to interpret the results.

Observed, Error, and True Scores

Consider the scores obtained at a recent blood pressure screening (table 6.1). Each of the 10 subjects has an observed recorded blood pressure; however, it is possible that errors of measurement may have entered into the recording system so that the observed score is not the person's true blood pressure value. For example, the observed score may be in error because of the amount of experience the tester has, how the actual measurement was taken, when it was taken, where it was taken, the type of instrument used, the time of day, what events took place prior to testing, and so forth.

Although it is unlikely that we can ever know exactly (without any error) what a person's blood pressure is, imagine that we can develop a means by which to measure it more accurately than is typically done in a laboratory or clinical setting. For example, we might place a pressure-sensitive apparatus directly in an artery to determine the pressure exerted during systole. (Obviously, we would have to ignore the fact that even thinking about such a procedure would undoubtedly alter a person's blood pressure reading.) Assume that we have done this for the subjects whose scores are reported in table 6.1. You will note that only 2 of the 10 subjects have observed blood pressures that are equal to their true blood pressures. The other blood pressure readings have various amounts of error associated with them. Some of the errors result in overestimating the true blood pressure, whereas others result in underestimating the true blood pressure.

Table 6.1 **Systolic Blood Pressure Recordings for 10 Subjects**

Subject	Observed blood pressure	= True blood pressure	+ Error score
1	103	105	-2
2	117	115	+2
3	116	120	-4
4	123	125	-2
5	127	125	+2
6	125	125	0
7	135	125	+10
8	126	130	-4
9	133	135	-2
10	145	145	0
Sum (Σ)	1250	1250	0
Mean *(M)*	125.0	125.0	0
Standard deviation *(s)*	11.6	10.8	4.1
Variance *(s²)*	133.6	116.7	16.9

Note: Units are mmHg.

Based on an example from Sax (1980).

A few key points can be seen in table 6.1:

▶ Each person's observed score is the sum of the true score and the error score. Your true score theoretically exists but is impossible to measure and can be thought of as what you actually know or how well you actually perform; it is without error. You can think of it as the average of an infinite number of administrations of the test in which you don't get any better because of practice nor do you get any worse because of fatigue. In a sense, your true score never changes for a specific point in time and it is perfectly reliable. Your error score results from anything that causes your observed score to be different than your true score; it is a value that theoretically exists but is impossible to measure. Sources of error include individual variability, instrument inaccuracy, cheating, testing conditions, and so forth.

▶ There is variation in the observed, true, and error scores (the standard deviations and variances are calculated for you).

▶ Error can be positive (increase the observed score) or negative (decrease the observed score).

▶ Error scores contribute relatively little to the observed variation.

▶ The error score mean is zero.

▶ The observed score variance (133.6) is equal to the sum of the true score variance (116.7) plus the error score variance (16.9).

Using observed (total), true, and error score variances, the reliability $r_{xx'}$ is defined as that proportion of observed score variance that is true score variance (i.e., true score variance divided by observed [total] score variance):

$$r_{xx'} = \frac{s_t^2}{s_o^2} = \frac{(s_o^2 - s_e^2)}{s_o^2} \qquad (6.1)$$

where s_t^2 is the true score variance, s_o^2 is the observed (total) score variance, and s_e^2 is the error score variance. In table 6.1, the reliability is $116.7/133.6 = .87$.

In theory, the true score is perfectly reliable, with a value of 1.00. (Certainly, the true score changes if there is a change in the phenomenon being measured, but at any given point in time, the true score is viewed as perfectly reliable and thus contains no error.) Thus, a test is said to be reliable to the extent that the observed score variation is made up of true score variation.

Using equation 6.1, we see that the limits on reliability are 0 and 1.00. If the observed score is made up of no true score variation, the reliability is 0.00, and if the observed score is made up of only true score variation, the reliability is 1.00. Neither of these two cases generally arises; however, for a test to be valid, it must be reliable, so it is important to report a test's reliability. *Depending on the nature of the decisions made from the test results, you generally want reliability to be .80 or higher.*

Return now to our original measurement and evaluation challenge. Kelly is interested in learning the reliability of the field test she will use because it will tell her whether the results she obtains are consistent from one testing period to another. If the test is any good, the results should vary little from one testing session to another. Additionally, the observed differences in $\dot{V}O_2$max obtained from the field test should reflect true differences in $\dot{V}O_2$max and not be simply a function of measurement errors.

The following practical implications derive from this presentation:

▶ There should be observed score variance. (If there is none, the reliability is undefined—because of division by zero.)

▶ The error score variance should be relatively small in relation to the total variance.

▶ Generally, longer tests are more reliable than shorter tests. This is true because as tests increase in length, there is an increase in observed score variance that is more likely a function of an increase in true score variance than in error score variance. (This assumes that, although both tests are made up of "good items," the longer test is made up of more good items than the shorter test.)

Mastery Item 6.3

Make a list of things that might cause a person's observed blood pressure to not reflect his or her true blood pressure.

You may be saying to yourself, This is all fine, but how does one ever know the person's true score? This is absolutely correct: One can never know the person's true score. However, the observed score is readily available, and there are ways of estimating the error score variation for a set of scores. Therefore, as noted on the right side of equation 6.1, we can estimate the reliability using the observed score variance and error score variance.

Mastery Item 6.4

Make a list of things that could cause a person's written test score to not reflect his or her true knowledge in the area tested.

Calculating the Reliability Coefficient

Let's turn to the actual calculation of a reliability coefficient. Reliability coefficients are classified into one of two broad types: interclass coefficients (based on the PPM correlation coefficient, presented in chapter 4) and intraclass coefficients (based on analysis of variance, or ANOVA, models, presented in chapter 5).

Interclass Reliability

Let us first look at three interclass reliability methods: test–retest reliability, equivalence reliability, and split-halves reliability.

Test–Retest Reliability   Consider the simplest way of determining if a test is reliable, or consistent. We could simply give the test to the subjects on two occasions (e.g., on the same day) and then correlate the two sets of observations using the PPM correlation coefficient and see if the correlation is high. This is exactly what is done with the test–retest reliability coefficient. Look at the two trials of sit-up performance presented in table 6.2. The PPM correlation coefficient is calculated to be .927, a correlation between scores from the two test administrations that is certainly high enough to call the test reliable. The coefficient suggests that 92.7% of the observed score variance is true score variance. If the time between testing occasions is longer (e.g., days or weeks), the test–retest reliability coefficient may be called **stability reliability.** That is, the measure is consistent or stable across time.

Mastery Item 6.5

Use SPSS to confirm the reliability of .927 reported in table 6.2. (Hint: Use SPSS to calculate a correlation coefficient as illustrated in chapter 4, p. 52.)

Table 6.2   **Sit-Up Performance for 10 Subjects**

Subject	Trial 1	Trial 2
1	45	49
2	38	36
3	54	50
4	38	38
5	47	49
6	39	38
7	39	43
8	42	43
9	29	30
10	42	42
Sum (Σ)	413	418
Mean *(M)*	41.3	41.8
Standard deviation *(s)*	6.6	6.5
Variance *(s^2)*	43.6	41.7
	$r_{xx'} = .927$	

Equivalence Reliability   A second way to determine interclass reliability is by using an equivalence reliability coefficient. Consider the teacher who is concerned about cheating during a written test. The teacher develops two parallel or equivalent forms of an exam and distributes the exams in the class so that no two adjacent students receive the same examination. However, how should this instructor now determine grades? Should there be two different grading procedures for the same class? Does student performance depend on which exam was completed? This teacher should first determine the equivalence of the two examinations. To do so, a test group is asked to take each of the examinations (both forms) under nearly identical conditions. Half of the subjects should take form 1 first and half should take form 2 first, so that no order effect causes scores on the test to be affected. An assumption must be made that the tests are parallel and that taking the first test neither hinders nor helps a student on the second test. The results from the two administrations are then correlated to determine if there is reliability, or consistency, between the two test forms.

Note again that this is simply a PPM calculation in which the two variables correlated are the test scores from the respective forms. This is an equivalence reliability coefficient.

You may be thinking that both of the interclass reliability examples are a bit far-fetched, because it is unlikely that teachers will administer tests on more than one occasion (which is a requisite for determining the reliability of a test). You are correct! Teachers will typically administer a test only once because time constraints and because examinee fatigue might otherwise come into play and negatively affect scores on subsequent trials. Additionally, practice can also affect the subsequent scores and thus the calculated reliability. However, there are ways to make minor adjustments in the equivalence methods and still reach a conclusion regarding a test's reliability. Consider how a teacher might create two "equivalent" forms of a single test. A teacher could create two equivalent forms of the test after the test is administered by assigning each person a score on two halves of the test (e.g., a score for the odd items and a score for the even items). Thus, the odd and even portions of the test can be perceived of as equivalent forms.

Mastery Item 6.6

When might it not be a good idea to split a written test into two equal parts consisting of the first half and the second half of the test?

Split-Halves Reliability The PPM correlation coefficient can be calculated between the scores on the halves of the test and used as an estimate of the reliability of the test. Table 6.3 has sample data for calculating the split-halves reliability. The split-halves reliability using the PPM correlation coefficient is .639. Because it was suggested earlier that a reliability of .80 or higher is desirable, you might be tempted to dismiss as unreliable the test whose scores are shown in table 6.3. However, an additional aspect of the data presented in the table needs to be considered. The reliability coefficient calculated from the data in table 6.3 is the correlation between two halves of the test (let's assume that each half consists of 13 items, giving a total test length of 26 items). We indicated that longer tests are generally more reliable than shorter ones. Because the .639 calculated in table 6.3 was based on a test of 13 items in length, we must now estimate the reliability of the original 26-item test. The Spearman–Brown prophecy formula, equation 6.2, is used to estimate the reliability of a test when the test length is changed:

$$r_{kk} = \frac{k \times r_{ll}}{1 + r_{ll}(k-1)} \tag{6.2}$$

where r_{kk} is the predicted (prophesied) reliability coefficient when the test length is changed k times, k is the number of times the test length is changed and is defined as

$$\frac{\text{the number of items for which an estimate of reliability is desired}}{\text{the number of items for which the reliability has been calculated}}$$

and r_{ll} is the reliability that has been previously calculated. Thus, to estimate the reliability for the 26-item test, we substitute into the Spearman–Brown prophecy formula and obtain the following:

$$r_{kk} = \frac{26 \div 13 \times .639}{1 + .639([26 \div 13] - 1)}$$

$$r_{kk} = \frac{1.278}{1.639}$$

$$r_{kk} = .78$$

Thus, the estimated reliability for the original 26-item test is .78. The reliability is said to have been adjusted with the Spearman-Brown prophecy formula.

Table 6.3   **Odd-Even Scores for Ten Subjects**

Subject	Odd score	Even score
1	12	13
2	9	11
3	10	8
4	9	6
5	11	8
6	7	10
7	9	9
8	12	10
9	5	4
10	8	7
Sum (Σ)	92	86
Mean (M)	9.2	8.6
Standard deviation (s)	2.2	2.6
Variance (s^2)	4.8	6.7
	$r_{xx'} = .639$	

This number can also be estimated from table 6.4, which shows values of r_{kk} calculated from equation 6.2 using numerous values of r_{11} (left column) and k (column headings). The number of times you desire to change the test length (k) is listed across the top of table 6.4 (0.25-5.0). You determine the predicted reliability (r_{kk}) by finding where the appropriate row and column intersect. For example, if your calculated reliability (r_{11}) is .40 and you increase the test length by a factor of five, the estimated reliability is .77.

You will note that there are values of k listed in table 6.4 that are less than 1. This indicates that the instructor can estimate the reliability for a test of shortened length. For example, assume that an instructor has a written test of 100 items with a reliability of .92. Randomly splitting the test into equal parts with 50 items each would result in two tests with predicted reliabilities of .85. This would reduce the time to administer and score the examination (and make the students happier!) and give the teacher the opportunity to have two forms of the test. The Spearman–Brown prophecy formula can be used to estimate the reliability of a test when the test length is changed. The formula can be used with the interclass reliability estimate or with the intraclass reliability, which is discussed next.

Mastery Item 6.7

Use table 6.4 to estimate the reliability of a 75-item test when the reliability you currently have is .50 for 25 items.

Intraclass Reliability

The interclass reliability, based on the correlation between two measures, is different from the intraclass reliability, based on ANOVA. Assume that you measured a certain skinfold on one group of subjects three times. You might want to estimate the reliability of the three measures. However, the interclass model permits you to correlate only two trials of skinfolds reputedly measuring the same thing because the PPM is used to correlate only two things at a time. The intraclass model, on the other hand, will allow you to estimate reliability for more than two trials. This is important because you might want to estimate the reliability for more than two trials, given that reliability generally increases as the number of trials increases.

Table 6.4 Values of r_{kk} From Spearman-Brown Prophecy Formula

r_{11}	k (change in test length)						
	0.25	0.5	1.5	2.0	3.0	4.0	5.0
.10	.03	.05	.14	.18	.25	.31	.36
.14	.04	.08	.20	.25	.33	.39	.45
.18	.05	.10	.25	.31	.40	.47	.52
.22	.07	.12	.30	.36	.46	.53	.59
.24	.07	.14	.32	.39	.49	.56	.61
.28	.09	.16	.37	.44	.54	.61	.66
.32	.11	.19	.41	.48	.59	.65	.70
.36	.12	.22	.46	.53	.63	.69	.74
.40	.14	.25	.50	.57	.67	.73	.77
.44	.16	.28	.54	.61	.70	.76	.80
.48	.19	.32	.58	.65	.73	.79	.82
.50	.20	.33	.60	.67	.75	.80	.83
.52	.21	.35	.62	.68	.76	.81	.84
.56	.24	.39	.66	.72	.79	.84	.86
.60	.27	.43	.69	.75	.82	.86	.88
.64	.31	.47	.73	.78	.84	.88	.90
.68	.35	.52	.76	.81	.86	.89	.91
.72	.39	.56	.79	.84	.89	.91	.93
.76	.44	.61	.83	.86	.90	.93	.94
.80	.50	.67	.86	.89	.92	.94	.95
.84	.57	.72	.89	.91	.94	.95	.96
.88	.65	.79	.92	.94	.96	.97	.97
.92	.74	.85	.95	.96	.97	.98	.98
.96	.86	.92	.97	.98	.99	.99	.99

Additionally, if there is a constant difference between two trials (i.e., each person's score goes up or down by the same amount), the interclass reliability would be 1.00, but from a theoretical perspective the results would not be consistent (something does seem wrong). For example, in testing skinfold fat, the measures could become smaller with each measurement if the subcutaneous fat is still compressed from the previous measure. Another example of constant change is demonstrated in table 6.5, where the PPM correlation is perfect ($r_{xx'} = 1.00$) yet the reliability (i.e., consistency of measurement) is lacking because each person's score increased by 10 on the second trial. The intraclass reliability model can address this issue. Significant mean differences across trials necessitate a deeper look into the changes across trials. It may be that subject learning or fatigue is affecting the reliability.

The most common names used for the intraclass reliability models are Cronbach's alpha coefficient, Kuder–Richardson formula 20 (KR_{20}), and ANOVA reliabilities. Each of these is calculated in essentially the same manner. The total variance in scores is partitioned into three sources of variation: people, trials, and people-by-trials. People variance is the observed score (total) variance between the subjects. Trial variance is based on the

Table 6.5 **Effect of a Constant Change in Measures**

Subject	Trial 1	Trial 2
1	15	25
2	17	27
3	10	20
4	20	30
5	23	33
6	26	36
7	27	37
8	30	40
9	32	42
10	33	43
Sum (Σ)	233	333
Mean (*M*)	23.3	33.3
Standard deviation (*s*)	7.7	7.7
Variance (*s²*)	59.1	59.1
	$r_{xx'} = 1.00$	

variance across the trials. People-by-trials variation is based on the fact that not all subjects perform equally differently across the trials. People variance is considered total variance. People-by-trials variance and trial variance are considered error variance. Reliability is estimated by subtracting error variance from total (observed) variance and dividing the result by the total (observed) variance.

Consider equation 6.1 (p. 84), in which reliability can be estimated from observed score and error score variance. People variance is observed variance. People-by-trials variance can be perceived as error variance, or all variance not attributable to people (i.e., trials and people-by-trials variances) can be perceived as error. Having estimates of observed and error variance permits you to use equation 6.1 to estimate the reliability of the scores.

The alpha coefficient is calculated as follows:

$$r_{xx'} = \text{alpha coefficient} = \left(\frac{k}{k-1}\right)\left(1 - \frac{\Sigma s^2_{trials}}{s^2_{total}}\right) \tag{6.3}$$

where *k* is the number of trials, Σs^2_{trials} is the sum of the variance of each trial, and s^2_{total} is the variance for the sum across all trials.

Table 6.6 presents an example of a calculation of the alpha coefficient. The variance calculations are identical to those you learned in chapter 3 (pp. 40 and 41). Note that the alpha reliability estimates the reliability for the total score (i.e., the sum across all trials). You can then use this result in the Spearman–Brown prophecy formula (equation 6.2) to estimate the change in the reliability coefficient if the number of trials is increased or reduced.

Mastery Item 6.8

Use the Spearman–Brown prophecy formula to confirm that the estimated reliability for two trials for the data in table 6.6 is .50.

Table 6.6   Calculating the Alpha Coefficient

Subject	Trial 1	Trial 2	Trial 3	Total
1	3	5	3	11
2	2	2	2	6
3	6	5	3	14
4	5	3	5	13
5	3	4	4	11
ΣX	19	19	17	55
ΣX^2	83	79	63	643
s^2	2.70	1.70	1.30	9.50

$k/(k\text{-}1) \times (1\text{-}(\Sigma s^2_{trials}/s^2_{total}))$

$3/(3\text{-}1) \times (1\text{-}(2.70 + 1.70 + 1.30)/9.50)$

$3/2 \times (1\text{-}5.7/9.50)$

$1.5 \times (1\text{-}.60)$

$1.5 \times .40 = .60 = $ coefficient alpha

Mastery Item 6.9

Use the Spearman–Brown prophecy formula to estimate the reliability of six trials from the data in table 6.6.

Mastery Item 6.10

Use SPSS to confirm the reliability estimate reported for the data in table 6.6.

Commands to calculate the alpha coefficient

1. Download table 6.6 from the text WWW site.
2. Start SPSS.
3. Click on the **Analyze** menu.
4. Scroll down to **Scale** and over to **Reliability Analysis** and click.
5. Highlight "trial1," "trial2," and "trial3" and use the arrow to place them in the **Item** box.
6. Click on **OK**.

Note: If you are using the student version of SPSS, you will not be able to complete mastery item 6.10 because reliability analysis is not an option. You will need to use SPSS to calculate the variance for each trial and for the sum of the trials. Then you can substitute these values into equation 6.3.

The alpha coefficient can also be used when data are scored "correct" (1) or "incorrect" (0). In this case, the alpha coefficient is referred to as the Kuder–Richardson formula 20 (KR_{20}). You should realize, however, that the alpha coefficient and KR_{20} are mathematically equivalent. Jackson, Jackson, and Bell (1980) provided an excellent discussion of the alpha coefficient.

Use the data in table 6.7 to estimate the alpha coefficient for the 10-item test presented. Note that the resulting alpha coefficient in this case is often called the Kuder–Richardson 20 (KR_{20}) reliability because the possible scores are 0 and 1. (KR_{20} will be further presented in chapter 10.) All the calculations, the interpretation, the use of the Spearman–Brown prophecy formula, and so on are the same as for the typical alpha coefficient.

Table 6.7 Student Scores on a 10-Item Multiple-Choice Quiz

Student	Item number									
	1	2	3	4	5	6	7	8	9	10
1	1	1	1	1	1	1	0	1	0	1
2	0	1	0	1	1	0	1	0	1	1
3	0	1	1	0	1	1	0	1	0	0
4	1	0	0	0	1	0	1	1	0	0
5	0	0	0	1	0	1	0	1	0	0

Go to the WWW site for chapter 6 and download the Mastery Item 6.12 data file. These are data from four trials of a tennis wall volley test where students had to hit a tennis ball against a wall for 30 s. They were given four trials. Use SPSS to calculate the interclass reliability between trial 1 and trial 2. Confirm that the result is .729. Now use the Spearman–Brown prophecy formula to estimate the reliability for four trials based on the reliability for a single trial obtained with the interclass correlation. The Spearman–Brown prophesied value should be .9150. Now use SPSS and calculate the alpha reliability across all four trials of the wall volley test. The actual alpha value obtained with the four trials is very close (.9136) to that which you predicted with the Spearman–Brown formula. Note that you will not always get an actual value that is this close to the predicted value.

Index of Reliability

Another statistic important to the interpretation of the reliability coefficient is the **index of reliability.** *The index of reliability is the theoretical correlation between observed scores and true scores and is calculated as the square root of the reliability coefficient (equation 6.5):*

$$\text{index of reliability} = \sqrt{r_{xx'}} \tag{6.4}$$

The square root of the percent of observed score variance accounted for by true score variance (i.e., the reliability, or r_{xx}) is the theoretical correlation between the observed and true scores. Thus, if the reliability of a test is .81, the theoretical correlation between observed and true scores is .90.

Standard Error of Measurement

Reliability obviously deals with a person's true score. Although true score can never be actually determined, as suggested earlier in this chapter, it can be thought of as the average of an infinite number of administrations of the test (where neither fatigue nor practice affects

The SEM represents how much a person's observed score would be expected to vary between tests of the same skill.

the subject's score). *Thus, for any given test administration, your best estimate of a person's true score is the obtained score.*

If the test is administered twice, you would average the test scores to get a better estimate of the true score. The theory is that random positive errors and negative errors will balance out in the long run. Regardless of the person's score, there will obviously be some error associated with it. In other words, it is unlikely in a real-life setting to have a score that is totally without error. Thus, a person's score is expected to change from test administration to test administration. *The standard error of measurement (SEM) reflects the degree to which a person's observed score fluctuates as a result of errors of measurement.* You should not confuse the **standard error of measurement (SEM)** with the standard error of estimate (SEE) presented in chapter 4 (p. 62). Although the two have similar interpretations (and look quite similar), they are different: The SEM relates to the reliability of measurement, whereas the SEE concerns the validity of an estimate.

The SEM is calculated as follows:

$$SEM = s\sqrt{1 - r_{xx'}} \qquad (6.5)$$

where s is the standard deviation of the test and $r_{xx'}$ is its reliability.

Assume a test has a standard deviation of 100 and a reliability of .84. The SEM is calculated to be

$$SEM = 100\sqrt{1 - .84}$$
$$SEM = 100(.4)$$
$$SEM = 40$$

If a person scored 500 on a test whose SEM was 40, one can place confidence limits around the person's observed score in an attempt to estimate what the true score is. *The SEM, just like the standard error of estimate, is interpreted as a standard deviation: The SEM is the standard deviation of the errors of measurement around an observed score.* It reflects how much a person's observed score would be expected to change from test administration to test administration as a result of measurement error. Because error scores are expected to be normally distributed, 68% of the scores would be expected to fall within ±1 SEM of the observed score. In our example, therefore, there is a 68% chance that the subject's true score is between 460 and 540 (i.e., 500 ± 40). Note that you could use table 3.3 to place confidence intervals around a particular observed score. You should be able to see that if you take the observed score and add and subtract 2 SEMs to the observed score, you have placed a 95% confidence interval around the observed score. This is because, as you learned in chapter 3, the mean score plus and minus 2 standard deviations results in capturing approximately 95% of the scores in a normal distribution.

Mastery Item 6.13

Verify that approximately 95% of true scores are within the range of 420 and 580 when the observed score is 500 and the SEM is 40 points (i.e., ±2 SEM).

Mastery Item 6.14

What are the chances that a person's true score is 600 or greater on a test when the observed score is 500 and SEM is 40?

A test does not necessarily have reliability in all settings. Said another way, the reliability of a test is situation specific. *A test is reliable under particular circumstances, administered in a particular way, and with a specific group of people.* It is not appropriate to assume that simply because a test is reliable for one group of subjects (e.g., females) it is automatically reliable for another group (e.g., males). The following list provides factors that can affect a test's reliability.

- **Fatigue**—Fatigue generally decreases reliability.
- **Practice**—Practice generally increases reliability. Thus, practice trials during teaching and training should be encouraged.
- **Subject variability**—Greater variability in the subjects being tested results in greater reliability.
- **Time between testing**—Reliability generally decreases as the time between test administrations increases.
- **Circumstances surrounding the testing periods**—Reliability generally increases with a greater similarity between the testing periods.
- **Appropriate level of difficulty for testing subjects**—The test should be neither too difficult nor too easy.
- **Precision of measurement**—Adequate accuracy must be assured with the measuring instrument. For example, one should measure the 50 yd (45.7 m) dash not to the nearest second but to the nearest hundredth of a second.
- **Environmental conditions**—Such factors as noise, excessive heat, and poor lighting can affect the measurement process.

Test users need to be sensitive to factors that could affect the reliability of the field test chosen.

VALIDITY

We have spent a great deal of time developing procedures for estimating a test's reliability because of the important role that reliability plays in determining a test's validity. *A test must first be reliable in order for it to be valid—to truthfully measure what it reports to measure.* Validity can be subdivided into several different types, of which we will discuss three: content-related validity, criterion-related validity, and construct-related validity. These are summarized here and greatly expanded on in the American Psychological Association's *Standards for Educational and Psychological Testing* (1999). Validity can also be broadly classified as logical or statistical in nature. Regardless of the type of validation procedure involved, a criterion of some sort exists. The criterion can be perceived of as the most truthful measure of what you are attempting to measure.

Types of Validity Evidence

Evidence Based on Content

Evidence based on content is evidence that the test characteristics are representative of the universe of potential items that could have been used. For example, the items on a written test at the end of a unit should reflect the material presented in the unit; the physical tests required for employment should reflect the types of tasks that would be performed on the job.

Evidence Based on Relations to Other Variables

Evidence based on relations to other variables demonstrates that test scores are systematically related to a criterion. A criterion measure is obtained, and test scores are correlated (often using the PPM correlation coefficient) with the criterion. For example, skinfolds validly measure percent fat (the criterion), and distance runs validly estimate $\dot{V}O_2$max (the criterion).

Evidence Based on Response Processes

Evidence based on response processes focuses on the test score as a measure of the unobservable characteristic of interest. Attitudes, **personality** characteristics, and unobservable yet theoretically existing traits are often validated with construct-related evidence. For example, attitudes toward physical activity theoretically exist; students can theoretically evaluate the effective teaching that is conducted within a classroom.

Content-Related Validity

Content-related validity is evidence of truthfulness based on logical decision making and interpretation. The terms *face validity* and *logical validity* are often used for content-related validity. The universe of interest, or the content universe, for a particular test needs to be well defined. For example, the items that appear on a particular cognitive unit test should reflect the knowledge content presented during the unit. A basketball skills test should theoretically include items that constitute the game of basketball (shooting, dribbling, passing, and jumping). That is, the test "looks" like it contains the material presented in class or the instructor desires to test students.

The fact that a test reflects content validity, however, does not necessarily make it valid. For example, consider someone who is taking skinfold measures to estimate percent body fat. Certainly skinfold measures have been shown to validly measure body fat. But if the person taking the measures is unqualified to do so (perhaps he or she is not well trained in the use of calipers) or measures at the wrong location (e.g., takes a posterior calf measure instead of a medial calf measure), the measurement may appear to be valid but is not. The criterion for content validity exists in the mind of the interpreter. Content experts, expert judges, colleagues, and textbook writers can serve as sources to content-validate instruments. Teachers developing cognitive tests are writing items that reflect the content of the course (so the items are content valid).

Criterion-Related Validity

Criterion-related validity is based on having a true criterion measure available. Validity is based on determining the systematic relationship between the criterion and other measures used to estimate the criterion. In short, criterion-related validity is evidence that a test has a statistical relationship with the trait being measured. Other terms for criterion-related validity are *statistical validity* and *correlational validity*; these terms are used because criterion-related evidence is based on the PPM correlation between a particular test and a criterion. For example, refer back to Kelly's situation in the measurement and evaluation challenge; she wants to measure the maximal oxygen uptake ($\dot{V}O_2$max) for a number of young adult subjects. Kelly knows that the best way to do so is to have each person complete a maximal exercise test on a treadmill, cycle ergometer, swimming flume, or other type of ergometer; however, Kelly does not have the equipment and resources for conducting such a maximal test on each subject. Therefore, Kelly is seeking alternative measures that can be used to estimate $\dot{V}O_2$max—submaximal tests, distance runs, and nonexercise models. These alternative measures must first be validated with the criterion measure. To do this, at some point in time, subjects must complete both the criterion test and the test (often called a field test) to be used to estimate the criterion. If a strong relationship is found between the criterion and the alternative test, future subjects need not complete the criterion measure but can have their value on the criterion estimated from the alternative (i.e., field) measure, or **surrogate measure.**

Criterion-related evidence is often further subdivided into **concurrent validity** and **predictive validity.** Both are based on the PPM correlation coefficient. The main difference between concurrent and predictive validities is the time at which the criterion is measured. For concurrent validity, the criterion is measured at approximately the same time as the alternative measure. Using a distance run to estimate $\dot{V}O_2max$ is an illustration of concurrent validity. The criterion is measured "in the future" with predictive validity. To establish predictive validity, the criterion might be assessed many weeks, months, or even years after the original test is conducted. The prediction of heart disease development in later life is based on predictive validity procedures; the criterion—development of heart disease—is not measured until many years later. However, it has been shown that lack of exercise, high body fat, smoking, high cholesterol, and hypertension are all predictive of future heart disease. (Of course, these same variables can be used to predict if one currently has heart disease. Thus, the time at which the criterion is measured and the interpretation of the correlation help to identify whether the criterion-related evidence is concurrent or predictive in nature.) The following list provides some examples of concurrent validity and predictive validity in exercise science, kinesiology, and education. The criteria are followed by a list of possible predictors.

Concurrent Validity

- $\dot{V}O_2max$ (criterion: oxygen consumption)
  - Distance runs (e.g., 1.0 mi, 1.5 mi; 2 km; 9 min, 12 min; 20 m shuttle)
  - Submaximal (e.g., cycle, treadmill, swimming)
  - Nonexercise models (e.g., self-reported physical activity)
- Body fat (criterion: hydrostatically determined body fat)
  - Skinfolds
  - Anthropometric measures (e.g., girths, widths, and lengths)
- Sport skills (criterion: game performance, expert ratings)
  - Sport skills tests (e.g., wall volley tests, accuracy tests, and total body movement tests)
  - Expert judges' evaluation at skill performance

Predictive Validity

- Heart disease (criterion: heart disease developed in later life)
  - Present diet, exercise behaviors, blood pressure, family history of heart disease or related health issues
- Success in graduate school (criterion: grade point average or graduation status)
  - Graduate Record Examination scores
  - Undergraduate grade point average
- Job capabilities (criterion: successful job performance)
  - Physical abilities
  - Cognitive abilities

Sport skills tests are good examples of criterion-related validation procedures. Green, East, and Hensley (1987); Hensley, East, and Stillwell (1979); Hensley (1989); and Hopkins, Schick, and Plack (1984) provide excellent examples of the procedures used to validate sport skills tests. A criterion measure must first be developed and then a variety of skills tests (i.e., a test battery) correlated with the criterion measure to determine which of these are most valid and most helpful in estimating the criterion. If a series of tests is used to estimate the criterion, multiple correlational procedures (see chapter 4) are used rather than the simple Pearson correlation coefficient. However, the logic is the same: An attempt is made to account for variation (i.e., increase the coefficient of determination) in the criterion measure from more than one measure. Consider a golf test. The criterion could be the average score from several rounds of golf. Thus, a study could be conducted in which everyone completes several rounds of golf to obtain the criterion measure. Each subject then completes a variety of skills tests (e.g., driving, long irons, short irons, chipping, and putting), which are then correlated with the criterion measure to determine which of the measures or combination of these measures best estimate the criterion measure. Note that there will always be some error in all of the measures (criterion and estimators).

Interpretation of the criterion-related validity coefficient depends on its absolute value. Because the criterion-related validity coefficient is simply a PPM correlation coefficient, it must range between –1.00 and +1.00. However, the closer the absolute value of the validity is to 1.00, the greater the validity. For example, look at table 6.8 where "Playing golf" is the criterion. The values under "Playing golf" are concurrent validity coefficients. The second highest concurrent validity coefficient listed in table 6.8 is –.65 (drive shot). The highest concurrent validity coefficient is .66 (middle-distance shot). The other values in table 6.8 are correlation coefficients between other golf skills test items.

Table 6.8   Correlation Matrix for Development of a Golf Skills Test

	Playing golf	Long putt	Chip shot	Pitch shot	Middle-distance shot	Drive shot
Playing golf	1.00					
Long putt	.59	1.00				
Chip shot	.58	.47	1.00			
Pitch shot	.54	.37	.35	1.00		
Middle-distance shot	.66	.55	.61	.40	1.00	
Drive shot	-.65	-.62	-.48	-.52	-.79	1.00

Reprinted from Green, East, and Hensley (1987).

Let's return to the standard error of estimate (SEE), which is often reported with concurrent validity coefficients. For example, assume that you have a submaximal test that estimates $\dot{V}O_2$max from a timed distance run of 1 mi in length and the SEE is 4 ml · kg^{-1} · min^{-1}. If someone has a predicted $\dot{V}O_2$max of 50 ml · kg^{-1} · min^{-1}, you can place confidence limits around the predicted score: You can be 68% confident that the actual $\dot{V}O_2$max is between 46 and 54 (i.e., 50 ± 4) ml · kg^{-1} · min^{-1}. Note that the standard error of estimate reflects the accuracy of estimating your score on the criterion measure; in other words, it is a validity statistic.

Development of the criterion measure is extremely important in criterion-related evidence of validity. Examples of how criterion measures can be obtained include the following:

▶ **Actual participation**—One can actually complete the criterion task (e.g., play golf, shoot archery, conduct job-related activities).

▶ **Known valid criterion**—One can use a criterion (e.g., run on a treadmill, obtain underwater weight) that has been previously shown to be valid.

▶ **Expert judges**—Experts judge the quality of the criterion performance. This is often used with team activities (e.g., volleyball) in which it is difficult or impossible to obtain a number that reflects performance on the task being measured.

> ▸ **Tournament participation**—Rankings of abilities can be determined when everyone participates with everyone else (best used when the skilled event is an individual sport).

> ▸ **Known valid test**—Subjects can complete a test that has been previously shown to be valid.

Construct-Related Validity

Construct-related procedures are often used to validate measures that are unobservable yet exist in theory. For example, one's intelligence quotient (IQ) exists in theory, but IQ is not something that can be readily measured. The same is true for attitude measures. Certainly, each of us has attitudes about various behaviors (e.g., exercise, dieting, physical activity), but it is difficult to measure these attitudes directly. This is where construct validity comes into play. *Construct-related validity evidence is essentially a marriage between logical (content) and statistical validity procedures.* To provide construct-related evidence for a particular measurement, you gather a variety of statistical information that, when viewed collectively, adds evidence for the existence of the theoretical construct being measured.

When you are collecting construct-related evidence, your working hypothesis would be worded like this: *If in theory the construct is valid, then such-and-such should occur.* Then test to see if it does occur. The logical part of construct validity is *what should occur.* The statistical part consists of the data that you gather. Continual gathering of information that supports the theory adds to the evidence for the existence of the construct. When what you think should occur is not supported by the data collection, there are two things to consider: It may be that the construct doesn't exist or that the if–then statement was inaccurate. Development of construct validity is highly related to the scientific method presented in chapter 5 (p. 64). A hypothesis is generated, a method is developed, data are collected and analyzed, and a decision is made based on the evidence obtained.

Kenyon (1968a, 1968b) developed a multidimensional instrument to measure Attitudes Toward Physical Activity (ATPA). Certainly attitudes toward physical activity exist. Some people like physical activity and others do not. But how can one measure these perceived attitudes? Kenyon provided evidence that there are a variety of reasons people like or dislike (or engage or don't engage in) physical activity (i.e., ATPA is a multidimensional construct). The unobservable, but theoretically existent, dimensions suggested by Kenyon include the following:

- ▸ Aesthetic experience
- ▸ Catharsis
- ▸ Health and fitness
- ▸ Social experience
- ▸ Pursuit of vertigo ("thrill of victory")
- ▸ Ascetic experience

Consider the aesthetic dimension, which indicates that some people like physical activity for the beauty of movement expressed in such activities as dance, ballet, gymnastics, diving, and skating. To provide construct-related evidence that this dimension exists, one could administer the ATPA to groups of people who engage in different types of behaviors. Your working hypothesis would be, *If the aesthetic dimension exists, people who attend dance concerts, ballet, and gymnastics events should score significantly differently on the aesthetic dimension than people who do not attend such functions.* This is exactly how construct-related evidence is obtained for such unobservable measures.

Construct-related evidence can be used to provide additional evidence for criterion-related validation evidence. Consider the golf test described earlier. A working hypothesis would be, *If this is a valid golf test, the following should occur: Students who have never played golf should score poorly on the test, beginning golfers should score better, experienced golfers should score higher, and members of the golf team should score the best.* This is referred to as the known group difference method of criterion validation. Conducting such a study and testing for differences in the group means (see ANOVA in chapter 5, pp. 74-76) could provide construct-related evidence for the golf test.

Mastery Item 6.15

Using the known group difference method, tell how you would provide construct-related evidence of a new field test you want to use to estimate aerobic capacity.

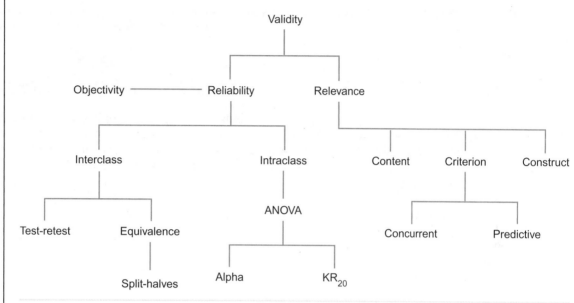

Figure 6.1   Diagram of validity and reliability terms.

Figure 6.1 illustrates the relationships among the various aspects of validity that we have introduced.

Much of the information presented in this chapter has related to the Pearson product-moment (PPM) correlation coefficient, introduced in chapter 4 (p. 52). In some cases the PPM is interpreted as a reliability coefficient. In other cases it could be an objectivity coefficient or a criterion-related (either concurrent or predictive) validity coefficient. In all cases the PPM correlation coefficient is calculated as you learned in chapter 4. The difference in the interpretation depends on what two things are correlated. This is described in figure 6.2. Essentially, if two trials of the same thing are correlated but measured at different times, then this PPM is interpreted as a reliability coefficient. If two different raters are correlated when they scored the same test, this PPM is interpreted as an objectivity coefficient. If two different forms of the same test are being correlated, this is an estimate of equivalence. If one of the measures being correlated is a criterion, then you are working with validity. Whether the PPM calculated is a concurrent or predictive validity coefficient depends on when the criterion was measured. This illustrates the generalized use of the PPM for estimating reliability, objectivity, and validity. It is important for you to be able to distinguish between these different correlations.

Correlation		Application/interpretation
X	*Y*	
Trial 1	Trial 2	→ Reliability estimate
Rater 1	Rater 2	→ Objectivity estimate
Form A	Form B	→ Equivalence estimate
Test	Concurrent criterion	→ Concurrent validity estimate
Test	Future criterion	→ Predictive validity estimate

Figure 6.2   Applications of the Pearson product-moment correlation in reliability and validity.

Mastery Item 6.16

Determine whether each of the following is a reliability coefficient, objectivity coefficient, or validity coefficient.

- A researcher uses the PPM to correlate an abdominal skinfold with hydrostatically determined percent body fat.
- A teacher correlates sit-up scores from students on Monday with sit-up scores from the same students on Tuesday.
- An instructor scores an essay test and then asks a colleague to also read and score the test. The instructor then uses the PPM to determine if the scores are similar from each reader.

APPLIED RELIABILITY AND VALIDITY MEASURES

Let's look at some examples of reliability and validity from exercise, sport science, and kinesiology. Recall that table 6.8 (p. 97) shows a correlation matrix used to develop a golf skills test. Table 6.9 presents various golf test batteries. Table 6.10 contains examples of reliability coefficients from various sport skills tests; remember that reliabilities are a function of the group being tested and are situation specific. Table 6.11 provides concurrent-validity coefficients for estimating $\dot{V}O_2$max from a variety of measures. Some of the authors used a single measure to estimate $\dot{V}O_2$max, whereas others used multiple regression.

Mastery Item 6.17

Comment in detail on the correlations presented in table 6.8.

Mastery Item 6.18

Which of the golf test batteries presented in table 6.9 would you use? Why?

Mastery Item 6.19

Why does the validity coefficient increase in value in the study by Murray and colleagues shown in table 6.11?

Table 6.9 Concurrent Validity Coefficients for Golf Test

2-item battery		3-item battery		4-item battery	
Middle-distance shot		Middlel-distance shot		Middle-distance shot	
Pitch shot	.72	Pitch shot		Pitch shot	
		Long putt	.76	Long putt	
				Chip shot	.77

Adapted from Green, East, and Hensley (1987).

Table 6.10   Reliability Measures From Sport Skills and Fitness Tests

Author	Test item	Reliability ($r_{xx'}$)
Engelman & Morrow (1991)	Traditional pull-up (boys)	.83 to .92
	Traditional pull-up (girls)	.91 to .92
	Modified pull-up (boys)	.68 to .83
	Modified pull-up (girls)	.77 to .83
Green, East, & Hensley (1987)	Golf—chip shot (males)	.85
	Golf—chip shot (females)	.86
	Golf—long putt (males)	.87
	Golf—long putt (females)	.93
	Golf—short putt (males)	.54
	Golf—short putt (females)	.46
Hensley, East, & Stillwell (1979)	Racquetball—short volley (males)	.77
	Racquetball—short volley (females)	.86
	Racquetball—long volley (males)	.85
	Racquetball—long volley (females)	.82
Hensley (1989)	Tennis—serve (males)	.86 & .95
	Tennis—serve (females)	.79 & .88
	Tennis—volley (males)	.70 & .72
	Tennis—volley (females)	.69 & .79
Hopkins, Schick, & Plack (1984)	Basketball—shot (males)	.84 to .95
	Basketball—shot (females)	.87 to .95
	Basketball—pass (males)	.88 to .96
	Basketball—pass (females)	.82 to .91
Nelson, Yoon, & Nelson (1991)	Modified push-up (boys)	.78 to .89
	Modified push-up (girls)	.77 to .91
Rikli, Petray, & Baumgartner (1992)	1/2 mile (boys)	.65 to .82
	1/2 mile (girls)	.32 to .77
	3/4 mile (boys)	.48 to .94
	3/4 mile (girls)	.58 to .83
	1 mile (boys)	.44 to .87
	1 mile (girls)	.34 to .90
Schick & Berg (1983)	Golf—5 iron	.90

Note: All reliabilities are intraclass.

Table 6.11 Concurrent Validity Measures for $\dot{V}O_2$

Author	Criterion	Predictor(s)	Validity (r)	Standard error of estimate (ml · kg^{-1} · min^{-1})
Getchell, Kirkendall, & Robbins (1977)	$\dot{V}O_2$max	1.5 mi run	.92	2.38
Jackson et al. (1990)	$\dot{V}O_2$peak	Gender		
		Activity code		
		Age		
		BMI	.78	5.70
Kline et al. (1987)	$\dot{V}O_2$max	1 mi walk		
		Gender		
		Age		
		Body weight	.88	5.00
Murray et al. (1993)	$\dot{V}O_2$peak	20 min steady-state run	.68	5.32
	$\dot{V}O_2$peak	20 min steady-state run		
		Gender	.73	4.96
	$\dot{V}O_2$peak	20 min steady-state run		
		Gender		
		Weight	.79	4.45

Note: BMI = body mass index.

Measurement and Evaluation Challenge

You should be able to determine the steps that Kelly must take to select and administer a reliable and valid field test of aerobic capacity to her college-age students. She must first determine if the test she has selected is reliable. That is, are the results consistent from one administration to another, assuming that the participants have not actually changed in training or activity levels? She must be sensitive to the standard error of measurement (SEM). Next, she must determine the concurrent validity between the proposed field test and the actual treadmill performance of the participants. Such information might be available in the research literature, or she may need to actually work with a researcher to obtain this vital information. She should concern herself with the types of participants she is testing and those used in the validation process. If they are similar, she should feel confident that the field test results will provide a fairly accurate estimate of the participants' aerobic capacities. The field test will not provide an exact measure of the participants' aerobic capacities. Thus, she must be concerned about the standard error of estimate (SEE) in estimating actual $\dot{V}O_2$max from the surrogate (i.e., field) measure.

SUMMARY

The issues of reliability, objectivity, and validity are the most important ones that you will encounter when you test and evaluate human performance, whether the performance is in the cognitive, affective, or psychomotor domain. Reliability coefficients represent the consistency of response and range from 0 (totally unreliable) to 1.00 (perfectly reliable). Likewise, objectivity (interrater reliability) values range from 0 to 1.00. The standard error of measurement (SEM), a reliability statistic, reflects the degree to which a person's score will change as a result of errors of measurement. A validity coefficient represents the degree to which a measure correlates with a criterion. Statistical validity coefficients range from −1.00 to +1.00. The absolute value of the validity coefficient is important. A value of 0 indicates no validity; 1.00 represents perfect correlation with the criterion. The standard error of estimate (SEE), a validity statistic, indicates the degree to which a person's predicted or estimated score will vary from the criterion score.

Last, you should realize that reliability and validity results are not typically generalizable. The reliability or validity obtained is specific to the group tested, the environment of testing, and the testing procedures. Whether the reliability and validity results that you obtain can be inferred to another population or setting must be studied before making such an inference.

Now that you are familiar with concepts related to reliable and valid assessment, you should be able to better evaluate the instruments that you might use in human performance testing.

Criterion-Referenced Measurement

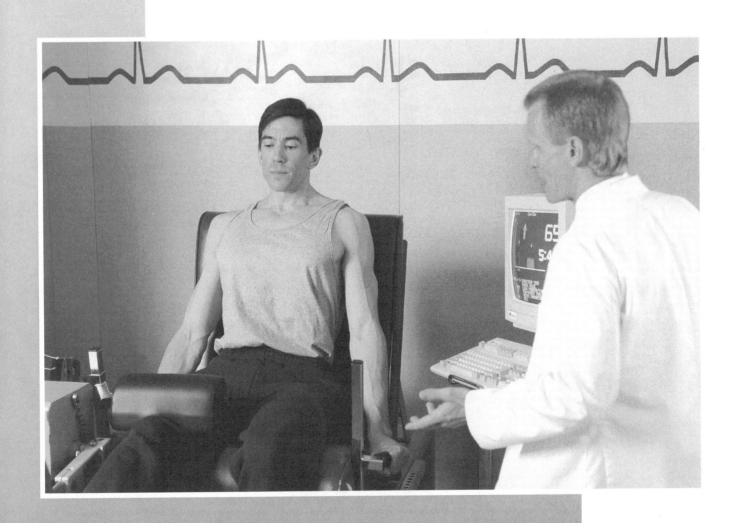

Measurement and Evaluation Challenge

Christina is an athletic trainer working primarily in sport rehabilitation. She recently has been faced with an increasing number of patients with pulled hamstrings. Currently there is speculation in the sport rehabilitation field that this increase in injuries could be related to creatine use. Christina would like to get more scientific information on the relationship between hamstring pulls and creatine use. However, when she received her professional preparation, the measurement techniques she learned were associated with only norm-referenced measurement. She is in a quandary about how to approach this question. She is interested in determining if creatine use is a valid predictor of hamstring pulls. Christina decides to go to the library and examine the literature to see if she can determine the techniques that will allow her to examine this question.

Objectives

After studying this chapter, you will be able to

▸ define a criterion-referenced test,

▸ explain the approaches for developing criterion-referenced standards,

▸ explain the advantages and limitations of criterion-referenced measurement,

▸ select appropriate statistical tests for the analysis of criterion-referenced tests,

▸ interpret statistics associated with criterion-referenced measurement,

▸ discuss and interpret epidemiologic statistics, and

▸ use SPSS to calculate criterion-referenced statistics.

In the area of human performance, we are blessed with many variables measured on the ratio scale (see chapter 3). The speed that a person runs and the distance that a person jumps are common ratio measures. Some variables may not be measurable on this scale and are instead reported as rankings and ratings. Players can be ranked on overall playing ability or rated on a specific skill. Other variables, such as gender and ethnicity, can be measured only categorically; recall from chapter 3 that these are termed nominal variables. Some variables can be measured in more than one way. For example, height is normally measured on a ratio scale and reported in feet and inches or in centimeters. However, let's assume that a teacher wants to equate teams in a basketball class on height. The teacher could rank the students from tallest to shortest and assign teams based on the player's rank. The teacher could likewise cluster students into groups based on height. The taller players could play against one another, whereas the shorter players would be matched against players of similar height.

This last example represents the establishment of **cutoff scores** to create forced categories. Cutoffs are important when comparisons of individual performances are not the primary concern but rather we want to know whether a certain performance (or minimum) level has been achieved. Conventional basic statistics do not apply when variables are measured in this manner; for example, calculating the mean and standard deviation for data that have been forced into categories would be inappropriate. Therefore, specific techniques applicable to nominal measurement must be used (as you should recall from chapter 5). Criterion-referenced tests (CRTs) are appropriate for this situation.

Technically, there is not a great deal of difference between what you learned in chapter 6 and what you will learn in chapter 7. The primary difference is in the level of measurement used to describe performance. In chapter 6, the variables were continuous in nature. The variables presented in chapter 7 are categorical in nature. However, beyond that, the concepts of reliability and validity presented in chapter 6 can be readily adapted to the variables presented in chapter 7. We will present information about criterion-referenced reliability and criterion-referenced validity. Our focus in chapter 6 was on the person's specific score. Our focus in chapter 7 is into which category the person is placed.

A **criterion-referenced test (CRT)** is one that is deliberately constructed to yield measurements that are directly interpretable in terms of specific performance standards. Performance standards are generally specified by defining a class or domain of tasks that should be performed by the individual (Nitko 1984, 12).

Mastery Item 7.1

What is a physical performance task that is well suited to criterion-referenced testing?

CRTs are used to make categorical decisions, such as passing or failing or meeting a standard versus not meeting the standard, or to classify subjects as masters or nonmasters. When specific, well-defined goals are identifiable, CRTs may best measure the reliability and validity of that particular item. CRTs are not limited to nominal measurement. Often, continuous variables can be used with criterion-referenced testing methods. For example, push-up or sit-up performance can be evaluated using criterion cutoffs rather than norm-referenced methods. Historically, programmed instruction that centered on **behavioral objectives**—specifically written goals with instructions on how they can be obtained—was well suited for this type of measurement approach. Mastery instruments based on behavioral objectives are best exemplified by tests involving licensing, such as the test you took to get a driver's license and, in the human performance area, Red Cross standards for lifesaving and swimming certification. It is easy to see in these examples that a minimum standard must be obtained before competency is proclaimed and a license is granted. In both cases, cutoff scores are referenced to a standard based on a theoretical minimal performance level.

DEVELOPING CRITERION-REFERENCED STANDARDS

Four basic approaches are used to develop criterion-referenced standards for tests of human performance (Safrit, Baumgartner, Jackson, and Stamm 1980):

▶ The *judgmental* approach is based on the experience or beliefs of experts. It reflects what they believe is an appropriate level based on their background and experience in testing and evaluating human performance. For example, many high school volleyball coaches require players to be able to serve overhand to play on the varsity team. The coach might set a cutoff level, such as placing 8 out of 10 overhand serves in the court.

▶ The *normative* approach uses norm-referenced data to set standards; some theoretically accepted criterion is chosen. The President's Challenge guidelines for qualifying for the youth fitness awards (see chapter 12) are an example of a criterion-referenced application of normative data. To qualify, a student has to attain the 50th or 85th percentile on all test items. This criterion was based not only on experts' opinions but also on the available norms.

▶ The *empirical* approach relies on the availability of an external measure of the criterion attribute. Cutoff scores are directly established based on the data available on this external attribute. This approach is the least arbitrary of the four. However, it is seldom used because of the lack of a directly measurable external criterion. An example is a firefighter having to scale a 5 ft (1.5 m) wall to perform his or her duties. This is a concrete example of a pass–fail item that is based on the empirical approach. Another excellent example of this approach is the work of Cureton and Warren (1990), presented later in this chapter (pp. 108-109).

▶ The *combination* method involves using all available sources: experts, prior experience, empirical data, and norms. Usually, experts' opinions and norms are the basis for making criterion-referenced decisions in human performance. The FITNESSGRAM Healthy Fitness Zone (see chapter 12) standards were established in this manner.

Mastery Item 7.2

Assume that you are conducting an elite basketball camp for high school players. What are some of the variables that you could use as a basis for camp eligibility?

Mastery Item 7.3

How would you go about establishing cutoff scores for one of the variables you selected in mastery item 7.2?

DEVELOPMENT OF CRITERION-REFERENCED TESTING

The specific use of the term criterion-referenced testing is generally traced to a 1962 article by Robert Glaser and D.J. Klaus. Glaser and Klaus developed this term because of a number of limitations they believed were inherent in norm-referenced tests; the primary shortcoming is that such tests are constructed to have content validity over a wide range of instructional goals and philosophies. Consequently, the more specific norm-referenced tests are, the less marketable they become. For this reason, norm-referenced tests are not well suited to the assessment of specific objectives. For example, if a norm-referenced approach is used to determine who receives a driver's license, then your ability to "pass" the test would be based on the group tested and not on your absolute ability to drive a car. The primary goal of a norm-referenced test is to establish a range of behavior to discriminate among levels of knowledge, ability, or performance. If a certain level of performance is necessary, then norm-referenced testing does not provide this information in the most efficient way. CRTs, on the other hand, are usually structured to assess far fewer objectives than a traditional norm-referenced test and therefore can be set up to identify specifically enumerated goals for behavioral items. For example, how many sit-ups should a 10-year-old boy be able to complete to be considered physically fit?

The primary difference between norm-referenced tests and CRTs is that CRTs are evaluated categorically. The traditional statistical techniques used to establish the reliability and validity of norm-referenced tests presented in chapter 6 cannot be used with CRTs. Therefore, you must choose specific techniques that best estimate the reliability and validity of criterion-referenced measures. Indexes of reliability associated with criterion-referenced tests are called indexes of dependability. Methods used to determine dependability are based on classical test theory or generalizability theory. The indexes allow you to determine not only **proportion of agreement** (P) (which refers to the consistency with which performances are categorized across methods or trials) but also the consistency with which decisions are made. Specific examples of indexes of dependability are presented later in this chapter (p. 113).

Cureton and Warren (1990) summarized the advantages and limitations of criterion-referenced measurement:

Advantages

▷ Criterion-referenced standards represent specific, desired performance levels that are explicitly linked to a criterion.

▷ Because they are absolute standards, they are independent of the proportion of the population that meets the standard.

▷ If standards are not met, then specific diagnostic evaluations can be made to improve performance to the criterion level.

▷ Because the degree of performance is not important, competition is based on reaching the standard, not on bettering someone else's performance level.

The following are other key advantages:

▷ Performance is linked to specific outcomes.

▷ Individuals know exactly what is expected of them.

Limitations

▷ Cutoff scores always involve some subjective judgment. Because few criteria are clear-cut, philosophical guidelines can drastically affect the selection of the performance criterion. Authorities often disagree on exact levels, so cutoffs are sometimes arbitrarily determined.

▷ Misclassifications can be severe. Consider a hypothetical situation in which a doctor is prescribing medication based on a criterion-referenced standard. Misclassification of patients could have severe consequences for their health.

▷ Because cutoffs must be set at some level, those individuals who attain the cutoff level may not be motivated to continue to improve. Conversely, individuals who never attain the cutoff could become discouraged and lose interest.

To examine some of these limitations, Cureton and Warren (1990) studied criterion-referenced standards for the 1 mi (1.6 km) run/walk test, for which the FITNESSGRAM (Cooper Institute for Aerobics Research 1987) and Physical Best (AAHPERD 1988) both provide criterion-referenced standards. To examine the validity of these standards, these authors developed an external criterion:

The criterion was defined as the lowest level of $\dot{V}O_2$max in children and adolescents consistent with good health, minimized disease risk, and adequate functional capacity for daily living. Because no empirical data specifically identifies [sic] the minimum level, the criterion $\dot{V}O_2$max was based primarily on indirect evidence relating aerobic capacity to health disease/risk. (p. 10)

Essentially, Cureton and Warren determined 1 mi run/walk speeds that corresponded to criterion levels of $\dot{V}O_2$max and converted these speeds to mile run times. The authors evaluated data on 581 boys and girls aged 7 to 14 against the FITNESSGRAM criterion and the Physical Best criterion. Their results are presented in figure 7.1. The figure indicates that 496 of the 581 cases (85%) were properly classified by the FITNESSGRAM standards, whereas only 357 (61%) were properly classified by the Physical Best standards. Fifteen percent (11% + 4%) were misclassified by the FITNESSGRAM, and 39% (35% + 4%) were misclassified by the Physical Best. This analysis highlights the importance of setting cutoff standards correctly.

FITNESSGRAM (a)

	Below the criterion VO_2	Above the criterion VO_2
Did NOT achieve the standard on the run/walk test	24 (4%)	21 (4%)
DID achieve the standard on the run/walk test	64 (11%)	472 (81%)

Physical Best (b)

	Below the criterion VO_2	Above the criterion VO_2
Did NOT achieve the standard on the run/walk test	130 (22%)	23 (4%)
DID achieve the standard on the run/walk test	201 (35%)	227 (39%)

Figure 7.1 Comparison of *(a)* FITNESSGRAM and *(b)* Physical Best standards for 1 mi run/walk times.

Another example of criterion-referenced standards is cholesterol levels set by professional associations. The American Heart Association (AHA) and the National Heart, Lung, and Blood Institute (NHLBI) have established cutoff values for blood cholesterol levels related to the risk of coronary heart disease. They are as follows:

▸ Low risk: <200 mg/dl

▸ Moderate risk: ≥200 mg/dl or ≤240 mg/dl

▸ High risk: >240 mg/dl

A physician who is counseling a patient about the risk of coronary heart disease would use the patient's blood test results and compare them with these standards. The physician might advise the following:

▸ No need for concern (patient's level = 180 mg/dl).

▸ Increase physical activity levels and eat a low-fat diet (patient's level = 215 mg/dl).

▸ Increase physical activity levels, eat a low-fat diet, and take prescription medication (patient's level = 300 mg/dl).

STATISTICAL ANALYSIS OF CRTs

Not only is the procedure for setting cutoffs critical, but so is the selection of statistical tests for examining the appropriateness of the cutoffs. Selection of the statistical tests to be used to analyze CRTs is based on the same principles as for selecting norm-referenced tests. The first factor to consider is the level of measurement of the variables involved. With CRTs, you categorize data into nominal variables; therefore, you must select statistical tests appropriate for this level of measurement. Remember that nominal variables are categorical in nature. For tests that are measured on a continuous scale to be evaluated with criterion-referenced instruments, the scores must first be categorized above and below the cutoff criterion. For criterion-referenced testing, the primary tool for analysis is a statistical technique using a **contingency table** (a 2×2 chi-square; see figure 7.2) for identifying those who score above the cutoff and those who score below the cutoff. Figure 7.2 illustrates the stability (dependability) of the CRT over 2 days. People classified as not meeting the standard (n_1) on both days or meeting the standard (n_4) on both days are consistently classified. Those classified as meeting the standard on one day and not meeting the standard the next (n_2) or vice versa (n_3) are misclassified. **Marginals** are the sum of observations for a specific row ($n_1 + n_2$ or $n_3 + n_4$) or column ($n_1 + n_3$ or $n_2 + n_4$) of a contingency table (see figure 7.2).

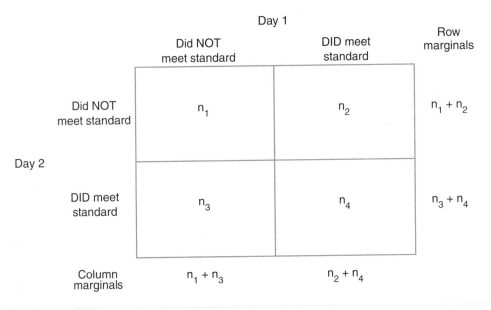

Figure 7.2   A 2×2 contingency table for a criterion-referenced test taken over 2 days.

The next factor to consider in analysis is the specific measurement situation. The measurement situations are the same as those associated with norm-referenced testing. To establish the reliability of a CRT, you first need to determine whether you're concerned with the equivalence or the stability of the test. To measure the validity, you must have a criterion measure. The criterion measure reflects the true state of circumstances regarding the test being investigated. Recall Christina's challenge from the beginning of this chapter. Her interest in the relationship between creatine use and muscle pulls is a validity study. The criterion is whether or not the person had a muscle pull and the predictor variable is whether or not the person was taking creatine.

STATISTICAL TECHNIQUES TO USE WITH CRTs

There are several statistics available and used to estimate the reliability and validity of CRTs. This text presents the techniques of chi-square (chapter 5), proportion of agreement *(P)*, the **Phi coefficient** (actually a Pearson product-moment correlation between two dichotomous variables), and Kappa (K). These are techniques that reflect association and agreement and can be used with data measured on a nominal scale.

As illustrated in chapter 5, the chi-square test is a test of association between nominally scaled variables. Logically, you would want there to be an association between how one does on the first attempt of a CRT and on the second attempt. This is an illustration of CRT reliability. Likewise, you would like there to be an association between how one does on a field test of a measure and how that person would do on a more truthful measure (i.e., the criterion) of the characteristic being measured. This is an illustration of CRT validity. Recall from chapter 5 that the null hypothesis in both of these tests is that there is *no* association but rejection of the null hypothesis results in deciding there *is* an association between the variables.

Note that the variables are scored 0 or 1 for both of the measures. You can calculate the Pearson product-moment correlation coefficient (chapter 4) between the dichotomously scored variables. This special case of the Pearson product-moment correlation coefficient is called the phi coefficient. The phi coefficient has limits of –1.00 and +1.00 with a value closer to 1.00 in absolute value indicating increased association and a value close to zero indicating no association.

SPSS produces the chi-square and phi coefficient as statistics options within the **Crosstabs** routine. This will be illustrated later in the chapter.

The proportion of agreement *(P)* value is established by adding the proportions in the cells that are correctly classified; thus *P* is equal to the number of agreements $(n_1 + n_4)$ divided by the total number $(n_1 + n_2 + n_3 + n_4)$. From figure 7.2 it is estimated by the following formula:

$$P = (n_1 + n_4) / (n_1 + n_2 + n_3 + n_4) \tag{7.1}$$

The *P* ranges from 0 to 1.00, and the higher the value, the more closely the data are correctly assigned to cells. The problem with *P* is that values up to .50 could happen simply by chance.

The **Kappa** (K) value is a widely used technique that allows for the correction for chance agreements. It is closely associated with the phi (Φ) coefficient, which is the Pearson product-moment (PPM) correlation calculated on nominal data. K is most appropriately used to assess interobserver agreement but can be used in test–retest situations or to examine the agreement between a predictor and a criterion that are nominally scaled. Although the proportion of agreement is a rough estimate of agreement or association between two nominal variables, the major problem with this statistic is that it does not consider the fact that some of these agreements could be expected purely because of chance. K takes chance agreement into account and therefore gives a more conservative estimate of the association between two nominal variables. The formula for K is

$$K = \left(P - P_c\right) / \left(1 - P_c\right) \tag{7.2}$$

where *P* is the proportion of observed agreement and P_c is the proportion of agreement due to chance. Consider the following example. Four hundred elementary students performed the 1 mi run on each of 2 days. The instructor wished to know if the test could consistently measure the ability of the students to achieve the cutoff established in the FITNESSGRAM. Table 7.1 presents these data.

Table 7.1 CRT Test-Retest Reliability Example

		Day 2		
		Did not achieve the standard	Did achieve the standard	Total
Day 1	Did not achieve the standard	80	20	100
	Did achieve the standard	50	250	300
	Total	130	270	400

$\chi^2 = 137.13$, $df = 1$, $p < .001$
Phi coefficient = .586
$P = (80 + 250)/400 = .825$
Kappa = .576

For this example, P is calculated to be:

$$(250 + 80)/400 = 330/400 \text{ or } .825$$

K is calculated to correct for chance. The P (.825) was estimated previously. The chance values (P_c) are as follows:

$$(130) \times (100)/(400 \times 400) = .081$$

and

$$(270) \times (300)/(400 \times 400) = .506$$

That is, multiply the marginals and divide by n^2. The sum of these properties is .587. Therefore, K = (.825 − .587)/(1 − .587) = .238/.413 = .576. This value is substantially lower than the P value of .825. Therefore, it is suggested that chi-square, phi, percent agreement, and Kappa values be calculated to give the most information about the association involved.

Thus, given a 2×2 table, determine the proportion of observed agreement (P) by summing the number of agreements that appear on the diagonal of the table and dividing by the total number of paired observations. Determine the proportion of chance agreement (P_c) for each cell on the diagonal by calculating the marginals for each row and column. When these marginals are cross-multiplied, the resulting values for each cell represent the values expected attributable to chance. Then obtain the proportion of chance (P_c) agreement by dividing the expected values attributable to chance by the total number of observations. Finally, sum these proportions across all the cells to obtain a total proportion of chance agreement.

Then substitute the proportion of agreement and the proportion of chance agreement into the Kappa formula. The values of K can theoretically range from −1.00 to +1.00; however, a negative value of K implies that the proportions of agreement resulting from chance are greater than those attributable to observed agreements. For that reason, K practically ranges from 0 to 1.00. The magnitude of the K is interpreted just as any other reliability or validity coefficient, with the higher the values, the better. However, because of the adjustment for chance agreement, values seldom exceed .75, which would be considered excellent. Values in the .60 to .75 range are usually considered to be good, whereas values ranging from .40 to .60 are often considered to be acceptable. K is an extremely useful statistic, and it can be used not only to evaluate interobserver agreement but also to measure stability in a test–retest situation and equivalence or test validity.

The equivalence reliability and stability reliability of CRTs such as FITNESSGRAM can be estimated using reliability coefficients.

A serious disadvantage of the K coefficient is that it is very sensitive to low values in the marginals and small contingency tables because chance values are so high. It is also limited to square contingency tables. Again, SPSS can provide the Kappa coefficient as one of the statistics options within the **Crosstabs** routine.

CRT Reliability

For the most part, the same types of reliability and validity situations exist for CRTs as with norm-referenced data. Equivalence reliability as well as stability reliability can be estimated (see chapter 6).

Equivalence Reliability

Mahar and colleagues (1997) examined the criterion-referenced and norm-referenced reliability of the 1 mi run/walk and the PACER test (both tests are used on the FITNESSGRAM). The sample consisted of 266 fourth- and fifth-grade children. They were administered two trials of the PACER test and one trial of the 1 mi run/walk. Equivalence reliability was examined between the 1 mi run/walk and each trial of the PACER test for the total sample and also by gender. Both P and K values were calculated for all cases. The results are presented in table 7.2.

Inspection of the results indicates that fairly high P values (.65 $\leq P \leq$ .83) are associated with varying levels of K (.30 $\leq$ K $\leq$ .65). Remember that you expect the K values to be more conservative than the P values. Whereas the equivalence reliability looks to be at least acceptable for the total sample and with boys alone, the values for the girls are much lower (P values of .66 and .65 and K values of .33 and .30). This points up not only the nature of CRT reliability estimates but also the importance of examining specific reliability situations.

Table 7.2   Criterion-Referenced Equivalence Reliability Between the 1 Mile Run/Walk and PACER

Tests	Total sample	Boys	Girls
Trial 1			
P	.76	.83	.66
K	.51	.65	.33
Trial 2			
P	.71	.76	.65
K	.43	.52	.30

Note: For trial 1, n = 126 boys, n = 95 girls, and total (both) n = 221; for trial 2, n = 122 boys, n = 91 girls, and total (both) n = 213.

Stability Reliability

Rikli, Petray, and Baumgartner (1992) examined the reliability of distance-run tests for children in kindergarten through grade 4. Test–retest reliability estimates using both the norm-referenced (intraclass reliability) and criterion-referenced (P) techniques were calculated. Data on the 1 mi, 3/4 mi, and 1/2 mi run/walk tests were gathered in the fall (on 1229 children—621 boys, 608 girls) and in the next spring (1050 children—543 boys, 507 girls). The P values for these data were calculated using the Physical Best and FITNESSGRAM cutoff values. The results are presented in table 7.3.

Table 7.3 **Criterion-Referenced Reliability Estimates**

Age

		5		6		7		8		9	
		F	S	F	S	F	S	F	S	F	S
Physical Best											
1/2 mi	M	.79	.86	.98	.95	.92	.86	.97	.83	.89	.90
	F	.88	.74	.98	.90	.89	.91	.96	.91	.92	.75
1 mi	M	.70	.70	.94	.89	.95	.92	.90	.94	.95	.93
	F	.75	.88	.88	.73	.81	.87	.95	.94	.92	.90
FITNESSGRAM											
1 mi	M	.75	.70	.76	.66	.85	.77	.91	.85	.86	.83
	F	.69	.51	.71	.45	.81	.85	.90	.84	.83	.94

Note F is fall semester, S is spring semester.
Reprinted from Rikly, Petray, and Baumgartner (1992).

Inspection of the results indicates that all reliability estimates fall in the acceptable range ($P \geq .70$) except the FITNESSGRAM standards for 5-year-old girls (fall = .69, spring = .51) and for 6-year-old boys ($P = .66$) and girls ($P = .45$) in the spring. These criterion-referenced values are consistently higher than the associated norm-referenced values. This is understandable because P values are not corrected for chance. Rikli and colleagues (1992) also explained this as follows: "The higher values for Physical Best are not surprising because P is always larger when there is a large percentage of scores that either meet or do not meet the standard on both the test and retest" (p. 274).

CRT Validity

The validity of CRTs is usually established with some type of criterion-related situation, either concurrent or predictive. Construct validity of sorts can be demonstrated by examining the overlap of two divergent groups measured on a continuum.

Criterion-Related Validity

An example of the criterion-related validity approach, in this case concurrent validity, can be seen in the work of Cureton and Warren (1990). Remember, Cureton and Warren studied criterion-referenced standards for the 1 mi run/walk test. The FITNESSGRAM (Institute for Aerobics Research, 1987) and Physical Best (AAHPERD 1988) tests were used. Both tests provided criterion-referenced standards. The data were presented in figure 7.1 (p. 109).

The results from these two CRT examples are presented in table 7.4. These results illustrate some of the problems of interpreting CRT results. Both tests have significant chi-square results, the phi coefficient is higher for the Physical Best standards, and the percent agreement and Kappa coefficient are higher for the FITNESSGRAM analysis.

Look again at figure 7.1a, which shows that 85% of the individuals were correctly classified for the FITNESSGRAM. Eleven percent achieved the standard on the run/walk test but were below the criterion $\dot{V}O_2$. These results are *false negatives*. That is, the participant is said to be OK on the field test (i.e., the run/walk), but in actuality (i.e., the criterion) he or she is below the standard. Notice too that 4% ($n = 21$) of the participants did not meet the standard on the field test but were above the criterion $\dot{V}O_2$. These individuals are referred

Table 7.4   Comparison of Two CRT Validities

	FITNESSGRAM	Physical Best
Chi-square result	$\chi^2 = 55.35$, $df = 1$, $p < .001$	$\chi^2 = 66.41$, $df = 1$, $p < .001$
Phi coefficient	.309	.338
Percent agreement *(P)*	.85	.61
Kappa	.288	.277

to as *false positives* because the field test results indicate they do not meet the standard but their performance on the criterion is above the standard. Compare the false negative and false positive results for the FITNESSGRAM and Physical Best results in figure 7.1. The impact of false negatives and false positives can be important in determining which field test you might use. To help you differentiate between false negatives and false positives, consider a field test of cholesterol that involves a simple finger stick to obtain a drop of blood. The criterion method for estimating cholesterol would be from drawing venous blood. The results of your finger stick (i.e., the field test) can be accurate (you have been correctly identified as having a healthy or unhealthy cholesterol level) or inaccurate. If the field test reports that your cholesterol level is healthy when in fact it is not, the results are a false negative. If the field test results indicate that your cholesterol level is too high when it is actually in the healthy range, the result is a false positive.

Construct-Related Validity

Setting cutoff scores is a difficult undertaking. The divergent group method can be used as a construct validation procedure. As illustrated in figure 7.3, the concept is to find two groups that are clearly different from each other. To establish a cutoff using this technique, we plot the distributions of scores for the divergent groups. The point in the curves where the scores overlap is used as the criterion cutoff score. This method was explained in more detail by Plowman (1992a). A theoretical application of this approach would be to select two groups of adults (or children). One of these groups would be physically active enough to obtain a health benefit, whereas the other group would not be active enough for a health benefit. Obtaining data on the amount of physical activity for each of these groups and then graphing it should help set a cutoff score for a minimum amount of physical activity needed for a health benefit.

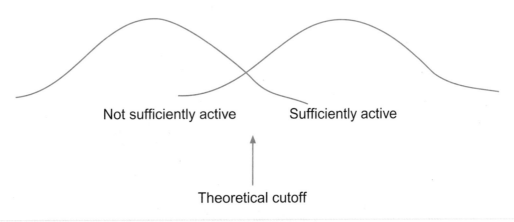

Not sufficiently active   Sufficiently active

Theoretical cutoff

Figure 7.3   A theoretical example of the divergent group method.

CRT EXAMPLES

The logic behind the use and interpretation of reliability and validity procedures with CRTs is similar to that with norm-referenced measurement as presented in chapter 6. If two trials of the same measure are administered, then stability reliability is being assessed. With CRTs this is the reliability of classification. If two different tests are being compared that are thought to measure the same thing, equivalence is being assessed. With CRTs, the equivalence is whether the two tests result in equivalent classifications for the individuals being assessed. If one of the measures is a criterion, the matter being investigated is validity. As we have pointed out several times in this chapter, determination of the criterion with CRTs is the most difficult aspect to establish. However, when the analysis is to determine if the measure is significantly related to a criterion, test validity is being investigated.

The following examples present specific applications of selected techniques for the assessment of reliability and validity using criterion-referenced testing. Try to calculate P and K for the following mastery items.

Mastery Item 7.4

Assume that two criterion-referenced physical fitness tests have been developed to establish the cutoff scores for physical performance on sit-ups. For test 1, the students perform the sit-ups with their hands clasped on their chests ("handches"). For test 2, they perform the sit-ups with their hands behind their heads ("handhead"). Are the tests equivalent? A sample of students is administered test 1 and test 2, and a 2 × 2 contingency table is developed to determine if there is equivalence in the classification on the two tests.

The data are presented in table 7.5. Use the following steps to obtain the chi-square, phi coefficient, and K statistics. SPSS does not calculate P, so you will have to do that by hand from the output you receive.

1. Download table 7.5 from the WWW site.
2. Start SPSS for Windows.
3. Click on the **Analyze** menu.
4. Scroll down to **Descriptive Statistics** and across to **Crosstabs** and click.
5. Put "handches" in the rows and "handhead" in the columns by clicking the arrow keys.
6. Click **Statistics.**
7. Check the **Chi-Square, phi,** and **Kappa** boxes.
8. Click **Continue.**
9. Click **OK.**

Table 7.5   **Example of Equivalence Reliablility**

Subject	handches	handhead
1	1	1
2	1	1
3	1	1
4	1	1
5	1	1
6	1	1
7	1	1
8	1	1

Subject	handches	handhead
9	1	1
10	1	1
11	1	1
12	1	1
13	1	1
14	1	0
15	1	0
16	1	0
17	1	0
18	1	0
19	1	0
20	1	0
21	0	0
22	0	0
23	0	0
24	0	0
25	0	0
26	0	0
27	0	0
28	0	0
29	0	0
30	0	0
31	0	0
32	0	0
33	0	0
34	0	0
35	0	1
36	0	1
37	0	1
38	0	1
39	0	1
40	0	1

Note: 0 = pass, 1 = fail.

Human performance specialists use test–retest reliability—the stability of a test over successive administrations—more frequently than equivalence to determine the reliability of CRTs. You can use an approach similar to that demonstrated in mastery item 7.4. Let's assume that we select one test and administer it on Friday (day 1) and then administer it to the same group of students on the following Monday (day 2). We are concerned with the consistency of classification across the two testing periods. The data are presented in table 7.6. Use the SPSS commands from mastery item 7.4 to calculate the statistics.

Table 7.6   **Example of Stability Reliability**

Subject	Friday	Monday
1	1	1
2	1	1
3	1	1
4	1	1
5	1	1
6	1	0
7	1	0
8	1	0
9	1	1
10	1	1
11	1	1
12	1	1
13	1	1
14	1	1
15	1	1
16	1	1
17	1	1
18	1	1
19	1	0
20	1	0
21	0	0
22	0	0
23	0	0
24	0	0
25	0	0
26	0	0

Subject	Friday	Monday
27	0	0
28	0	0
29	0	0
30	0	0
31	0	1
32	0	1
33	0	0
34	0	0
35	0	0
36	0	0
37	0	0
38	0	1
39	0	1
40	0	1

Note: 0 = failed to meet criterion, 1 = met the criterion.

Mastery Item 7.6

From a validity standpoint (whether it be predictive validity, concurrent validity, or construct validity), the application of the 2 × 2 contingency table is appropriate. For example, let's assume that we have a standard for flexibility fitness and that we suspect that if people attain a certain degree of flexibility, they can reduce their incidence of low back injury. Therefore, we want to test our sit-and-reach test to determine if we can properly classify those who have a history of low back problems as opposed to those who do not. The data are presented in table 7.7. Use the SPSS commands from mastery item 7.4, and interpret the validity of the sit-and-reach test to predict low back pain.

Table 7.7   **Example of Statistical Validity**

Subject	lowback	sitrch
1	1	1
2	1	1
3	1	1
4	1	1
5	1	1
6	1	1
7	1	1
8	1	1

(continued)

Table 7.7 (continued)

Subject	lowback	sitrch
9	1	1
10	1	1
11	1	1
12	1	1
13	1	1
14	1	1
15	1	1
16	1	1
17	1	1
18	1	1
19	1	1
20	1	1
21	1	0
22	1	0
23	1	0
24	1	0
25	1	0
26	1	0
27	1	0
28	1	0
29	1	0
30	1	0
31	0	0
32	0	0
33	0	0
34	0	0
35	0	0
36	0	0
37	0	0
38	0	0
39	0	0
40	0	0
41	0	0
42	0	0

Subject	lowback	sitrch
43	0	0
44	0	0
45	0	0
46	0	0
47	0	0
48	0	1
49	0	1
50	0	1
51	0	1
52	0	1
53	0	1
54	0	1
55	0	1
56	0	1
57	0	1
58	0	1
59	0	1
60	0	1

Note: For lowback 0 = no back pain, 1 = back pain; for sitrch 0 = passed, 1 = failed.

APPLYING CRITERION-REFERENCED STANDARDS TO EPIDEMIOLOGY

Epidemiologic research is a tool that is becoming increasingly popular in human performance measurement. It is closely related to CRT because the variables are often nominal in nature and some of the statistics used are those calculated from a 2×2 contingency table. The criterion measure is categorical: for example, alive or dead; has a disease or does not have a disease. The "predictor" variables can be nominal (e.g., gets sufficient physical activity or does not get sufficient physical activity) or continuous (e.g., weight). It is when the predictor and criterion variables are both nominal that epidemiologic statistics are most like those of CRT (and can even be calculated with SPSS **Crosstabs**).

Epidemiology is the study of the distribution and determinants of health-related states and events in populations and the applications of this study to the control of health problems (Last 1992). Epidemiology is the fundamental science of public health that uses hypothesis testing, statistics, and research methods to develop an understanding of the frequency and distribution of mortality (death) and morbidity (disease or injury) and more importantly the risk factors that are causally related to mortality and morbidity (Stone, Armstrong, Macrina, and Pankau 1996). In our fields, modern epidemiologic research has clearly discovered the increased risk for a variety of chronic diseases related to a sedentary or physically inactive lifestyle (Ainsworth and Matthews 2001; Caspersen 1989).

Descriptive epidemiology seeks to describe the frequency and distribution of mortality and morbidity according to time, place, and person. For instance, what was the rate of breast cancer in adult women in the United States during the 1990s? Epidemiology may help identify risk factors of mortality and morbidity. Analytical epidemiology pursues the causes and prevention of mortality and morbidity. For example, does obesity increase the risk of breast cancer in women? In women who are obese, does moving into a healthy weight range lower the risk of breast cancer?

Epidemiology uses both prospective research approaches, tracking a study group into the future, and retrospective research approaches, looking back at a database of previously collected data. It uses a variety of research designs, some of which are depicted in table 7.8.

Table 7.8   **Research Designs in Epidemiology**

Type	Description
Experimental	
Randomized clinical trial	Randomly assigns subjects to treatments or exposures
Community trial	Randomly assigns whole communities to treatments or exposures
Observational	
Cases series	Notes cases at a particular time or place
Cross-sectional	Takes a snapshot of identifiable groups at one point in time
Proportionate mortality or morbidity study	Compares results of a study group to the population
Case-control	Compares known cases of mortality or morbidity with matched noncases
Cohort	Longitudinal, generally tracks populations in the long term

Epidemiology is a science that requires the use of advanced statistics and complicated multivariate models to understand the relationships between risk factors and mortality and morbidity while controlling for confounding factors or extraneous variables. Complicated statistical models such as logistic regression and proportional hazards are used to test those relations. Those types of analyses are beyond the scope of this text and are not necessary for us to know at present. But we do need to know some basic procedures and statistics to understand how criterion-referenced standards play a role in epidemiology.

Two basic statistics are the calculations of incidence and prevalence:

▶ **Incidence**—the number, proportion, rate, or percentage of new cases of mortality and morbidity. Incidence could be calculated in a randomized clinical trial or a prospective, longitudinal cohort study.

▶ **Prevalence**—the number, proportion, rate, or percentage of total cases of mortality and morbidity. Prevalence would be calculated in a cross-sectional study.

Values of incidence and prevalence are often expressed as a rate, which is the number of cases per unit of the population. An example would be 10 cases per 1,000 in the population or 100 deaths per 100,000 in the population. The value of expressing incidence and prevalence as a rate is that two populations of very different sizes can be compared. For example, the rate of mortality in Dallas, Texas, can be compared with the rate of death in New York City.

In analytical epidemiology, we convert our measures of incidence or prevalence into estimates of risk:

- **Absolute risk**—the risk (proportion, percentage, rate) of mortality or morbidity in a population that is exposed or not exposed to a risk factor.
- **Relative risk**—the ratio of risks between the exposed or unexposed populations. This statistic is calculated with incidence measures.
- **Odds ratio**—an estimate of relative risk used in prevalence studies.
- **Attributable risk**—the risk of mortality and morbidity directly related to a risk factor. It can be thought of as the reduction in risk related to removing a risk factor.

Let's combine criterion-referenced standards with an example of a simple analysis in epidemiology. High cholesterol is defined by the American Heart Association and the National Heart, Lung, and Blood Institute as a value of 240 mg/dl or above. Thus, the criterion reference standard for total cholesterol is 240 mg/dl. Let's examine the results of a theoretical epidemiologic study about the relationship of cholesterol and mortality attributable to heart attack. Examine table 7.9, which is a 2 × 2 contingency table. We have conveniently labeled each cell as A, B, C, or D. This will make all descriptive and analytical calculations quite simple. We also conduct our analyses on incidence and prevalence bases. In this study, 56 subjects with high cholesterol and 44 subjects with cholesterol below that criterion are compared. All subjects had a genetic history of early coronary heart disease. Note that both variables are categorical in this example.

Table 7.9   Results of a Hypothetical Study Relating Cholesterol and Heart Attack Mortality

Exposure	Outcome	
	Heart attack deaths	No heart attack deaths
High cholesterol	A 25	B 31
No high cholesterol	C 7	D 37

If you examine all the results in figure 7.4 you can observe the following:

- All calculations can be made from the easy-to-follow formulas using the A, B, C, and D cell identifiers.
- The absolute risk for heart attack death was 32% for all subjects, 45% for subjects with high cholesterol, and 16% for subjects without high cholesterol.
- If a subject had high cholesterol, the relative risk of 2.81 indicated that high cholesterol elevated the risk of heart attack mortality by a multiplier of 2.81.
- If a subject had high cholesterol, the odds ratio indicated elevated odds of heart attack mortality by a multiplier of 4.26.
- The attributable risk indicated that high cholesterol contributed to 64% of the heart attack mortality. Thus, heart attack mortality could be reduced by 64% if high cholesterol were no longer present in people of this population.

$$\text{Total} = \frac{A + C}{A + B + C + D} = \frac{25 + 7}{25 + 31 + 7 + 37} = \frac{32}{100} = .32 \text{ or } 32\%$$

$$\text{High} = \frac{A}{A + B} = \frac{25}{25 + 31} = \frac{25}{56} = .45 \text{ or } 45\%$$

$$\text{Not high} = \frac{C}{C + D} = \frac{7}{7 + 37} = \frac{7}{44} = .16 \text{ or } 16\%$$

→ Absolute risk

$$\text{RR} = \frac{A/(A + B)}{C/(C + D)} = \frac{.45}{.16} = 2.81$$

→ Relative risk

$$\text{OR} = \frac{AD}{BC} = \frac{25 * 37}{7 * 31} = \frac{925}{217} = 4.26$$

→ Odds ratio

$$\text{AR} = \frac{[A/(A + B)] - [C/(C + D)]}{A/(A + B)} = \frac{.45 - .16}{.45} = .64 \text{ or } 64\%$$

→ Attributable risk

Figure 7.4 Statistical analysis of epidemiological data in table 7.9.

The example used in table 7.9 and figure 7.4 was contrived to serve as a simple demonstration of some basic concepts and analyses in epidemiology. However, research studies using epidemiologic methods have demonstrated very strong relationships between levels of physical activity and fitness and a variety of mortality and morbidity outcomes from chronic diseases. Chapter 11 will discuss some of those specific findings in more detail.

Mastery Item 7.7

1. Go to the WWW site for chapter 7 and download the data from table 7.9.
2. Confirm that you can calculate the odds ratio and relative risk by using the **Crosstabs** routine.
3. Do so by running **Analyze → Descriptive Statistics → Crosstabs** and placing "cholesterol" in the rows and "heart attack" in the columns.
4. Then go to **Statistics** and click on **Risk**.
5. When you review the SPSS results, you should see that the odds ratio and relative risk values presented previously are presented in the SPSS output.

Mastery Item 7.8

In table 7.10, a 2 × 2 contingency table, are results from a study conducted by Bungum, Peaslee, Jackson, and Perez (2000). The study examined the relation of physical activity during pregnancy and the risk of Cesarean birth compared with a normal vaginal birth. Perform the analysis presented in figure 7.4 with these data.

Table 7.10   **Results of a Study by Bungum et al. (2000)**

Exposure	Outcome	
	Cesarean section birth	Vaginal birth
Sedentary	A 26	B 67
Active	C 7	D 37

Measurement and Evaluation Challenge

When Christina arrived at the library, she perused *Measurement and Evaluation in Human Performance*, third edition, and found that she needed to select criterion-referenced measurement tools to assess the relationship between pulled hamstrings and creatine use. She decided to ask the athletes she was treating two simple questions:

Question 1: In the last 12 months, have you sustained a hamstring injury?

Question 2: During the past 12 months, have you taken creatine?

The answers would be yes or no. The criterion measure is hamstring pulls, and the predictor is creatine usage. Notice that both variables are nominal (with two categories: yes or no).

From her readings, Christina believed that she could study the validity of creatine use as a predictor of hamstring injuries by examining the proportion of agreement *(P)* and Kappa (K) values. She would ask all of her patients the two questions (not just the ones who had hamstring injuries) and set up a 2 × 2 contingency table. She would use each of these statistics and epidemiologic statistics to investigate hamstring pull risks associated with use of creatine. She hopes, as a result of this study, to obtain some information that will help her in counseling athletes on creatine use.

SUMMARY

There are specific measurement situations in the area of human performance well suited for criterion-referenced measurement; moreover, there are specific statistical techniques that must be used with these CRTs. The primary problem associated with criterion-referenced testing in human performance is in establishing a criterion or cutoff score. Because few measurement problems in human performance have concrete criterion scores associated with them, cutoffs have to be established from experts' opinions or normative data. Cutoffs can often be arbitrary, thus affecting empirical validity. The establishment of these scores also affects the reliability and the statistical validity of the test. Therefore, criterion scores must be set with a high degree of caution.

In the area of youth fitness testing, criterion-referenced standards have been established by test developers (e.g., FITNESSGRAM). In other areas of human performance, such as sport skills testing, such standards have typically not been set. In epidemiological research and practice, many cutoffs have been established that are directly related to health risks. See Morrow and Falls (2003) for a summary of criterion-referenced testing use with the FITNESSGRAM.

Criterion-referenced statistical techniques are used to analyze data. Criterion-referenced measurement can be a valuable tool for you to examine measurement in human performance. Criterion-referenced testing is the method of choice when variables are categorized and where an obvious level of proficiency must be achieved before proceeding to the next level (e.g., flotation and treading water skills need to be mastered before one enters the deep end of the pool).

The typical statistics used with criterion-referenced testing reliability and validity are chi-square, phi coefficient, proportion of agreement *(P)*, and Kappa (K), which adjusts the proportion of agreement for chance.

Last, you learned how epidemiologic statistics are closely related to criterion-referenced testing procedures. Epidemiology is a powerful method for identifying risk factors for various disease outcomes.

chapter 8

Alternative Assessment

Larry D. Hensley, University of Northern Iowa

Outline

Measurement and Evaluation Challenge

Mariko Brown is a second-grade physical education teacher who has been working with her students on various locomotor patterns. She has provided instruction, given demonstrations, and allowed students opportunities to practice a variety of movement patterns, including skipping, hopping, galloping, and sliding. Furthermore, students are taught how to adjust and vary speed in the several movement activities and how to make smooth transitions from one form of locomotion to another. Ms. Brown is explicit in describing and demonstrating the various skills, even giving the students a handout that illustrates how to perform each of the movement patterns. She has allowed her students time to practice the various movement patterns and now wishes to determine how well they are doing. She wants not only to see the progress of each student but to be able to diagnose any problem a student has and to be able to make appropriate recommendations for improvement. How can Ms. Brown assess each student's understanding and performance of the various movement patterns?

Objectives

After studying this chapter, you will be able to

- define alternative, performance-based assessment and distinguish it from traditional, standardized testing;
- discuss recent trends in assessment practices in education;
- explain the advantages and disadvantages of alternative assessment;
- identify criteria for judging the quality of alternative assessments;
- identify various alternative assessment techniques;
- create an alternative assessment, complete with scoring criteria; and
- identify methods for improving alternative assessment.

I t is difficult to find a hotter issue in education today than assessment and accountability. Since the early 1990s there has been a renewed interest in using assessment practices as a way to enhance student learning. This coincides with the educational reform movement based on standards-based instruction. Furthermore, according to many authorities, this increased interest in assessment has resulted in a paradigm shift, so that the focus is now on what some call alternative, or performance-based, assessment techniques rather than on more traditional, standardized testing techniques. Other authors also use the term authentic assessment to refer to this "new" approach to assessing student progress. We defined measurement and evaluation in chapter 1; however, the term **assessment** has been popularized by educators during the last decade or so and now is used more often in education than either measurement or evaluation. *Assessment is the process of collecting information. Evaluation results when a judgment or interpretation of the assessment is made.* In other words, assessment includes both measurement and evaluation.

But what is alternative assessment, and how does it affect assessment in physical education, particularly in (but not limited to) the psychomotor domain, where skill development is such an important objective?

Unfortunately, there is not universal agreement among educators as to what constitutes alternative assessment, although education authorities generally refer to it as any type of assessment task other than the traditional, standardized paper-and-pencil, multiple-choice type of test. The issue is complicated by the fact that the terms *performance assessment* and *authentic assessment* are often used synonymously with *alternative assessment*.

The following definitions should help you gain a better understanding of these various types of assessment:

▶ **Alternative assessment** is considered different from the conventional form of standardized testing (i.e., multiple-choice or true–false test, sport skills test, or physical fitness test). Alternative assessment involves a wide variety of nontraditional techniques for assessing student achievement, such as

- ▶ portfolios,
- ▶ role-playing,
- ▶ interviews,
- ▶ event tasks,
- ▶ individual or group projects,
- ▶ open-ended questions,
- ▶ student logs or journals,
- ▶ exhibitions, and
- ▶ observation.

Furthermore, alternative assessments lend themselves more to criterion-referenced standards than norm-referenced standards, and the scoring is largely based on subjective judgments.

▶ **Performance assessment** includes "testing methods that require students to create an answer or product that demonstrates their knowledge or skills" (U.S. Congress 1992). Students are asked to conduct an activity or create a product, rather than simply choose an answer from a list of alternatives provided by the teacher. Testing has traditionally focused on whether students get the right answer, with little regard for whether they truly understand the problem or whether they can apply their understanding. In performance assessment, the distinction is made between being able to describe how a skill should be performed (knowledge) and being able to actually perform it (performance—certainly an important feature in sport, fitness, and physical education).

▶ **Authentic assessment** is "designed to take place in a real-life setting rather than in an artificial or contrived setting which typifies traditional forms of assessment . . . frequently [it involves] directly observable behavior" (National Association for Sport and Physical

Education 1995, vii). For example, authentic assessment may involve observing a student make strategic choices and perform skills in a game of tennis rather than simply asking questions about strategy and measuring the skill out of the context of the game, as is often done with a sport skills test. Moreover, authentic assessment is multidimensional or holistic in that it requires the student to incorporate higher-level thinking into the behavior required by the assessment task. Playing a game of tennis not only involves executing various psychomotor skills but also requires an understanding of the rules as well as some elements of strategy.

Although some theorists argue that it is important to differentiate among performance assessment, authentic assessment, and alternative assessment, it is our view that for all practical purposes the terms are synonymous and may be used interchangeably. Lund and Kirk (2002, p. 7) reached the same conclusion, although they chose to use the term *performance-based assessments*. We prefer to use the term *alternative assessment* because it implies that we are doing something other than the more traditional, standardized testing. We concur, however, that the assessments should be authentic in nature and that the assessment tasks should be performance-based.

Before we examine the nature of alternative assessment and what has prompted many educators, including physical educators, to jump on this bandwagon, we will first provide an example from a sport setting.

Mastery Item 8.1

Search the Internet to find the latest information about alternative assessment. What is the context in which it is most likely to be used?

Observation of performance is one method of alternative assessment commonly used by coaches in making training decisions about their players.

Tina Gonzales, an eighth-grade girls basketball coach, is meeting with prospective team members during the first day of practice. She welcomes the students to the tryouts for the basketball team, discusses pertinent rules and regulations, and then explains that she will be assessing each student's performance over the next five practice sessions to determine which students actually make the team. Coach Gonzales is faced with the reality that although 40 students are trying out for the basketball team, by rules she may have only 15 players on the team. She must do something that will enable her to select the best players or those players who have the greatest potential for contributing to team success. The coach is about to embark on the process of assessment, which will provide her with information about the abilities of the prospective players that she will then use to make these decisions. During the next five practice sessions, Coach Gonzales has the student-athletes participate in a variety of basketball drills and simulated game situations, as well as actual game play. She carefully observes each student's performance on the various activities, making notes daily to document what she sees. As the week progresses, she begins to synthesize his notes as well as her recollection of the performance of each student, mentally comparing her observations with her predetermined standards of performance. She does all of this to help her make a judgment about which students should be retained on the team. At the end of the week, Coach Gonzales announces that she has systematically assessed each prospective player's performance, and she proceeds to identify the 15 players who have made the cut for the eighth-grade basketball team based on the results of this assessment. Although the coach has now determined the membership of the team, as practice continues and actual competition ensues she continues her systematic observation of each player's performance to determine a "starting five" and to diagnose each player's strengths and weaknesses. The events described here take place regularly, as coaches at all levels, from youth sports to elite, professional sports, carefully observe their athletes to make a variety of judgments about who plays, how they play, and how to improve their performance.

This example illustrates the use of alternative assessment; the coach assessed student-athletes on their progress toward defined behaviors without using formal, standardized tests of any type. Rather, observation and professional judgment served as the primary basis for his assessment. *Although this example represents the most widely practiced form of alternative assessment—observation of performance—other alternative assessments include (but are not limited to) portfolios, exhibitions, student logs or journals, individual or group projects, and role-playing.* Now that you have an idea of what alternative assessment is, let's review the development and evolution of alternative assessment in U.S. education, which has been the main site of its use.

RECENT TRENDS IN EDUCATIONAL ASSESSMENT

Since the mid-1980s there has been widespread concern among parents, business leaders, government officials, and educators about the effectiveness of schools. This concern has prompted many calls for educational reform, as both teachers and schools are now being held more accountable for what students learn. *At the heart of the most recent wave of educational reform is the identification of learner outcomes, framed in the context of standards that define what students should know and be able to do.* Content standards specify what should be learned in various subject areas. The National Association for Sport and Physical Education (NASPE) first published content standards for physical education in 1995 in the publication *Moving Into the Future—National Standards for Physical Education: A Guide to Content and Assessment.* The standards were revised slightly and updated in 2004. Performance standards, meanwhile, define an acceptable level of achievement or performance. They are supposed to answer the question, How good is good enough? We currently do not have nationally recommended performance standards in physical education, leaving it up to individual teachers and schools to define acceptable performance. Furthermore, a cornerstone of this educational reform movement is assessment—not only to provide indexes of accountability but also to perform as an integral part of the instructional process. If used appropriately, assessment should provide both the teacher and the student with information about progress toward desired instructional goals.

At the same time, however, the increased demands for accountability and the heightened emphasis on assessment come at a time of growing dissatisfaction with the traditional forms of assessment, such as multiple-choice, machine-scored tests and standardized sport skills and physical fitness tests. Many acknowledged assessment authorities in education tout alternative assessment as the solution to this dissatisfaction. *This movement toward alternative assessment is motivated, at least in part, by the belief that alternative assessment methods facilitate teaching, enhance learning, and result in greater student achievement.* Perhaps the fundamental question is whether the new assessment models that embrace some form of alternative assessment improve education. Unfortunately, there is little empirical evidence that provides a clear-cut answer to this simple question.

But what is alternative assessment and how does it affect assessment in physical education, not only in the psychomotor domain, where skill development is such an important objective, but also in the cognitive and affective domains? In a study of public school physical education teachers, Mintah (2003) reported that teacher observation, self-observation, checklists, peer observation, and event-tasks were the most commonly used forms of alternative assessment. In fact, 100% of the teachers studied reported using teacher observation for assessment. This finding should not really surprise us given that much of physical education is based on overt, observable behavior or performance. Mintah went on to report that the physical education teachers perceived authentic assessment (alternative assessment) to have a positive impact on student self-concept, motivation, and skill achievement. Although anecdotal information such as this provides valuable support for the use of alternative assessment techniques, direct evidence regarding the efficacy of alternative assessment is still lacking.

Mastery Item 8.2

Have you experienced alternative assessment in some setting? How does alternative assessment differ from traditional, standardized testing?

Mastery Item 8.3

Why is assessment of a physical performance task, a sport skill, for example, well suited to alternative assessment?

ALTERNATIVE ASSESSMENT OF PSYCHOMOTOR SKILLS

According to Wood (1996), "Traditional psychometric assessment devices (e.g., multiple choice tests, sport skills tests) may no longer be sufficient for assessment in the quickly changing educational landscape characterized by emphasis on learning outcomes, higher order cognitive skills, and integrated learning" (p. 213). Furthermore, traditional assessment instruments and techniques tend to measure narrowly defined characteristics, do not facilitate integration of skills or processes, and are frequently artificial in nature. This has resulted in an emerging shift to increase the use of alternative forms of assessment and to validate techniques that many physical educators have used for years. Clearly, the landscape of educational assessment in general and specifically in kinesiology and physical education is undergoing change (Wood 2003).

Authenticity of Alternative Assessment

Assessment is said to be authentic when the assessment task is designed to take place in a real-life setting, one that is less contrived and artificial than traditional forms of testing. In other words, the assessment task has contextual significance. Authenticity is not a dichotomous characteristic, being either present or absent; rather, authenticity is a multifaceted characteristic that exists in varying degrees. For example, playing volleyball in practice is not quite the same as playing volleyball in competition. Thus, some assessments are more authentic than others.

The idea of performance-based, authentic assessment is not new. This is especially true in the field of physical education where the content manifests itself in directly observable behavior, particularly when considering psychomotor skills such as fundamental movement patterns, sport skills, and selected components of physical fitness. In fact, advocates for alternative assessment methods, performance assessment in particular, have often used examples from sport to explain their positions.

Mastery Item 8.4

Use the Internet as well as other reference sources available to you to locate three examples of alternative assessment tasks appropriate for physical education or a clinical setting.

Mastery Item 8.5

Explain why alternative assessments lend themselves to the use of criterion-referenced standards rather than norm-referenced standards.

For instance, students have been asked to perform a myriad of straightforward tasks designed to assess skills such as throwing, catching, putting, serving a tennis ball, and passing a volleyball. Whereas the majority of published motor ability and sport skills tests are performance based, requiring the student to complete a task or demonstrate a prescribed skill, relatively few are high in authenticity. That is, the context in which the assessment is conducted and the task is performed is artificial in nature and bears little resemblance to a real-life situation. The following directions from the AAHPERD Tennis Skills Test illustrate a contrived assessment context:

> *Directions for the ground stroke (forehand–backhand) test:* The student taking the test is positioned at the center mark of the baseline while the teacher or test administrator is located along the center line on the other side of the net. The teacher then tosses, using an overhand throwing motion, 10 balls to the forehand side of the student and then 10 balls to the backhand side of the student. Directions specify that the tossed ball should land near the service line within approximately 6 ft [1.8 m] of the student. The student, meanwhile, attempts to return each tossed ball using either a forehand or backhand stroke into the designated scoring area on the other side of the net. Each attempt is scored according to placement (balls landing deeper in the court receive more points) and power (as determined by the distance the ball bounces). [The actual test directions provide more detailed instructions for scoring.] (Hensley 1989, 13)

Whereas the ground stroke test just described was shown to have moderately high concurrent validity when compared with expert judges' ratings ($r = .76$ to $r = .86$), this assessment technique is contrived in nature and is relatively low in authenticity. The fact that the test requires the students to actually execute a tennis ground stroke does make the assessment performance based, albeit limited to a single task; however, this does not ensure its authenticity. Whereas this limitation does not negate the usefulness of this particular test in situations in which good psychometric properties are highly valued, it becomes problematic in the day-to-day formative assessment designed to facilitate teaching and enhance student learning.

Mastery Item 8.6

Educators have shown considerable interest in moving toward the use of alternative assessments. What reasons are frequently given for this shift in assessment practices?

To be most useful in an instructional setting, an alternative assessment technique should consider both context (situation–task) and performance (construct–skill). That is, the assessment task should represent a completed performance having contextualized meaning that is directly related to the eventual use of the skill (Siedentop 1996). The previously described tennis ground stroke test is fairly representative of most assessments in the psychomotor domain, and of sport skills tests in particular, in that the assessment task typically involves a performance of some type but rarely provides contextual meaning. Students are taught specific skills, are given frequent practice of these skills in regimented drills, and then are tested on their ability to perform these isolated skills using standardized tests. For example, students learn to dig, to forearm pass, and to overhead pass in volleyball skill drills, but they often cannot perform these same skills in a game situation or play a satisfactory game of volleyball. Because basic individual skills may serve as the foundation for future activity or performance in a game, it is necessary to assess both individual skills as well as game performance.

Mastery Item 8.7

Explain why many sport skills tests used in physical education are considered to be relatively low in authenticity.

Technical Quality of Alternative Assessment

Whereas there is a certain intuitive appeal to many forms of alternative assessment, there is little substantive evidence to suggest that these techniques are superior to the more traditional forms of standardized testing. Yet, the popularity of alternative assessment methods has risen dramatically in recent years. Before we rush to judgment about the usefulness and appropriateness of so-called alternative assessments, it is important to consider the technical quality of the assessments. *As with any form of assessment, the traditional psychometric properties of reliability and validity are important.* The use of portfolios, exhibitions, or direct observation will not in itself ensure quality assessment information, correct conclusions, and appropriate inferences. Evaluators and other decision makers need to understand issues of reliability and validity as they relate to these different forms of assessment.

Generally, the overall purpose of assessment is to provide valid information for decision making. Good assessment information provides an accurate indicator of an individual's performance and enables the teacher, coach, or other decision maker to make appropriate decisions. But what constitutes good assessment information? What determines the quality of an assessment? According to Herman, Aschbacher, and Winters (1992), when we talk about the quality of an assessment, we are really asking the following:

1. Does an assessment provide accurate information for decision making?
2. Do the results permit accurate and fair conclusions about student or athlete performance?
3. Does using the results contribute to sound decisions?

To be able to answer yes to these questions, three criteria must be met: reliability, validity, and fairness.

Reliability

As previously defined, reliability relates to the consistency of scores or observations. An unreliable test score is essentially useless because it does not provide meaningful information to the user. Inasmuch as alternative assessment depends heavily on the subjective judgment of the teacher (or other assessor) to score and interpret the performance or product of the assessment task, you need to be particularly concerned about interrater reliability or objectivity. Too often teachers or other assessors use their own idiosyncratic criteria for judging a performance. Consider the earlier example of Coach Gonzales' assessment of those players trying out for the basketball team. What if Coach Gonzales' assessment of one of the players, say Patricia, was different from the assessment by her assistant coach? Which assessment is accurate? What does this really mean? For Patricia it means a great deal: whether she makes or fails to make the basketball team. Considering the subjective nature of alternative assessments, the user must strive to minimize inconsistency in scoring to have confidence that the judgment is a result of the actual performance, not some extraneous aspect of the situation. Decisions about an individual cannot be valid unless based on reliable information.

To ensure reliable scoring of alternative assessments, three conditions must be met:

▶ Well-defined, explicit criteria for judging the performance or product have been established.

▶ Those making the judgments—peers, teachers, coaches, or exercise specialists—thoroughly understand the criteria.

▶ Those making the judgments have learned how to apply the scoring criteria in a consistent manner. The uniform application of scoring criteria can best be achieved through training and practice.

Consider Ms. Brown's situation presented at the beginning of this chapter. She must be certain that bias does not interfere with her observation and recording of her students' performance of the locomotor skills. It is important that she and other observers evaluate the same performance in the same way by devising criteria by which to score the students.

Validity

As you know from chapter 6, validity is an indication of how well an assessment actually measures what it is supposed to measure. Although reliability is necessary, it is not a sufficient condition for validity. An assessment could be perfectly reliable but not relevant to the decision for which it is intended. If the assessment result is not related to the characteristic reportedly being measured, it may jeopardize accurate conclusions about an individual's performance and subsequent decisions. Broadly speaking, validity relates to the meaning and consequences attached to the test scores (Messick 1995).

The magnitude, or consequence, of the decisions made from the assessment information is what we call the stakes of the assessment. For instance, if a teacher is conducting a classroom-based assessment of children's throwing, with the primary purpose of measuring and evaluating whether they understand and can apply the concept of opposition to the throwing motion, we would call this a low-stakes assessment. The information gleaned from the assessment would be useful to the teacher in diagnosing the students' performance and providing constructive feedback about how to improve their throwing. On the other hand, if the decisions to be made from the assessment information have many significant consequences (such as student retention or promotion, a final course grade, or graduation), where misinterpretation could lead to a dire aftereffect, we would call this high-stakes assessment. The **stakes** of the assessment depend on the consequences of the decisions made from the assessment information.

Unfortunately, we have little empirical information about the validity of alternative assessments. The primary form of validity reported for alternative assessments, particularly those developed by teachers for classroom use, is face (content-related) validity, based on the assumed relationship between instruction and assessment. Although this linkage is important, content-related (face) validity alone should not be accepted as sufficient evidence for the use of an assessment method. We need to have corroborating evidence that alternative assessments yield valid conclusions. This is particularly true with high-stakes assessments, where the consequences of the decisions are great.

However, inasmuch as most alternative assessments are substantially different from standardized testing, traditional validation procedures may be inappropriate (Miller and Legg 1993). In this light, measurement specialists have proposed different frameworks and criteria for the validation of alternative assessments. Linn, Baker, and Dunbar (1991), for example, suggest the following criteria for determining the validity of alternative assessments:

▶ **Consequences**—Does the assessment lead to the intended consequences? What are the unintended consequences?

▶ **Fairness**—Does the assessment enable all students to demonstrate their true capabilities, or does it unfairly disadvantage some students?

▶ **Transfer and generalizability**—Do the results of the assessment generalize to other problems or situations in the domain of interest?

▶ **Cognitive complexity**—Does the assessment adequately consider higher levels of understanding and complex thinking?

▸ **Content quality**—Is the assessment task selected to measure a given content area worth the time and effort of students and assessors?

▸ **Content coverage**—Does the assessment adequately sample the breadth of content of the domain of interest?

▸ **Meaningfulness**—Is the assessment task meaningful to students and does it motivate them to perform their best?

▸ **Cost and efficiency**—Is the cost of data collection and scoring of the assessment reasonable and efficient?

Even though these criteria may seem more applicable to classroom assessments, careful consideration of each in the design and use of alternative assessments in physical education and human performance will undoubtedly enhance the quality or validity of the assessment.

Fairness

Although fairness is not a psychometric property in the same sense as reliability and validity, it is critically important in all forms of assessment, whether traditional or alternative. **Fairness** simply means that an assessment allows all students, regardless of gender, ethnicity, or background, equal opportunity to do well. Although there is tremendous diversity within our society and students do not come to school with the same background, exposure, motivation, or values, all students should have an equal opportunity to demonstrate the skills and knowledge being assessed. *Fairness should be evident in the development or selection of the assessment task as well as in the criteria used for judging the performance or product.*

Is the assessment task fair and free from bias? Does the assessment task favor either boys or girls, students from a particular ethnic group, students who have lived in a particular location, or those whose families have greater financial resources? To be fair, the task should reflect the knowledge, values, and experiences that are familiar to and appropriate for all students and should seek to measure the knowledge and skills that all students have had adequate time to acquire. Furthermore, it is important that the scoring procedures and criteria for judging a performance or the product created are free from bias. This helps to ensure that the ratings of a performance reflect the examinee's true capability and are not a function of the biases and perceptions of the person judging the performance. See chapter 9 for more detail about assigning student grades.

Despite the fact that authentic alternative assessments are closely related to the instructional process, they will compare unfavorably with traditional, standardized assessments on many psychometric criteria normally used. Traditional evaluation procedures tend to demonstrate greater reliability, objectivity, and validity than alternative assessment, largely because of the more objective nature of assessment. Because of this, we need to rethink the technical quality used to judge the validity of alternative assessments. *Technical quality is just as important in alternative assessment as it is in traditional, standardized assessment.* Furthermore, the higher the stakes associated with the assessment, the greater the attention that should be given to the technical quality of the assessment—its reliability, validity, and fairness (Herman, Aschbacher, and Winters 1992). Although various organizations and measurement experts work to construct suitable standards for alternative assessments, the challenge for the practitioner who uses these assessments is to carefully consider the issues of validity, reliability, and fairness to ensure that any assessment you might use leads to correct judgments and appropriate decisions.

Mastery Item 8.8

Alternative assessment is not without its critics. What are some major points of criticism levied against its use?

Mastery Item 8.9

Assume you are interested in evaluating an individual's ability to kick a soccer ball. Create an alternative assessment task that could be used for assessing kicking performance of elementary school children.

ESTABLISHING CRITERIA FOR JUDGING ALTERNATIVE ASSESSMENTS

Alternative, performance-based assessments are backed by criterion-referenced standards. Regardless of the type of assessment task (e.g., portfolio development, public exhibition, student log, or teacher observation), the individual's performance or product is evaluated according to a prescribed standard or criterion. Figure 8.1 illustrates the fundamental components of an alternative assessment.

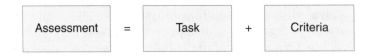

Figure 8.1 Assessment components.

Some authorities claim that a unique feature of alternative assessment is that students are provided with the standards beforehand. Yet others claim that this is not unique to alternative assessment; rather, it is a practice that good teachers use as they continually assess how their students are doing and adjust their instructional plans accordingly. Although the uniqueness of this can be debated, it is important to understand that the assessment task must be accompanied by scoring standards—that is, **performance criteria.** Without some type of scoring standard, the assessment task is essentially nothing more than an instructional activity and should not be called an assessment at all. Unfortunately, as teachers have jumped on the alternative-assessment bandwagon, many of the assessment activities that have been created in the name of alternative assessment are inadequate in that they simply do not contain an evaluative component. This is the biggest mistake teachers and schools make when it comes to alternative assessment. If there are no developed standards for judging the students' performance or products and the desired outcome has not been satisfactorily defined, then there is no evaluative component. The lack of an evaluative component does not diminish the usefulness of the task as an instructional activity, but let's not call it an assessment.

Mastery Item 8.10

Develop a scoring system, in the form of a simple checklist, that could be used for judging performance on the assessment task for kicking created in mastery item 8.9.

The National Center for Research on Evaluation, Standards, and Student Testing (CRESST) defines performance criteria as guidelines, rules, characteristics, or dimensions that are used to judge the quality of student performance. Criteria indicate what we value in student responses, products, or performances. In alternative assessments, students must clearly understand what criteria will be used to judge their performance. Because of the complexity of many alternative assessments, educators must explicitly describe the responses or performances they want to elicit in the assessment task.

Criteria for judging student performance have been called many things: scoring guidelines, rating scales, checklists, and, more recently, rubrics or scoring rubrics. For our purposes, the term **scoring rubric** will be used, but regardless of the name, these criteria represent the backbone of an effective performance assessment system. *According to Marzano, Pickering, and McTighe (1993, 29), a scoring rubric consists of a fixed scale and a list of characteristics describing performance for each of the points on the scale.* Thus, a scoring rubric is essentially comparable to rating scales that have been used in physical education for many years. Although the form may be similar, its use may differ from scales in that scoring rubrics facilitate instruction and should be provided to students, parents, administrators, and so on rather than being a secretive system on which an instructor mysteriously bases grades.

Scoring rubrics can take several forms. The simplest scoring rubric, a checklist, is designed to judge whether specific behaviors or characteristics are present or absent in students' performances or the products of their work. This scoring system uses only two levels of scoring for rating the performance. There is no attempt to judge how well each of the behaviors is displayed, only that they are present or not. A checklist type of scoring rubric has frequently been used to judge students' ability to perform various tumbling and gymnastic stunts. Table 8.1 illustrates a simple checklist that could be used for assessing the throwing performance of elementary school children.

Table 8.1   Sample Checklist for Throwing Performance

Behavior	Check if observed	Comments
Nonthrowing side to target		
Throwing hand starts behind ear		
Steps toward target with opposite foot		
Elbow leads hand forward		
Forearm extension prior to release		
Release point just past the head		
Trunk rotation		
Follow through toward target		

More complex scoring rubrics often include multiple dimensions or behaviors scored using some type of numerical scale, qualitative scale, or a combination of both. Basically, a numerical scale uses numbers to describe performance levels along a continuum, whereas a qualitative scale uses adjectives rather than numbers to describe performance levels. Although the number of levels or points along the continuum can vary—usually somewhere between 3 and 7 points—some authorities advocate an even number of scale points, because this forces the assessor to choose something other than the middle score. Be cautious, however, about including too many levels or score points on the scale because it can become difficult to clearly differentiate among the various levels. Lund (2000) recommended that the number of levels to include on a scale should match the number of levels of performance that the teacher wishes to specify. That is, if the teacher wanted to designate three levels of performance—high, medium, or low—then a 3-point scale should be used. Table 8.2 shows a scoring rubric for the forehand and backhand drives in tennis that uses both a numeric and a qualitative scale. This has been adapted from the rating scale used for validation purposes with the AAHPERD Tennis Skills Test noted previously (Hensley 1989).

Obviously, the sample scoring rubric illustrated in table 8.2 is designed to assess a single dimension or discrete skill, either the forehand or backhand drive in tennis. This represents a task-specific scoring rubric. The teacher simply observes the student's performance and

Table 8.2   Sample Scoring Rubric for Forehand/Backhand Drive in Tennis

5—Excellent	Proper grip, good balance and footwork, and near-perfect form. Consistently demonstrates correct stroke mechanics. Shots are hit with power and consistently placed appropriately.
4—Good	Proper grip, good balance, adequate footwork, and acceptable, but not perfect form. Demonstrates above average consistency of stroke mechanics. Moderate power and consistent placement within court area.
3—Satisfactory	Proper grip, acceptable balance, but footwork is poor. Form is somewhat erratic and inefficient, resulting in inconsistent shot placement. Style of stroke is more defensive in nature, but can sustain a short rally.
2—Fair	Uses improper grip at times, poor footwork, and basically incorrect form. Inconsistent stroke mechanics. Defensive style of play, merely trying to get the ball over the net. Unable to sustain a rally.
1—Poor	Incorrect grip, off balance, with poor footwork. Form is very poor and erratic. Virtually no control of ball placement. Experiences difficulty in getting ball over net.

then attempts to match his or her impression of the performance to the specified criteria on the scoring rubric. The observation and accompanying assessment could be accomplished while the student is performing specific skill drills, practicing with a partner, or playing an actual game. The authenticity of the assessment will certainly be enhanced if it is conducted during a tennis game or match, thus giving contextualized meaning to the performance.

Assessing a student's performance of a particular skill, such as a forehand or backhand drive, in the manner just described may be most useful when conducted for diagnostic purposes related specifically to the skill or task in question. This is termed **analytic assessment.** However, it does little to indicate if the student can actually play a game of tennis. We suggest that, rather than merely assessing an isolated skill such as the forehand or backhand drive in tennis, teachers focus on **holistic assessment,** in which the overall quality of actually playing the game (tennis in this case) is judged. Whereas analytic assessment may provide useful feedback about the individual's strengths and weaknesses in separate skills, holistic assessment is both more efficient and more powerful than analytic assessment, requiring the individual to integrate knowledge of the game, strategy, and a variety of skills into a single, authentic performance. Such an approach is likely to be faster and simpler than performing analytic assessments for a number of independent skills. Keeping with the tennis example, table 8.3 illustrates a scoring rubric designed for the holistic assessment of tennis-playing ability.

Mastery Item 8.11

Explain the difference between analytic assessment and holistic assessment, and describe when one approach would be preferred over the other.

Designing scoring rubrics can be a difficult and time-consuming task, particularly if you follow an analytical approach to assessment and attempt to assess individual performance on the multitude of skills presented in a typical physical education class using a task-specific scoring rubric. It is particularly challenging to find just the right language to accurately describe the various gradations in performance. To guide you in designing scoring rubrics for the holistic assessment of sport skills, table 8.4 provides a generalized scoring rubric. Such a rubric may be used to assess a variety of sport skills. You can adapt this generalized scoring rubric to the specific elements of a particular sport or activity. Note the similarity between the scoring rubrics in tables 8.3 and 8.4.

Table 8.3 Sample Scoring Rubric for Holistic Assessment of Tennis-Playing Ability

5—Excellent	Demonstrates mastery of tennis skills and ability to consistently execute all strokes with little or no conscious effort, resulting in few unforced errors. Extensive knowledge base. Anticipates opponent's shots and employs effective strategy specific to the task or situation.
4—Good	Demonstrates competency and ability to perform basic tennis skills without making many errors. Complete understanding of rules and strategies of tennis. Usually selects appropriate strategy and shot selection for situation and generally displays consistent performance.
3—Satisfactory	Displays basic understanding of tennis and is able to perform fundamental skills adequately to be able to play game. Performance is frequently inconsistent, with numerous errors being made. Understands basic strategies, but lacks ability to effectively employ them.
2—Fair	Demonstrates inability to perform more than the basic skills. Has difficulty in executing even the basic skills, makes frequent errors, some critical, during performance. Generally inconsistent performance, displays only a minimal understanding of strategies and rules.
1—Poor	Rarely, if ever, performs skills well enough to be able to play an effective game of tennis. Demonstrates little understanding of tennis and inability to execute skills without making significant and frequent errors. Makes little attempt to adjust performance.

Table 8.4 Generic Scoring Rubric for Holistic Assessment of Sport Skills

5—Excellent	Demonstrates mastery of sport-specific skills and ability to consistently perform with little or no conscious effort, resulting in few errors. Extensive knowledge base and understanding of sports activity. Employs effective strategy specific to the task or situation.
4—Good	Demonstrates competency and ability to perform basic skills without making many errors. Complete understanding of rules and strategies of the specific sport or activity. Usually selects appropriate strategy and skill for situation and generally displays consistent performance.
3—Satisfactory	Displays basic knowledge of sport or activity and ability to adequately perform fundamental skills required to play game. Performance is frequently inconsistent, with numerous errors being made. Understands basic strategies, but lacks ability to effectively employ them.
2—Fair	Demonstrates inability to perform more than the basic skills. Has difficulty in executing even the basic skills, makes frequent errors, some critical, during performance. Generally inconsistent performance, displays only a minimal understanding of strategies and rules.
1—Poor	Rarely, if ever, performs skills well enough to be able to play an effective game. Demonstrates little understanding of sport or activity and inability to execute skills without making significant and frequent errors. Makes little attempt to adjust performance.

Mastery Item 8.12

A high school coach is evaluating tennis players at a summer camp for the purpose of grouping them for drills. She and two of her staff watch each camper play a Pro set. They evaluate the players' abilities using the scoring rubric in table 8.3. The scores for 12 campers from each of the three raters are presented in the MI 8.12 data set located on the text's WWW site.

1. Use SPSS to calculate the alpha coefficient to estimate the objectivity (rater reliability) of the three raters.
2. To do this, use Analyze → Scale Reliability Analysis and enter the rating from each of the raters in the "Items" box.
3. Confirm that the interrater reliability is .9066. How would you estimate the reliability for a single rater? (Hint: See chapter 6 and methods to estimate the reliability when you change the test length.)
4. Confirm that the estimated reliability for a single rater is .76 (you will have to do this by hand).

The previous examples illustrate a combined numeric–qualitative scoring scale. In contrast, some public schools have developed developmental scoring rubrics that use a numerical scale containing several levels (table 8.5). Similar rubrics have been designed for most fundamental movement patterns as well as for student outcomes associated with each of the content standards for physical education recommended by NASPE. Because of the developmental nature of these scoring rubrics, you can use them with a variety of assessment tasks designed for students from kindergarten through secondary school.

Table 8.6 illustrates a scoring rubric designed by a physical education teacher in Iowa City, Iowa, for assessing responsibility of her elementary school students. Note that three dimensions of responsibility are assessed using this rubric. Students also use this rubric for self-assessment. This is an example of alternative assessment in the affective domain.

Table 8.7 provides a generalized numerical scoring rubric for assessing content knowledge (cognitive domain). This rubric can be easily modified, of course, to identify the specific content being assessed.

Table 8.5 Developmental Scoring Rubric for Prekindergarten Through Age 9

Developmental level	Catching	Throwing
6	Can catch an object thrown with increased velocity or catch an object while moving	Can throw with increased velocity and accuracy
5	Can transfer catching skills to a game situation	Can transfer throwing skills to a game situation
4	Can catch a variety of objects at different levels with a partner	Shows trunk rotation and accuracy
3	Can catch a variety of self-tossed objects	Follows through toward target
2	Can catch a bounced ball from a partner	Shows opposition
1	Arms extended toward thrower, shows avoidance reaction	Limited body movement; arm dominated

Adapted from Wichita Public Schools, Kansas.

Table 8.6   **Responsibility Scoring Rubric for Elementary Physical Education**

Level	Listening	Effort	Social interaction
4—Caring	Helps others to follow directions.	Supports coaches and classmate without being asked. Encourages classmates to participate in all physical activities both in and out of school.	Supports all classmates with positive comments most of the time. Willingly works with and accepts anyone in the class.
3—Self-responsibility	Follows directions promptly.	Has a "can do" attitude toward participating and learning. Willingly participates in vigorous physical activity outside of class.	Makes positive comments and "put ups" most of the time. Willingly works in a group.
2—Involvement	Hears and sees.	Willingly participates in all activities with enthusiasm.	Makes positive comments and "put ups" most of the time. Willingly works with a partner.
1—Self-control	Allows others to hear and see.	Participates most of the time.	Keeps self from calling others names or making negative comments or "put downs." Keeps self from striking or pushing others.
0—Irresponsibility	Keeps others from hearing or seeing.	Does not participate most of the time.	Calls other students names. Strikes or pushes others. Makes negative comments or "put downs."

Adapted from Karen Nagle, Iowa City, Iowa Community Schools.

Table 8.7   **Generic Scoring Rubric for Content Knowledge**

Level	Knowledge
4	Demonstrates a thorough understanding of the important concepts or generalizations and provides new insights into some aspect of that information
3	Displays a complete and accurate understanding of the important concepts or generalizations
2	Displays an incomplete understanding of the important concepts and generalizations and has notable misconceptions
1	Demonstrates little understanding of the concepts and generalizations and has several misconceptions

Adapted from the Mid-continent Regional Education Laboratory (McREL), Aurora, CO.

As these examples show, you can develop a wide range of scoring rubrics to guide your judgment of psychomotor performances and affective behaviors, as well as cognitive knowledge and understanding. *The purpose of the assessment will go a long way toward determining the kind of scoring criteria needed, but whatever the purpose, the criteria should reflect the valued qualities of the desired performance or product.* Although you may be able to find scoring rubrics that have been published or have been developed by others, it is likely that none of these will exactly meet your needs. However, as you identify the criteria and develop your own scoring rubrics, do take advantage of the work of others. You will probably be able to adapt many of the existing rubrics to fit your needs. Look at the criteria used by others. Note the scoring scale and the way in which the descriptions are written. Scoring rubrics used in other subject areas may be modified to meet the assessment needs in physical education and human performance. Lund and Kirk (2002) identified seven basic steps to follow when writing rubrics:

1. Envision the desired student performance.
2. Determine the criteria.
3. Pilot the assessment.
4. Write levels for the rubric.
5. Create a rubric for students (so the students understand what is expected of them).
6. Administer the assessment.
7. Revise the rubric.

See the book by Lund and Kirk (2002) for a detailed discussion of each of these steps. Additional information about designing scoring rubrics for physical education is also provided in the NASPE Assessment Series volume, *Creating Rubrics for Physical Education* (Lund 2000).

In all cases, when you are using alternative, performance-based assessment methods, the assessment task must be accompanied by scoring standards. To do otherwise simply raises questions about the credibility of the assessment and negates the usefulness of the information obtained.

Mastery Item 8.13

A teacher has specified "social responsibility" as a major class objective. Develop a scoring rubric that could be used by the teacher for assessing social responsibility through periodic observation of students.

ALTERNATIVE ASSESSMENT TECHNIQUES

As previously mentioned, there are many types of alternative assessment techniques. Each type of alternative assessment brings with it different strengths and weaknesses relative to credible and dependable information. Because it is virtually impossible for a single assessment tool to adequately assess student performance, the real challenge comes in selecting or developing alternative, performance-based assessments that complement both each other and more traditional assessments to equitably assess students in physical education and human performance. We can group most alternative assessments used in physical activity settings into five categories: observation, projects, portfolios, exhibitions, and logs or journals. We next present differentiations and examples of alternative assessments in each of these categories.

Observation

Human performance provides many opportunities for the student to exhibit behaviors that may be directly observed by others, a unique advantage of working in the psychomotor domain. *The nature of performing a motor skill makes assessment through observational analysis and subjective judgment a logical choice for many physical education teachers.* In fact, investigations of measurement practices of physical educators have consistently shown a reliance on observation and subjective assessment methods (Hensley et al. 1989; Matanin and Tannehill 1994; Mintah 2003). Assessment practices that depend on the judgment of observers have frequently been criticized for questionable validity and reliability, being susceptible to personal bias, generosity error (the tendency to overrate), lack of objective scoring, being conducted unsystematically, not being judged against well-defined criteria, and not being recorded. These criticisms notwithstanding, observation and informed judgment may provide the best hope for alternative assessment in physical education that is user-friendly and authentic and yet provides meaningful and relevant information. In fact, we propose that sound professional judgment is inseparable from good assessment and represents the foundation for all types of assessment.

When we talk of assessment of psychomotor skills or activities common in sport and physical education, often there is not a single correct response or performance; there are a variety of ways to satisfactorily perform the designated activity or skill. This is particularly important to recognize when we consider observational assessment techniques. Consequently, students' performance of psychomotor skills must be judged against well-defined criteria that direct attention to the dimensions of the performance itself. *Criteria not only guide the assessment process but promote learning by offering clear performance targets to students.*

Individual or Group Projects

Projects have long been used in education to assess a student's understanding of a subject or a particular topic, although this practice is not as common in physical education or human performance. Although there are an infinite number of tasks that could be used

for this purpose, projects typically require students to apply their knowledge and skills in completion of the prescribed task. Often projects require creativity, critical thinking, analysis, and synthesis on the part of the student. Examples of student projects used in physical education and human performance include the following: writing a research report on a specified topic, interviewing a local athlete or sports official, designing one's own personal fitness program, developing a video or audiotape, preparing a poster presentation, performing a skills analysis of oneself or another, creating a new game or dance, participating in a role-playing scenario, and others.

Group projects enable a number of students to work together on a complex problem that requires planning, research, internal discussion, and presentation. Although it is often difficult to ascertain the contribution of each member of the group to the final product, group projects are often attractive because they encourage and facilitate cooperation, a valued outcome in most physical activity settings. An unfortunate aspect that results in decreased validity of the evaluation is the fact that you may not be able to accurately ascertain whether a single individual, or only some members, did all the work or if all group members participated and contributed equally. *Projects are judged in accordance with established standards of excellence (scoring criteria) that are known to the participants ahead of time.*

The following example of a project designed for middle school or high school students involves a research component, analysis and synthesis of information, problem solving, and effective communication.

Students working on a group project.

Directions to students: Some people are calling for mandatory drug testing of interscholastic athletes. Yet mandatory testing is rare. What is behind the call for mandatory drug testing? That is, what benefits are there to mandatory drug testing? What are the disadvantages?

Imagine that you have been hired by the local school board to investigate the issue of drug testing. You have been charged with developing a report to be presented to the school board describing the pros and cons of mandatory drug testing in the schools and providing a recommendation to the board with supporting rationale.

Portfolios

Portfolios are systematic, purposeful, and meaningful collections of individuals' work that have been assembled over time. Portfolios have been used by artists, photographers, and fashion models for many years to display their best work. In many respects, the items included in a portfolio may be similar to the products of student projects that have been purposefully collected over a longer time. Portfolio collections may also include input provided by teachers, parents, peers, administrators, or others.

Although there are no definitive guidelines as to the format of a portfolio, the following are two basic types:

- ▶ **Showcase or model portfolio**—A portfolio consisting of work samples chosen by the student that document a student's best work. The student has consciously evaluated his or her work and selected only those products that are of the highest quality.

- ▶ **Descriptive or representative portfolio**—A portfolio consisting of representative work of the student; the student makes no attempt to evaluate the products. It may also include other types of process information, such as drafts of student work or records of student achievement or progress over time.

The type of portfolio, its format, and the general contents are usually prescribed by the teacher. It's a good idea to limit the portfolio to a certain number of pieces of work, to prevent the portfolio from becoming a massive collection that has little meaning to the student and presents a monumental evaluation task for the teacher. This also requires students to exercise some judgment about which items best fulfill the requirements of the portfolio task and demonstrate their level of achievement. The portfolio itself is usually a file or folder that contains the student's collected work. The contents could include items such as the following: training log, student journal or diary, written reports, photographs or sketches, letters, charts or graphs, maps, copies of certificates, computer disks or computer-generated products, completed rating scales, fitness test results, game statistics, training plans, report of dietary analyses, and even video or audiotapes. The potential projects that could result in portfolio items are almost limitless. However, Kirk (1997) provided the following list of possible portfolio projects that may be useful for physical activity settings.

Portfolio Projects for Physical Activity

- Write a self-evaluation of current skill level and playing ability, with individual goals for improvement.
- Conduct ongoing self- and peer evaluation of skill performance and playing performance (process–product checklists, rating scales, criterion-referenced tasks, task sheets, game play statistics).
- Prepare a graph or chart that shows and explains performance of particular skills or strategies across time.
- Analyze your game-playing performance (application of skills and strategies) by collecting and studying your individual game statistics (i.e., shooting percentage, assists, successful passes, tackles, and steals).
- Create and perform an aerobic dance, step aerobic, or gymnastic routine, with application knowledge and skills (evidence: a routine script or videotape of performance).
- Document participation in practice, informal game play, or organized competition outside of class.
- Keep a daily physical education journal in which you set daily goals; record successes, setbacks, and progress; and analyze the situation to make recommendations for present and future work.

- Using a self-analysis or preassessment, select or design an appropriate practice program and complete schedule. Record results.
- Set up, conduct, and participate in a class tournament for assigned group. Keep group and individual records and statistics (as an individual or as a part of a group).
- Write a newspaper article as if you were a sports reporter reporting on the class tournament or a game (must demonstrate knowledge of the game).
- Develop and edit a class sports or fitness magazine.
- Complete and record a play-by-play and color commentary of a class tournament game, as if you were a radio (audiotape) or television (videotape) announcer.
- Interview a successful competitor about his or her process of development as an athlete and his or her current training techniques and schedule (audiotape or videotape).
- Interview an athlete with a disability about his or her experience of overcoming adversity. Apply what you have learned to your situation (audiotape, videotape, or article).
- Write an essay on the subject of, "What I learned and accomplished during gymnastics [or any activity unit] and what I learned about myself in the process."

A scoring rubric should be used to evaluate portfolios in much the same manner as any other product or performance. Moreover, providing this rubric to the students in advance enables them to become better attuned to the characteristics of quality work—and better positioned to produce a portfolio of high quality. Determine whether the portfolio will be judged as a whole or whether the individual items or projects will be judged separately. A holistic evaluation of a portfolio, usually along several dimensions, is probably more common and economical, but it does not provide the detailed feedback to the student that an analytic evaluation of each item or task would. Table 8.8 illustrates a holistic scoring rubric for judging a portfolio along three dimensions.

Table 8.8   Scoring Rubric for Complete Portfolio

	Format/design	Content items for each section	Reactions to portfolio tasks
Bull's-eye	Follows prescribed format with no errors. Design is attractive. Shows creativity.	Appropriate, well-conceived items. Accurate information. Highest quality work.	All reflections are descriptive, insightful, and capture the student's feelings.
On target	Follows prescribed format with few errors. Design is neat but has little imagination.	Appropriate and generally useful items. Mostly accurate. Acceptable work.	Reflections clearly describe student's feeling about the tasks.
Getting close	Follows prescribed format for most part. Design distracts and could be neater.	Some items show little thought in selection and are inappropriate. Marginal work.	Some reflections include personal reactions, but they are often vague.
Missed the mark	Does not follow prescribed format. Numerous errors. Inappropriate design.	Many inappropriate items, indicating a lack of basic knowledge. Poor quality work.	Little evidence of personal reflection on the tasks. Vague and repetitive.

Example of a Portfolio Project

The following is an example of a portfolio project and accompanying **analytic scoring** rubric for a middle school or high school activity unit. This portfolio project could be used with a variety of sports, such as volleyball, basketball, tennis, badminton, or soccer.

Directions to students: In addition to participating in the class tournament, select one of the three following activities. In each of the activities, you must demonstrate and apply your knowledge of the skills, rules, traditions, and strategies of a selected sport from class and your ability to analyze the game and communicate your analysis of the game.

1. Write an article reporting the class tournament game that you observed, including game statistics and an analysis of the game based on team or individual performances. Include action photographs for emphasis.
2. Complete and record on audiotape a play-by-play commentary of a class tournament game or other live game as if you were a radio sports announcer. You may complete the project with a partner.
3. Complete and record on videotape a play-by-play commentary of a class tournament game or other live game as if you were a TV sports broadcaster. You may complete this project with a partner or small group.

Table 8.9 presents a suggested scoring scale for a portfolio (Kirk 1997). Melograno (1998) provided an extensive presentation of portfolio development in physical education, including a large number of examples of various portfolio possibilities.

Table 8.9 Scoring Rubric for Portfolio Task

Outstanding	• Provides a clear demonstration of an accurate working knowledge of rules, strategies, skill, and an understanding of the game • Provides an accurate, quick, and colorful analysis of the game for the audience • Demonstrates the ability to communicate ideas well • Provides outstanding evidence that the student researched the role, responsibilities, and techniques of the announcer/reporter • Provides an overall product of the highest quality
Proficient	• Provides a demonstration of a mostly accurate knowledge of the areas of the game • Provides a mostly accurate analysis and interesting report of the game • Provides acceptable evidence of the completed research on the role and responsibilities of the announcer/reporter • Provides an overall product that is acceptable
Novice	• Performance demonstrates a lack of essential knowledge about the game; many inaccuracies and errors occur • Analysis and report of the game is slow, inaccurate, and boring • Does not communicate ideas clearly • No evidence of research on the roles, techniques, and responsibilities of the announcer/reporter • Provides an overall product of poor quality

Adapted from Kirk 1997.

Exhibitions

An **exhibition** is a public display or performance during which a student showcases learning and competence in particular areas. The exhibition would be evaluated using a scoring rubric specifically developed for the type of performance or display.

Possible exhibitions include the following:

- ▶ Performing an aerobics routine for a school assembly
- ▶ Organizing and performing a jump rope show at halftime of a basketball game
- ▶ Performing in a folk dance festival at the county fair
- ▶ Demonstrating wu shu (a Chinese martial art) at the local shopping mall
- ▶ Designing a poster about the benefits of physical activity for display at the annual fitness fair
- ▶ Drawing and coloring pictures of various sport activities for display at a local restaurant or business

In addition to being evaluated by the primary teacher, exhibitions may also be judged by an expert panel of adults or peers (e.g., teachers, parents, coaches, community members, employers, or students).

Student Logs or Journals

Both logs and journals represent a type of student project but are included as a separate category of alternative assessment because of their widespread use in physical activity settings. A log records behaviors over a period of time. Often the information recorded shows changes in behavior, trends in performance, results of participation, progress, and the regularity of physical activity. Sometimes logs indicate the choices or decisions made by individuals as well as their feelings about their behaviors. Journals are very similar, often recording student participation, but generally include more information about feelings, thoughts, perceptions, or reflections about actual events or results. The entries in journals often report social or psychological perspectives, both positive and negative, and may be used to document the personal meaning associated with one's participation. As with all types of alternative assessment, the log or journal is judged according to prescribed criteria specified in advance.

Mastery Item 8.14

The U.S. Surgeon General's report on physical activity and health as well as the national standards for physical education published by NASPE proclaim the importance of regular physical activity. Create an alternative assessment task that would enable you to assess the physical activity behavior of high school students outside of a classroom setting.

Mastery Item 8.15

Develop a scoring rubric that could be used with the assessment task created in mastery item 8.14.

Mastery Item 8.16

How might you go about providing evidence that the task you created for assessing physical activity behavior is valid?

GUIDELINES FOR DEVELOPING ALTERNATIVE ASSESSMENTS

It should be obvious by now that the nature of alternative assessments precludes the publication of assessments that have universal appeal and usefulness. Clearly, one size does not fit all when it comes to alternative assessments. As a result, most professionals generate their own assessments. This does not mean that you cannot adapt and enhance ideas that you have obtained from other sources, but in many instances you will need to create your own alternative assessment that meets your specific needs. *A guiding requirement in the development of any type of alternative assessment is that it consist of both a task and performance criteria.* That is, to create an assessment, you need to create both a task and criteria.

Thus, the process of developing meaningful alternative assessments is similar to the process of developing traditional assessments.

The basic steps include the following:

1. Determine the purpose of the assessment.
2. Determine the target or desired student outcome.
3. Select or create an appropriate assessment task.
4. Establish criteria for judging the performance or product.
5. Determine the quality of the assessment.

Determining Purpose

The critical first step in the process is determining the purpose of the assessment. For instance, is the purpose of the assessment to diagnose or determine deficiencies in an individual's performance or the product he or she has created? Or is the purpose to ascertain individual achievement of specified objectives to award a grade? Or is the purpose of the assessment to obtain information for evaluating your program or to comply with state reporting requirements? *Furthermore, it is important to recognize the importance attached to the assessment and to the decisions that will be made based on the results. That is, what are the stakes of the assessment and who will be using the results?*

Defining the Target

You need to know what you are expecting before you start. In other words, can you clearly define the desired outcomes for your students? What should students know and be able to do? In 1995 NASPE first published national standards for physical education. This document identifies what students should know and be able to do as a result of participation in school physical education. It has been used by school districts throughout the nation to guide their curriculum development and to help articulate the desired outcomes for students. Moreover, the document provides examples of alternative assessments linked to the recommended standards. The national physical education standards were updated in 2004, with additional work in progress designed to articulate performance standards as well. If you become a physical educator, whether you use the national physical education standards, recommendations of the U.S. Surgeon General, or ACSM guidelines, you must clearly identify the goals, objectives, or targets that are to be covered on the assessment.

Selecting the Appropriate Assessment Task

A key to good assessment is matching the assessment task to the target. If the task does not match the intended outcomes, then it serves no useful purpose for assessment and the results are not meaningful for decision making. Ask yourself if the assessment task elicits the desired performance or work. Have the students had the opportunity to acquire the knowledge and skills needed for the task? If not, consider a different task. And keep in mind that not all targets or desired outcomes are best measured using alternative assessments.

The following questions summarize important concerns to consider about the usefulness of alternative assessment tasks:

> ▶ Does the task match the outcome goals?
> ▶ Is the task meaningful? To whom—students, parents, teachers, others?
> ▶ Is the task fair and free of bias?
> ▶ Is the task authentic?
> ▶ Is the performance of the task measurable?
> ▶ Is the task feasible, given available resources, time, and costs?
> ▶ Can criteria be established to judge performance of the task?

Setting Performance Criteria

We have already discussed extensively the need for and importance of performance criteria. *A task alone does not constitute an assessment—we must also have criteria to be able to judge the performance or product.* Some teachers and professionals, particularly those just starting out with alternative assessment, tend to focus only on the task, to the exclusion of the criteria. How will you know if an individual has accomplished the goal or target that was set? How will you know when the performance is good enough? How will you know if the portfolio is acceptable? How will you be able to diagnose a student's performance and provide meaningful feedback? Each of these questions is answered by the criteria, not the task.

Determining the Quality of the Assessment

Recall that when we talk about the quality of an assessment, we are most interested in reliability, validity, and fairness. We emphasize throughout this textbook the importance of reliability and validity in measurement and evaluation practices of all types.

For alternative assessments, be particularly careful to avoid the pitfalls that threaten reliability, validity, and fairness and that can lead to improper decisions and unintended consequences. Moreover, for high-stakes assessment, there is a greater requirement to provide evidence of the reliability and validity of the assessments. Evidence of interrater reliability, as well as the stability of assessment results, can be determined using the statistical methods described for criterion-referenced tests (CRTs) in chapter 7. As discussed earlier in this chapter, providing evidence of the validity of alternative assessments is more problematic, although using the techniques available for CRTs may be helpful. The difficulty arises because the criterion cutoff score is often very arbitrary, thus affecting the validity of the evaluation. Until more appropriate techniques are developed, your affirmative answers to the following questions adapted from Herman, Aschbacher, and Winters (1992) may provide some confidence in the validity of your alternative assessment.

> ▶ Can the scores be used to describe what individuals have learned?
> ▶ Are the scores useful for making generalizations about an individual's performance?
> ▶ Can the scores be used to diagnose individuals' strengths and weaknesses?
> ▶ Are the scores unbiased?
> ▶ Is there corroborating evidence that the assessment serves its intended purpose?
> ▶ Does the assessment have positive consequences for learning and instruction?

IMPROVING ALTERNATIVE ASSESSMENT

As physical educators and instructors strive to improve the quality of their programs, many will heed the reform initiatives of education experts who propose that standards-based education and alternative, performance-based assessment offer great promise for enhancing the education system. To improve the preparation of physical education teachers and other physical activity specialists so that they are able to conduct meaningful assessments, our profession needs to embrace a new way of thinking about assessment. *Although we are proposing that physical education teachers include alternative assessment techniques in their repertoire of assessment methods, we are not suggesting that teachers completely abandon traditional, standardized testing techniques.* There is a need for both, depending on the purpose of the assessment. Regardless of the approach taken, we advocate that teachers use observational assessment techniques or other forms of alternative assessment combined with well-conceived scoring rubrics for meaningful assessment in physical education class or other physical activity settings. The design and incorporation of clear, developmentally appropriate, and explicitly defined scoring rubrics with subjective judgment of student performance are essential to ensure validity, consistency, and fairness.

Unfortunately, many of the procedures and tasks being touted under the alternative assessment banner are merely student-centered activities with no basis for judging their quality. Stiggins (1987) suggested that the most important element in designing performance-based (alternative) assessments is the explicit definition of the performance criteria. Moreover, Herman, Aschbacher, and Winters (1992) stated that criteria for judging student performance lie at the heart of alternative assessment. *If we expect alternative assessments to realize their promise and live up to expectations, then we must demand high-quality assessments accompanied by clear, meaningful, and credible scoring criteria.*

The following guidelines (adapted from Gronlund 1993) provide ways to improve the credibility and usefulness of alternative assessment in physical education.

Tips for Improving Alternative Assessment

1. Ensure that assessments are congruent with the intended outcomes and instructional practices of the class.
2. Recognize that, together, observation and informed judgment comprise a legitimate and meaningful method of assessment.
3. Use an assessment procedure (i.e., holistic, analytic) that is appropriate for the use to be made of the results.
4. Use authentic tasks in a realistic setting; thus, provide contextualized meaning to the assessment.
5. Design and incorporate clear, explicitly defined scoring rubrics with the assessment.
6. Provide scoring rubrics and evaluative criteria to students and other interested persons.
7. Be as objective as possible in observing, judging, and recording the performance.
8. Record assessment results as soon as possible after the observation.
9. Use multiple observations whenever possible.
10. Supplement observational assessment with other evidence of achievement.

A balanced approach to assessment is the prudent path to follow. The issue is not whether one form of assessment is intrinsically better than another. No assessment model is suited for every purpose. The real issue is determining what type of performance indicator best serves the purpose of the assessment and then choosing an appropriate assessment method that is suitable for providing this type of information.

Mastery Item 8.17

A meaningful alternative assessment should consist of a task as well as criteria. What is the most difficult part of creating alternative assessments? Why?

Measurement and Evaluation Challenge

To judge students' understanding and competence of the movement skills presented to them, Ms. Brown asks the students in her class to create and demonstrate a movement sequence that combines each of the four locomotor patterns into a continuous movement. Furthermore, she explains that the movement should use different speeds of traveling and should flow continuously from one basic movement form to another. The students proceed to practice the four locomotor patterns, trying to combine them into a continuous movement, until they believe they are ready to demonstrate their movement sequence for Ms. Brown. The complete task requires approximately 1 min per student as Ms. Brown diligently watches each student perform, recording on a checklist the satisfactory completion of each element of the activity. In this assessment activity, students were asked to show that they knew the various locomotor patterns and to exhibit the skills required to perform the movements. Assessment activities similar to this are frequently seen in elementary school physical education classes, particularly as instruction relates to motor skill development.

SUMMARY

Alternative, performance-based assessment conducted in authentic settings has become a foundation of current educational reform taking place in schools. This approach to assessment is supposed to enhance learning and achievement, yet there is sparse empirical evidence to validate this claim. Nevertheless, there is a certain intuitive appeal to alternative assessment as educators and others continue to jump on this bandwagon. Alternative assessments take many forms, but those typically used in physical education and human performance generally include observation, individual or group projects, portfolios, exhibitions, and student logs or journals. A guiding principle for the development and use of any type of alternative assessment is that it consists of both a task and performance criteria. Furthermore, we have emphasized that reliability, validity, and fairness are just as important with alternative assessment as they are with traditional, standardized testing. For those of you who are interested in more information about alternative, performance-based assessments, an excellent resource text is provided by Lund and Kirk (2002). In addition, the several volumes published in the NASPE Physical Education Assessment Series are designed to provide a collection of current, appropriate, and realistic assessment tools for physical educators. A variety of Web sites, such as PE-Central (www.pecentral.org), provide alternative assessment suggestions and sample scoring rubrics applicable for physical education.

Grading:
A Summative Evaluation

Outline

Measurement and Evaluation Challenge

Tom Jones is getting ready to teach a unit on tennis to his high school physical education class. He wants to weight the cognitive and the psychomotor aspects of the unit equally. Drawing on his experience, Tom plans to give two written tests during the unit to measure cognitive objectives; the first examination will count for 20% of the final grade, and the final examination will count for 30% of the final grade. He plans to use two skills tests, serving and the drop shot, to measure the psychomotor objectives of the unit. The serving test will be weighted as 35% of the grade, and the drop shot test will be 15% of the final grade. What steps should Tom follow to ensure that the final **composite score** for each student will reflect these criteria?

Objectives

After studying this chapter, you will be able to

- ▸ list the proper criteria for grade assignment;
- ▸ illustrate the methods for assigning grades to one test or performance; and
- ▸ use the methods for assigning final grades.

You will also understand that there are many uses for grades, including the following:

- ▸ Motivating and guiding the learning, educational, and vocational plans of students as well as their personal development;
- ▸ Communicating to students and parents about students' progress;
- ▸ Condensing information for use in determining college admissions, graduate school admissions, and scholarship recipients;
- ▸ Communicating to possible employers a student's strengths and limitations; and
- ▸ Helping the school tailor education to student needs and interests and to evaluate the effectiveness of teaching methods.

This chapter is directed to students who plan to become teachers of physical education, a profession that requires the assessment of student performance. The assessment of student performance in public and private schools generally culminates with the report of a grade. The process of grading can be a challenging one that many beginning, and even veteran, teachers do not look forward to. Although teachers and students enjoy giving and receiving good grades, they do not enjoy giving and receiving poor grades. However, prospective teachers must realize that assigning grades is a professional obligation and an important component of teaching physical education.

Besides dealing with the many minor difficulties that arise during the process of determining grades, you also need to consider that grades, which are assessments of one person by another, are based on subjective evidence. Sometimes the amount of subjectivity included in a set of grades is large. Grades based on an assessment of a student's answers to an essay examination and participation in class discussions are examples of such situations. Even grades arrived at through seemingly precise and objective methods often involve more subjectivity than you might think. "Objective testing" (true–false, multiple-choice, and others) requires somewhat subjective decisions, such as determining what items to include on the test and what response is most correct for each item. Often instructors devise a precise formula for combining certain achievements of the student, but this formula is generally based on a subjective weighting of the importance of the course objectives, similar to those Tom Jones selected in the measurement and evaluation challenge described at the beginning of this chapter. *Thus, although it is true that some grades involve a greater amount of subjectivity than others, it is also true that all grades involve some degree of subjectivity.*

Generally, the greater the degree of subjectivity involved in the grading system, the greater the unreliability. In other words, a subjective system, if repeated, would probably not assign the same grades to the same students. This lack of objectivity and consequent lowering of reliability lead to a second, more serious, concern about grades: the lack of a common understanding of what a particular grade represents. *Grades are affected by the type of marking system used, the particular instructor who assigns the grade, the makeup of the class, the institution where the grades are earned, and many other related factors.* For example, an E might stand for excellent in one grading system but for failing in another; a grade of B from an instructor who rarely gives B's has a meaning different from that of a B given by an instructor who seldom gives anything lower; an A obtained in a class of superior students may represent a more significant achievement than an A received in a class with less competition; and a certain level of performance might be given an A at one institution but a lower grade at another.

Whereas the use of insufficient objective evidence reduces the *reliability* of grades, the lack of a clear and generally accepted definition of what a particular grade represents affects the *validity* of grades. Grades, which should reflect the degree to which an individual has achieved the course objectives or goals, are often contaminated by many other factors.

No specific rules for grading can be firmly established because any situation in which grades are assigned is different in some respects from any other. Differences exist in areas such as teaching techniques, course objectives, equipment and facilities available, and type of students in the class. However, an effective, consistent grading process allows an instructor and his or her students to have confidence in the validity of the assigned grades. If a student understands and accepts the teacher's grading system, he or she may not be happy with a poor grade but can accept it as being fair. Were you ever unhappy when you received a grade different from the one you expected? Was it easier to accept a poor grade if you felt the teacher's grading method was fair to all students? The challenge of this chapter is to provide you with the skills and knowledge required for developing and using good grading practices in your instructional programs, whether you will use these practices in an academic setting or any similar situation that requires a final assessment of the degree to which objectives have been met.

Mastery Item 9.1

Recall a past physical activity course you took. What were the objectives and the methods used for assigning grades? Were they fair to all students? If you received a grade below your expectations, did you accept it as fair?

EVALUATIONS AND STANDARDS

In chapter 1 we defined several terms important to the measurement and evaluation process. We differentiated between a **formative evaluation,** conducted during an instruction or training program, and a **summative evaluation,** a final, comprehensive judgment conducted near the end of an instruction or training program. Evaluations result from a decision-making process that places a judgment of quality on a measurement. In physical activity instruction, teachers make formative evaluations at the beginning of and during the instructional process and use them to detect weaknesses in student achievement and to direct future learning activities. A formative evaluation can be a formal measurement activity, such as a pretest prior to instruction, or a very informal, subjective evaluation given by the instructor (such as verbal feedback during tennis practice). This chapter focuses on formal summative evaluations, including the proper steps for conducting such evaluations, which result in a final grade being determined for the student's overall performance for the entire instructional unit or for the subject's final achievement of the objectives presented.

For a judgment to be made about the quality of performance, the performance must be compared with a standard. In chapters 1, 6, and 7 we discuss norm-referenced and criterion-referenced standards for evaluation. To review briefly, a norm-referenced standard is established by comparing an individual's performance to performances of others of the same gender and age. The establishment of this standard usually requires some type of data analysis. A criterion-referenced standard, on the other hand, is a specific predetermined level of performance that has been established—from past databases or expert opinion—before the individual performs. In a criterion-referenced evaluation, the individual either achieves the standard (passes) or does not achieve the standard (fails). In a norm-referenced evaluation, the individuals are ranked from excellent to poor based on their position among the comparison individuals' scores.

Mastery Item 9.2

Before continuing with this chapter, return to chapters 1, 6, and 7 to review formative and summative evaluations and norm- and criterion-referenced standards. Give an example of a formative and summative evaluation in a swimming class. How would normative- and criterion-referenced standards apply to a gymnastics unit?

PROCESS OF GRADING

On what should students or subjects be graded? As explained in chapter 1, in physical education and human performance, there are three large domains of potential objectives:

▶ The **psychomotor domain** concerns physical performance.
▶ The **cognitive domain** concerns mental performance.
▶ The **affective domain** concerns attitudes and psychological traits (also called the psychological domain).

Teachers of such subjects as mathematics, science, English, and history have instructional objectives limited mostly to the cognitive domain. Thus, in one sense their evaluation process is simpler than that of a physical education instructor. To evaluate students effectively, an instructor must have a clear understanding of the instructional objectives of the unit he or she is teaching. He or she must select and administer tests and measurements that are relevant to these objectives. The instructor then must compare the resulting test scores to appropriate standards and, finally, determine grades. Figure 9.1 illustrates this continuum.

Figure 9.1   The grading process.

Mastery Item 9.3

Consider what types of assessment strategies and instruments you might use to measure objectives in each of the three domains.

An effective and successful grading process requires that students understand the course objectives, know the tests and measurements used for grading, and know the method with which test scores will be combined to determine final grades. Inform your students of these factors at the beginning of instruction, and use formative evaluation techniques to update them on their personal progress throughout the course.

DETERMINING INSTRUCTIONAL OBJECTIVES

There are three questions to consider when determining whether a potential objective should be part of your instructional unit and thus also a part of your grading process:

1. Is the objective defensible as an important educational outcome?
2. Does every student have an equal chance to demonstrate his or her ability on the objective?
3. Can the objective be measured objectively, reliably, relevantly, and validly?

What Not to Grade On

Often students in physical education are graded on such objectives as attendance, correct uniform, shower taking, leadership, attitude, sportsmanship, participation, team rank, or improvement. Figure 9.2 summarizes the reported attributes used for grading; as you ask the preceding three questions of each of the most graded attributes in figure 9.2, you can see that they are largely inappropriate as bases for students' grades and thus should be eliminated. Rarely is it defensible to grade on one or more of these elements. Research indicates that many of these parameters are widely overused as objectives in physical education (Hensley and East 1989). Let us briefly analyze why they are inappropriate.

Although attendance, correct uniform, and shower taking are obviously worthwhile and necessary for appropriate physical education instruction, they fail to meet the first criterion: having an important educational outcome. Students in mathematics may have to attend class and bring their textbooks, but they are seldom graded on these required factors, nor should they be.

Leadership, attitude, sportsmanship, and participation are worthwhile objectives of any physical education or athletic program. However, to grade these factors reliably and validly, an instructor would need to implement a formal and systematic program of measurement

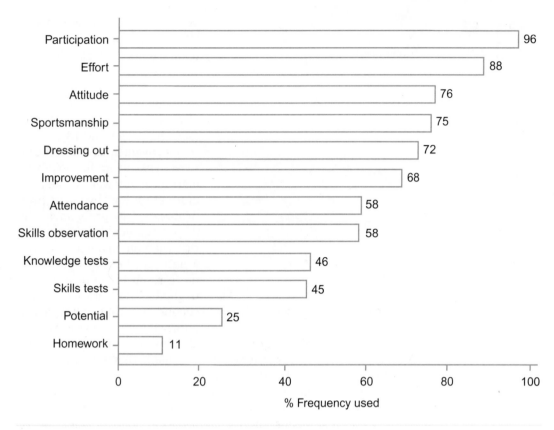

Figure 9.2 Attributes used for grading physical education.

and evaluation (third question). Implementing such a program takes time and expertise. Generally, the physical education instructor simply does not have the time or the expertise to perform this task. When these factors are used for grading (unless some of the techniques that are explained in chapter 8 are used), they are usually based on random teacher observations, which tend to be subjective and possibly biased. The result is grades that lack objectivity, reliability, and validity and are thus unfair to students.

Grading on team rank in team sport classes fails to meet the second criterion—provision of an equal opportunity to demonstrate ability. A poor performer might be placed on a good team and receive an A, whereas a good performer could be placed on a poor team and receive a low grade. This strategy is not fair to all students because their grades depend on the performance of other students, over which they have no control. *From a measurement and evaluation prospective, grading students in team sport classes may be one of the most problematic tasks a teacher faces.* Sport skills tests

Physical educators and fitness programmers can keep individuals motivated to improve by providing opportunities to succeed.

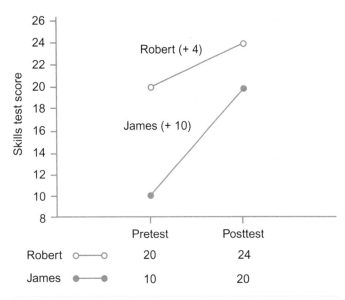

	Pretest	Posttest
Robert ○——○	20	24
James ●——●	10	20

Figure 9.3 The problem with grading on improvement.

are often used to address this problem. However, an individual sport skills test that is isolated from the team game may lack relevance and validity. Sport skills testing is examined in chapter 13. Authentic assessment (subjectively evaluating individuals as they participate in game conditions, chapter 8) is also used but requires well-developed rating scales (see chapters 8 and 13).

Improvement is one of the most attractive objectives to use in grading in physical education. As an instructor you want your students to improve, but grading on improvement presents some difficult problems. One problem is illustrated in figure 9.3. A low beginner (James) improves at a much greater rate than an advanced beginner (Robert). This is a natural phenomenon; the lower beginner has more room for improvement than the advanced beginner. The marathon runner running his or her second marathon will typically show a much more improved time than the experienced runner whose improvement is based on comparing the 20th to the 19th marathon time. Another problem is that improvement scores tend to be less reliable than either the pretest or posttest from which they are computed. A third problem is that some students, knowing that they are to be graded on improvement, may provide false initial performance scores that will inflate their improvement scores. Some teachers attempt to support improvement grading by suggesting that a student's improvement on final performance be compared with his or her potential. However, the means for validly determining potential are often unreliable or simply don't exist.

Why do you suppose educators, and especially physical educators, like to grade on improvement? It is probably because this practice gives more students a chance to earn high grades, as both those students who score high and those who show great improvement receive A's. A great number of educators are reluctant to give low grades because they believe low grades discourage effort, which in turn increases the probability of low grades, resulting in a cycle that continues until the student dislikes the subject. The concern for physical educators is that when students become "turned off" by the subject, they are less likely to continue physical activity during their lifetimes.

However, little, if any, evidence exists to support this contention. It is doubtful that students feel rewarded when given high grades for improvement when their actual performance level is lower than most of their peers. For example, poor swimmers know they are poor swimmers (without being told so) and so do their peers; an honest student knows that in the long run it is actual level of achievement that is important, not the rate of improvement. Suppose it is true that low grades cause a student to dislike school, or, more specifically, that low physical education grades destroy incentive and reduce the probability of a student engaging in physical activity. The solution of not giving low grades is like a physician treating the symptoms rather than the cause of the illness. Instead, instructors need to determine why an individual is performing poorly and provide opportunities for success. This might apply to other fitness situations in clinical or community settings as well; you do not want clients to be discouraged from attending fitness classes or programs.

Mastery Item 9.4

Who would you grade higher on improvement—a student who improved from 17.0 s to 14.0 s in the 100 m dash or one who improved from 14.0 s to 13.0 s?

First, an instructor may want to increase the number of opportunities for achieving success by broadening the curriculum from, for example, touch football, basketball, and softball to a larger variety of activities. Because the skills required to succeed in physical education are relatively heterogeneous compared with those required in other subjects, varying the physical education curriculum will probably result in a higher percentage of students achieving success than in the other areas. For example, learning how to swim a particular stroke is not nearly as dependent on knowing how to do a forward roll as, say, learning to derive a square root is dependent on knowing how to divide.

A second means of encouraging success lies in ability grouping and enrolling students in courses suitable to their skill levels. If differences in initial ability are small, grading on improvement becomes unimportant (because it becomes almost the same as grading on achievement). Other subject areas, such as mathematics, English, and the sciences, are ahead in regard to ability grouping; to adopt this approach, physical educators need to develop valid and reliable measuring devices to classify students by ability.

A third approach involves educating individuals that a grade is simply an expression of the level of achievement of the course objectives attained by the student; it is not a reward or a punishment. Furthermore, not every student can excel in every course. *One way to help ensure that grades remain a clear indication of a student's performance is to think of grades as measurements rather than as evaluations.* In grading, a measurement is a quantitative description of a student's achievement, whereas an evaluation is a qualitative judgment of a student's achievement. In other words, a grade, as a measurement, represents the degree to which the student has achieved the course objectives. If a grade is considered an evaluation, it should indicate to some degree how adequate a particular student's level of achievement is.

There are several advantages to treating grades as measurements rather than evaluations. Judgments of the adequacy of a student's achievement depend not only on how much he or she achieved but also on the opportunity for achievement and effort. This makes it difficult to report evaluations precisely in a standard grading system. Additionally, it is probably of more value to a future teacher or employer to know that a student was outstanding, mediocre, or poor in a particular course of study than to know that the student did as well as could be expected or that the student failed to live up to the expectations of a teacher. Given valid and reliable measurement of a student's achievement in various areas, the future teacher or employer can make his or her own evaluations in light of current circumstances, which would probably be more valid than having evaluations made by others under totally different conditions. *The worth of physical education must be conveyed to the student through the teaching procedures and the curriculum, not through the awarding of high grades regardless of the level of proficiency.*

Remember that Tom Jones has decided to award one half of the total grade based on the cognitive aspects of the basketball unit. Thus, he is broadening the opportunities for students to succeed, and his grading method does not reflect the less defensible objectives of attendance, improvement, and so forth.

What to Grade On

Base grades on reliable and valid tests and measurements that are representative of important instructional objectives. In individual or team sport classes, sport skills tests, rating scales, and round-robin tournament performance can be effective criteria for grading. In individual sports, such as bowling, golf, and archery, you can evaluate the student's achievement of objectives with the student's performance. In fitness classes, field tests of physical fitness provide reliable and valid data for evaluation. Because the cognitive domain is important in physical education, knowledge testing should be part of the overall grade. Chapters 10 to 14 provide you with a selection of reliable and valid tests and testing procedures for use in human performance.

In summary, use the following procedures to make grading fair, reliable, and valid:

▸ Carefully determine defensible objectives for each course before it begins.

▸ Group students according to ability in the physical skills necessary for the course.

▸ Construct tests and measurements as objectively as possible, realizing that all tests are subjective to some degree.

▸ Remember that no matter how well constructed a test is, no test is perfectly reliable.

▸ Realize that the distribution of grades for any one class does not necessarily fit any particular curve, but that over the long run physical skills are probably fairly normally distributed.

▸ Determine grades that reflect only the level of achievement in meeting the course objectives and not other factors.

▸ Establish grades on the basis of status, not improvement.

▸ Avoid using grades to reward the good effort of a low achiever or to punish the poor effort of a high achiever.

▸ Consider grades as measurements, not evaluations.

CONSISTENCY IN GRADING

A goal of any grading system should be to achieve consistency in grade determination. Theoretically, a student's grade should not depend on any of the following:

▸ A particular section of a course. If students are taking PE 100–Beginning Tennis, their grades should not be affected by whether they are in the 9:00 A.M. or the 2:00 P.M. section.

▸ A particular semester of the class. A student's level of performance should warrant the same grade whether the course is taken in the fall semester or the spring semester.

▸ Other students in a class. A student's grade depends on his or her performance only and should not be influenced by the performance of other students in the class.

▸ A particular instructor. Instructors X and Y should give the same grade to the same level of performance.

The goal of consistency is extremely hard to achieve because of student and teacher differences, which are natural phenomena of any instructional setting. As you study further in this chapter, you will discover that lack of consistency in grading is one of the weaknesses of several grading schemes.

Mastery Item 9.5

You are going to teach a general physical fitness course to students in ninth grade. What objectives would you establish for that course? What tests would you select for grading purposes?

GRADING MECHANICS

Four steps are involved in the mechanical process of arriving at a grade to represent each student's level of achievement of course objectives.

1. Determine the course objectives and their relative weights.
2. Measure each student's achievement of the course objectives.

3. Combine measurement results to obtain a composite score for each student.

4. Convert the composite scores to grades.

Each step may be accomplished in a variety of ways, depending on different situations and philosophies.

Step 1: Determine Course Objectives and Their Relative Weights

Determining course objectives and their significance is the most important of the four steps and requires a considerable amount of thought. In fact, it is basic to every aspect of teaching a course, not just grading. Determining course objectives precedes planning the sequence of presentation of material, necessary equipment, teaching procedures, assignments, and grading procedures, and it should be based on as much knowledge of the abilities of the prospective students as you can obtain, the general objectives of physical education, and such practical considerations as the number of students, the limitations of the facility, the duration of the course or unit, and the number, frequency, and duration of meeting times.

The objectives that you finally decide on for a physical education course can be classified into the psychomotor, cognitive, and affective (or psychological) domains (see also chapter 1, pp. 9-10). Stating the objectives in behavioral terms will help make the second step (measuring achievement) easier. In other words, list the actual performance levels, the specific knowledge, and the social conduct expected to be achieved by students. Sample objectives for a badminton unit are as follows:

Sample Badminton Objectives

▶ Cognitive objective: Know the setting rules when the game becomes tied at certain scores.

▶ Psychomotor objective: Be able to place at least four out of five low, short serves into the proper portion of the court.

▶ Affective objective: Be aware of proper etiquette when handing the shuttlecock to the opponent at the end of a rally.

How you weight the course objectives will vary greatly from one situation to another, depending on your philosophy, the age and ability of the students, and so on. For example, you would likely place less emphasis on the affective domain objectives for a physical fitness unit for 10th-grade boys or girls (single gender) than for a coeducational volleyball unit at the same grade level. *The actual weight you give to each course objective should result in a balanced and defensible list of goals to be achieved by each class member.*

Informing students of what is expected of them at the outset facilitates the planning of student experiences, teaching methods, and grading procedures and reduces student anxiety.

Inform students of what is expected of them.

Prepare a list of course objectives for a 5-week soccer unit for a coeducational high school class. Remember to state the objectives in behavioral terms. If necessary, review chapter 7 (p. 104) for information about behavioral objectives.

Step 2: Measure the Degree of Attainment of Course Objectives

Recall that measurement is the procedure of assigning a number to each member of a group based on some characteristic. In this case, the characteristic involved is the degree of achievement of each course objective. Unlike step 1, steps 2 through 4 occur after the teaching and, presumably, the learning have taken place. This is not to imply that all measuring should be of a summative nature. On the contrary, there is merit in obtaining measures throughout a unit (formative evaluation) because doing so leads to increased awareness on the part of the student, as well as the teacher, of progress toward the course objectives. However, the testing or measuring used for grading students' degree of achievement should obviously not occur before the completion of instruction and learning. Chapters 10, 13, and 14 provide ideas for constructing, evaluating, selecting, and administering tests and other devices to arrive at the numerical value to assign to each student that most accurately represents his or her level of achievement of the course objective being measured.

Step 3: Obtain a Composite Score

Seldom does a course have a single objective. Often, more than one measurement is made to determine the amount of achievement for each objective. *For these reasons, it is usually necessary to combine several scores to arrive at a single value representing a student's overall level of achievement of the course objectives.* This composite score is then normally converted into whatever grade format (e.g., A-B-C-D-F; pass–fail) is being used. The correct method to obtain the composite score depends on several factors, the most important of which is the precision of the scores that are to be combined.

In the case of performance scores, it is apparent why a composite score cannot be obtained by simply totaling the various raw scores. Usually units differ: Feet cannot be added to seconds, or the number of exercise repetitions cannot be added to a distance recorded in inches. As was mentioned briefly in chapter 3 under the discussion of standard scores, scores from distributions bearing differing amounts of variability cannot simply be added together because the variability affects the weight each score contributes to the composite score. The following example illustrates why adding raw scores on two tests that do have the same units (such as written tests) may not result in the desired composite score: Imagine that the scores from a class of 25 students were distributed as shown in table 9.1 on three tests worth 9 points, 9 points, and 27 points. This rather extreme example was chosen to make a particular point: that a score from a set of scores having a large variability will have more weight in the composite score than a score from a set of scores with little variability, regardless of the absolute values of the scores. The variability of the scores on the first two tests is greater than the variability for test 3. It would seem that, because the total number of points possible on test 3 is triple that of tests 1 or 2, the score achieved on test 3 would have the most influence on the student's composite score. Notice, however, some possibilities:

Student A was one of the seven students achieving the highest score made on test 3; his scores on the first two tests were average. Student A received a composite raw score of 34. The composite raw score for student B, who was above average on the first two tests but was among the lowest scorers on test 3, is 37, higher than that for student A. Student C scored at the average on two tests and above average on one test, as did student A. However, because student C's above-average performance came on test 1 (on which the scores were more variable than the test on which student A achieved above average, test 3), student C's composite raw score, 37, is also higher than student A's. Finally, even though student D scores as high as anyone in the class on test 3, the low scores made on the first two tests lowered the composite raw score, 30, below the others.

Unless two (or more) sets of scores are similar in variability, summing a student's raw scores to arrive at a composite score may lead to some incorrect conclusions.

Assume that, instead of representing the distributions of scores on three tests, the values in table 9.1 describe the distributions of the measures of how well psychomotor, cognitive, and affective objectives were met by the students. Further assume that you had decided that achievement of the cognitive objectives, the affective objectives, and the psychomotor objectives would represent 20%, 20%, and 60% of the final grade, respectively. To achieve this weighting, you simply made the number of points possible for each of the three objectives the desired percentage of the total number of possible points (i.e., 9, 9, and 27). As concluded previously, unless the variability of the three sets of scores is quite similar, the actual weighting will be different from that originally planned.

The solution is to do the weighting after a score for each objective is obtained rather than to attempt to build the weighting factor into the point system, unless you can assume that equal or nearly equal variability among the sets of scores will occur.

Table 9.1 **Test Score Distributions From a Class of 25 Students**

Test 1 (9 points)		Test 2 (9 points)		Test 3 (27 points)	
Score	Frequency	Score	Frequency	Score	Frequency
9	1	9	1	27	
8	2	8	2	26	
7	3	7	3	25	
6	4	6	4	24	7
5	5	5	5	23	11
4	4	4	4	22	7
3	3	3	3	21	
2	2	2	2	20	
1	1	1	1	19	

Student	Score on test 1	Score on test 2	Score on test 3	Total raw score
A	5	5	24	34
B	8	7	22	37
C	9	5	23	37
D	3	3	24	30

If calculating composite scores by combining raw scores is not feasible because of different units of measurement or a lack of equal variability among the sets of scores, how can you form composite scores? As described in chapter 3 (p. 43), convert each set of scores into the same standard distribution so that a common basis is established to make comparing, contrasting, weighting, and summing of scores from several sets of scores possible. There are several methods of doing this (three of which are discussed shortly); selecting the best method depends on the precision of the measurement involved and the assumption of normality.

In the case of the precision of scores, determine whether the scores are on an ordinal scale or an interval or ratio scale. As outlined in chapter 3, when you use an ordinal scale of measurement (such as the rankings from a round-robin tournament), it is only possible to say that A is greater than B. However, with interval and ratio scales (such as the number of push-ups performed), it is possible to state how much greater A is than B, because these scales have equal-sized units.

The second consideration involves determining whether the distribution of scores approximates a normal distribution. If it does not, is it because the trait being measured is itself not normally distributed, or is it because, even though the trait is normally distributed, the sample at hand for some reason does not reflect this?

There are five possible situations involving these two considerations (table 9.2). The reason there are not six possible situations is that if the scores are ordinal measures, the distribution of scores—a simple ranking of the students—cannot approximate a normal distribution. Each situation is associated with one of the three methods of converting scores to a standard distribution: rank, normalizing, and standard score. Although statistical tests are available to determine whether a distribution of interval or ratio scores is significantly different from a normal distribution, such tests are beyond the scope of this book. However, a visual inspection of the frequency distribution of the sample is usually sufficient to reveal its closeness to a normal distribution. If you are still uncertain whether a distribution of scores approximates the normal distribution closely enough, you can use the normalizing method (see p. 165).

Table 9.2 Methods of Obtaining a Composite Score Based on Shape of Distribution and Scale of Measure

	Scale of measurement	
Shape of distribution of sample	Ordinal	Interval or ratio
Nonnormal	Rank method	Rank method
Nonnormal but trait is normally distributed	Normalizing method	Normalizing method
Approximately normal	(Not possible)	Standard score method

The end result of the rank method is simply a ranking of the students, whereas the end result of the other two methods is a set of standard scores. *It is not possible to arrive at a composite score if some of the sets of scores are converted to ranks and some are converted to standard scores. Therefore, if one of the several scores being summed to obtain a composite score must be expressed as a rank, then all the scores must be expressed as ranks.* For this reason, you should plan in advance the type of measuring that you will use during a unit of instruction.

Rank Method

The simplest method, requiring the least precise measurement, is numerical ranking of the performance of the students on each test taken. However, this lack of precision also makes this method less reliable than others, and thus this method should be avoided, if possible. In the rank method, the best performance is given a 1, the second-best performance a 2, and so on until the worst performance is given a rank equal to the number of students being measured. *To obtain a composite score for each student in the numerical ranking system, simply sum the ranks for each student. The lowest total represents the overall best achievement.* In the event that one or more ranks are missing for a student, you may use a mean rank for the student, which is obtained by dividing the sum of the ranks by the number of values contributing to that sum. If you wish to weight the rankings, add in those considered most important more than once to arrive at the total. For example, assume that three rankings were obtained: the first to count 10%, the second 40%, and the third 50% of the final grade. For each student, a composite score would be obtained by summing the first rank, four times the second rank, and five times the third rank. As before, the lowest total would represent the best overall achievement. This may seem to contradict what you studied in chapter 3, page 32; however, the robustness of ranks in this situation permits performing mathematical operations on them even though they represent ordinal data.

A variation of the numerical ranking system is to rate each student as belonging to one of a set number of categories. For example, the best five achievements might be rated as 1, the next five as 2, and so on. Another variation goes one step further by giving the categories letter grades rather than numerals. This latter procedure does not require a particular number of students to be classified into each category. It has some merit in that it is more informative to the students than pure numerical rankings, but the principles used to arrive at a composite score are the same. In fact, a composite score is obtained in this system by changing the letter grades to numerals and proceeding in a fashion similar to that described for the numerical ranking system. For example, the categories A+, A, A–, B+, B, B–, C+, C, C–, D+, D, D–, and F are given the values 12, 11, 10, 9, 8, 7, 6, 5, 4, 3, 2, 1, and 0, respectively. (Note that in this method, a higher number is better than a lower one.) A student received an A+ on the affective objectives (10%), a B– on the cognitive objectives (40%), and a C– on the psychomotor objectives (50%). To arrive at a composite score for the student, the letter grades are converted to their numerical equivalents, multiplied by the weight of the corresponding objective, and summed:

$$\left(12\times 1\right)+\left(7\times 4\right)+\left(4\times 5\right)=12+28+20=60$$

Dividing this sum by 10 (the sum of the weights: 1 + 4 + 5 = 10) and comparing the resulting value to the categories converts this student's performance to a C+ (60/10 = 6 = C+).

Mastery Item 9.7

Using the system and weighting scheme just shown, what grade would you assign a student receiving grades of B+, C, and A-, respectively?

Normalizing Method

Use the normalizing method when the scores obtained in measuring a trait that is known or believed to be normally distributed do not appear to result in an approximation of the normal curve. For example, you would expect a distribution of the number of basketball free throws made in a given time period for males in the 10th grade to approach normality as the number of data points increases. If the measurement takes the form of ranks, the normal curve will not be approximated. Other reasons, such as not obtaining a representative sample, may cause a distribution to be other than normal even though the trait being measured is normally distributed. The normalizing method is much like converting a set of raw scores into a distribution of some standard score such as the *T* score. The difference, however, is that in the present case the raw scores are first converted to percentiles, which in turn are converted into a standard score scale (in most cases *T* scores are used).

A description of the procedures for converting a set of raw scores into percentiles appears in chapter 3 (p. 37), and the resulting percentiles can be converted into T scores using table 9.3. To illustrate how this procedure converts a nonnormal distribution into a normal distribution, notice that a score, regardless of its raw-score standard deviation distance from the mean, that falls 34.13% above the mean (1 standard deviation above the mean in a normal curve) is equivalent to a T score of 60, which is 1 standard deviation above the mean in the T score scale. An example of the normalizing method resulting in a T score corresponding to each raw score is shown in table 9.4. Once each set of scores is converted to T scores, the T scores can be weighted as desired and then combined to arrive at a composite score for each student. These composite scores can then be converted to the appropriate grade of the particular format being used.

Table 9.3   Conversion of Percentiles to T Scores

Percentile	T score	Percentile	T score	Percentile	T score
0.02	15	13.57	39	90.32	63
0.03	16	15.87	40	91.92	64
0.05	17	18.41	41	93.32	65
0.07	18	21.19	42	94.52	66
0.10	19	24.20	43	95.54	67
0.13	20	27.43	44	96.41	68
0.19	21	30.85	45	97.13	69
0.26	22	34.46	46	97.72	70
0.35	23	38.21	47	98.21	71
0.47	24	42.07	48	98.61	72
0.60	25	46.02	49	98.93	73
0.82	26	50.00	50	99.18	74
1.07	27	53.98	51	99.38	75
1.39	28	57.93	52	99.53	76
1.79	29	61.79	53	99.65	77
2.28	30	65.54	54	99.74	78
2.87	31	69.15	55	99.81	79
3.59	32	72.57	56	99.87	80
4.46	33	75.80	57	99.90	81
5.48	34	78.81	58	99.93	82
6.68	35	81.59	59	99.95	83
8.08	36	84.13	60	99.97	84
9.68	37	86.43	61	99.98	85
11.51	38	88.49	62		

Note: Although the T score scale theoretically extends from 0 to 100, in practice T scores lower than 15 or higher than 85 are rare and thus are not included in the table.

Table 9.4   **Example of Normalizing Method**

Raw score	f	cf	cfm	Percentile	T score from table 8.3
85	1	6	5.5	91.7	64
74	1	5	4.5	75.0	57
63	1	4	3.5	58.4	52
59	1	3	2.5	41.7	48
53	1	2	1.5	25.0	43
47	1	1	0.5	8.4	36
	6				

Note: In this table, the *f* column is simply the frequency of occurrence of each score. The *cf* column is the cumulative frequency associated with each score (beginning with the lowest score). To obtain the values in the *cfm* (cumulative frequency of the midpoint) column, add one half of the *f* value for each interval to the *cf* value from the next lowest score. For example, the value of 3.5 in the *cfm* column for the interval called 63 is obtained by adding 0.5 (1/2 of the frequency of 1 for this interval) to 3.0 (the *cf* of the interval below the interval of 63). Finally, to express the values as percentiles, divide 100 by *n* (in this case, 6) and multiply each *cfm* value by the resulting quotient.

Standard Score Method

If you judge it safe to assume that the raw-score distribution closely approximates a normal distribution, convert the raw scores to a standard score scale such as the T score scale as described in chapter 3. The net result of this procedure is, as in the normalizing method, a *T* score corresponding to each raw score. If you convert all sets of scores to *T* scores (or some other standard scale), a common basis exists for comparing scores made on two different tests even though the two raw scores were expressed in different units or had differing variability. Furthermore, you can now add *T* scores representing various achievements to obtain meaningful composite scores that can be used in determining final grades. As with the rank method, you can also weight the various tests (i.e., the respective objectives on which they are based) by multiplying the *T* score by the appropriate weights you determined previously.

Once you obtain a composite score for each student, the final step is to change this score to the appropriate grade of the particular format being used. As in each of the other three steps, various factors will affect how you do this. The decision of what procedures to follow is based on the form of the composite score, policies of the school system or department, and your individual philosophy.

Step 4: Convert Composite Scores to a Grade

At the end of step 3, the composite scores will be in one of two forms: Each student will have a total or average ranking or a total or average standard score. In effect, both forms are an ordering of the students, although in the rank method the lowest total (or lowest average) usually represents the best achievement, whereas in the normalizing and standard score methods the highest total (or highest average) represents the best achievement. The procedures for converting a set of composite scores to grades are the same regardless of the form of the composite scores and involve answering two related questions:

1. Is the class to be graded below, at, or above average in achievement of the course objectives in comparison to similar classes?
2. What percentage of students should receive each grade?

If tests and measuring devices were absolutely reliable and valid and if course objectives remained constant over time, teachers would not need to answer these questions before converting composite scores to grades. However, measurements are not perfect, objectives change, unexpected or unplanned events occur, facilities and equipment change over time, and several other factors make it impossible to compare the achievement of the current class to previous classes on a strictly objective basis. *Grading is especially difficult for new teachers because they lack experience on which to base their answers to these questions.* After arriving at some subjective answers, you can use several methods to convert the composite scores to grades.

Observation

Observation is one of the simplest methods for determining grades. List the scores from best to worst. Examine the list for natural gaps or breaks in the scores. Table 9.5 lists the scores for 15 high school boys. As you see, two gaps appear in the data. Those observed gaps are used for cutoffs for the letter grades A, B, and C.

This system usually works well for a small number of scores, because gaps in the data often appear. However, observation is not useful for a large number of scores, because observable gaps may not be present. *Furthermore, this method does not ensure consistency.* The natural gaps could be quite different in two classes; a grade of A in the fall semester class might fall in the B group in the spring.

Table 9.5   **Scores Graded by the Observation Method**

Scores	Frequency	Grade
150	1	
140	2	A
110	3	
100	2	
90	2	
80	1	
70	1	B
40	1	
30	1	
20	1	C

Predetermined Percentages

The predetermined-percentages method may be used with composite scores in the form of rankings or standard scores because it is not the value of the score that matters but its position in the distribution. Once you decide on the percentage of students to whom each grade will be assigned, you need only multiply the percentage values by the number of students in the class, as follows:

$$k = n \times (P / 100) \tag{9.1}$$

where k is the number of students to receive a particular grade, n is the total number of students, and P is the percentage of particular grade. The resulting product is the number of each letter grade to be assigned. Table 9.6 shows this procedure for a class of 45 students,

Table 9.6 Grades Determined by Preset Percentages

Grade to be given	Percentage, preset by teacher, of students to receive	Number of students to receive grade (k)	Scores	Frequency	Total actual number of students receiving grade
A	15%	45 × .15 = 6.75 → 7	120	2	6
			110	2	
			100	2	
B	25%	45 × .25 = 11.25 → 11	90	3	12
			80	4	
			70	5	
C	45%	45 × .45 = 20.25 → 20	60	8	21
			50	5	
			40	5	
			30	3	
D	10%	45 × .10 = 4.5 → 5	20	4	4
F	5%	45 × .05 = 2.25 → 2	10	2	2

which, on the basis of test scores and other evidence, has achieved substantially above the average of other similar classes. The teacher decides to allot 15% A's, 25% B's, 45% C's, 10% D's, and 5% F's. If the composite scores are in the form of standard scores or the sum (or average) of several rankings (but not if the composite scores are a single rank in which the best student has a rank of 1, the second-best student has a rank of 2, and so on), you can modify the predetermined percentage method slightly by using the natural breaks in the distribution of the composite scores. As the last column of the table indicates, the actual number of scores for a grade may be slightly different from the calculated k, but the cutoff points are selected to result in numbers of students close to the predetermined number.

As with the observation method, the predetermined-percentages method of grading does not ensure consistency in grade assignment from class to class or semester to semester. Even if the predetermined percentages remain consistent, because scores vary from class to class and semester to semester, the actual cutoff point (in terms of scores) will vary. Thus, a score that "translates" into an A in the fall semester might be a B in the spring. You can adjust for this, however, by changing the percentages of each grade to be given.

Grading on the Curve

Although *grading on the curve* is a phrase often heard in educational settings, the process is not well understood. Actually, grading on the curve is a variation of the predetermined-percentages method in which the assumption is made that the differences in student abilities in a class are normally or at least approximately normally distributed and therefore that the percentages of each grade assigned can be determined through use of the normal curve.

Some confusion about grading on the curve stems from the fact that it is possible to use the normal curve in two ways. The practical limits of the normal curve (±3 standard deviations) can be equally divided into the number of different symbols to be assigned, or certain standard deviation distances from the mean may simply be selected as the limits for each symbol assigned. Refer to table 9.7 for data used to illustrate each approach. In this example, the composite scores of 65 students have been compiled, and you desire to use the normal

Table 9.7   **Composite Scores for 65 Students**

98	78	71	64	60	52	40
93	78	70	64	59	51	38
91	77	69	63	57	50	37
88	76	68	63	57	48	36
86	75	67	63	56	47	26
85	74	67	63	56	47	
83	73	66	62	55	46	
81	73	65	62	55	45	
81	72	65	61	54	44	
79	71	65	61	53	41	

curve to determine the cutoff points for assigning the grades of A, B, C, D, and F. The mean and standard deviation of the composite scores are 63.4 and 15.09, respectively.

First Approach   For practical purposes, the normal curve may be considered to extend ±3 standard deviation units above and below the mean. (Recall that 99.74% of the area under the normal curve is found between these two points.) This total width of 6 standard deviation units is divided equally into the same number of parts as there are different grades to be assigned—in this case, five (A, B, C, D, F). Each grade thus encompasses a width of 1.2 standard deviation units (6/5 = 1.2) (figure 9.4). Because the grading format used in this illustration has an uneven number of grades, one half of the middle grade (C) falls on either side of the mean. A grade of C will be assigned to those students whose composite scores lie between 0.6 standard deviation units above and 0.6 standard deviation units below the mean. A grade of B will be assigned to students whose composite scores are between 0.6 and 1.8 standard deviation units above the mean; D will be assigned to those whose composite scores lie between –0.6 and –1.8 standard deviation units below the mean. If this process is continued, it will result in the limits of 1.8 and 3.0 standard deviation units above the mean for grades of A and –1.8 and –3.0 standard deviation units below the mean for grades of F. To account for very extreme scores, you can specify that composite scores more than 3.0

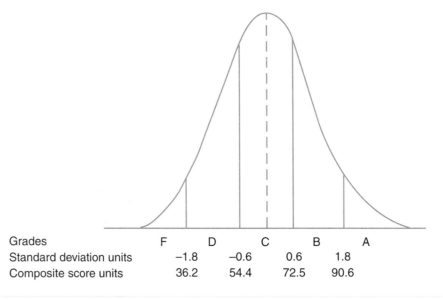

Grades		F		D		C		B		A
Standard deviation units			–1.8		–0.6		0.6		1.8	
Composite score units			36.2		54.4		72.5		90.6	

Figure 9.4   Relationship among grades, standard deviation units, and composite score units.

standard deviations above the mean also correspond to A's and that composite scores more than –3.0 standard deviations below the mean also correspond to F's.

The final procedure involves expressing the standard deviation units in terms of the composite score values. Because in this example, 1 standard deviation is equivalent to 15.09 composite score units, the cutoff point between C and B and between C and D must lie 9.05 composite score units above and below the mean, respectively. (The value of 9.05 results from multiplying 15.09 by 0.6.) The two cutoff points are thus 72.5 and 54.4 (63.4 ± 9.05). The cutoff points between B and A and between D and F are obtained by adding to and subtracting from the mean the product of 15.09 and 1.8. The result is 27.16, and thus the cutoff points are 90.56 and 36.24 (see figure 9.4). To simplify, you would probably round the cutoff points; you could also establish a conversion chart for assigning grades to the composite scores (table 9.8).

Table 9.8   Conversion of Composite Scores to Grades: Example of the First Approach

Composite score	Grade
91 and over	A
73-90	B
54-72	C
36-53	D
35 and below	F

If a grading format involving five grades is used in conjunction with the normal curve, the teacher is, in effect, deciding to assign approximately 3.5% of the class A's, 24% B's, 45% C's, 24% D's, and 3.5% F's. (To verify this, review how to use table 3.3, p. 44.) Applying the conversion chart in table 9.8 to the 65 composite scores results in the assignment of 3 A's (4.6% of the class), 15 B's (23.1%), 30 C's (46.1%), 14 D's (21.6%), and 3 F's (4.6%). Slight differences occur between the normal-curve percentages and the actual percentages because of the rounding in establishing the conversion chart and the fact that the set of 65 scores is not exactly normally distributed.

Second Approach   The second approach to grading on the curve involves selecting certain standard deviation distances from the mean as the limits for each grade. In this approach, instead of dividing the 6 standard deviations (the practical limit of the normal curve) into equal units for each grade, select distances that conform to your notion of the percentage of each symbol to be assigned. One possibility is depicted in table 9.9.

Table 9.9   Curve Grading: Example of the Second Approach

	Based on table 9.7			
Grade	Standard deviation distance for the grade (predetermined by instructor)	Percentage of students to receive grade (use table 3.3)	Number of students to receive the grade (n = 65)	Composite scores corresponding to the grade
A	Above +1.48	7%	.07 × 65 = 5	≥86
B	+0.47 to +1.47	25%	.25 × 65 = 16	71-85
C	-0.47 to +0.46	36%	.36 × 65 = 23	57-70
D	-1.47 to –0.46	25%	.25 × 65 = 16	41-56
F	Below –1.48	7%	.07 × 65 = 5	≤40

Note: The first and last columns would constitute a simple conversion table.

The selection of these particular standard deviation distances would result in assigning approximately 7% A's, 25% B's, 36% C's, 25% D's, and 7% F's to a set of normally distributed scores. As in the first approach, these standard deviations must be converted to the units of the composite scores by multiplying the selected constant by the standard deviation of the composite scores. For the data found in table 9.7 (mean = 63.4, standard deviation = 15.09), the resulting cutoff points are displayed in table 9.9. (As before, slight differences between the normal-curve percentages and the actual percentages occur because of rounding and the fact that the 65 scores are not exactly normally distributed.)

Although any set of standard deviation distances may be chosen, if the assumption regarding normality is generally true, the selected values will necessarily result in symmetrical assignment of grades. If a nonsymmetrical distribution of grades is desired, the second approach to curve grading actually becomes the predetermined-percentages method described previously. *The normal-curve method does not ensure consistency in grading from class to class because the mean and standard deviation may change from class to class.*

Mastery Item 9.8

The mean and standard deviation of a set of composite scores are 38.5 and 6.6, respectively. Calculate the normal-curve grading scale if you are to have a grade distribution of 10% of the students receiving A's, 20% B's, 40% C's, 20% D's, and 10% F's.

Mastery Item 9.9

Convert the data in table 9.7 to grades using the following standard deviation distances: A, above +1.35; B, +0.40 to +1.35; C, +0.39 to -0.40; D, -1.34 to -0.39; F, below -1.34.

Norms

Norms are derived from a large number of scores on tests and measurements from a specifically defined population. The scores are statistically analyzed to produce a descriptive statistical analysis, which allows the production of percentile norms or standard score norms. Norms are available for many physical fitness and sport skills tests (see chapters 11-14). The U.S. government performs large-scale surveys of various health-related variables such as cholesterol and blood pressure. These normative data are used to develop national health profiles, and the variables are related to morbidity and mortality. The National Children and Youth Fitness Studies I and II produced percentile norms on various physical fitness tests for American youth (Pate, Ross, Dotson, and Gilbert 1985; Ross et al. 1987).

Mastery Item 9.10

Review how to perform percentile and standard score calculations (chapter 3, pp. 37 and 43). Convert the scores in table 9.7 to percentiles and to *T* scores.

Norms can be used both for assigning grades on individual test items and for making your own norms for composite scores to arrive at final grades. For example, table 9.10 provides the 75th and 25th percentile norms for 1 mi (1.6 km) walk test scores for 400 males and 426

Table 9.10 **Percentile Norms for the 1 Mile Walk Test (min:sec)**

Percentile	Males	Females
75	11:42	12:49
25	13:38	14:12

Adapted from Jackson, Solomon, and Stusek 1992.

females aged 18 to 30 years. You might assign an A to the top quartile (above P75), a B to the two middle quartiles (P25-P75), and a C to the last quartile (below P25). *National and other published norms can be used to establish grade scales. However, the fairest norms for grade determination are local norms.* To be used in grading, norms must be representative of the students' gender, age, training, and instructional situation. Norms developed at the instructional location on students similar to those who are to be graded will ensure representativeness.

You can develop local norms for composite scores at your facility by following these steps:

1. Determine the objectives of your instructional or training program.
2. Select tests and measurements to assess these objectives.
3. Administer the same tests with standardized procedures for several years.
4. Collect enough data to have at least 200 scores for each gender and age you intend to evaluate.
5. Conduct a statistical analysis of the data and establish percentile or standard score norms.

If you follow these steps, you will be able to establish norms that are representative of your students and their learning situation. The grading standards will provide consistency in grade assignment between classes, semesters, and instructors if all instructors are involved in the development and use of the norms. However, until you have sufficient data to establish norms, you will have to use some other grading technique. *Ultimately you should work toward establishing local norms, because this method reduces most of the inconsistencies in grading.*

Arbitrary Standards

The previous techniques for converting scores to grades have involved data analysis of the observed measurements. Using arbitrary standards for grade assignment does not require data analysis of test scores. With this process, criterion-referenced standards are established for each grade. Table 9.11 provides an example of such standards on a 100-point knowledge test.

Table 9.11   Arbitrary Standards for Grading a 100-Point Knowledge Test

Point range	Grade
100-90	A
89-80	B
79-70	C
69-60	D
59-0	F

The biggest advantage of this type of grading system is that it provides consistency in grading. A 90 on the knowledge test will be an A in whatever class or semester the student takes the test. The system is simple and easy to understand. However, setting the standards in physical performance tests without knowledge of expected student performance levels results in a "best guess" process that may result in an undesirable grade distribution. If the standards you use are accurate reflections of student achievement levels, the arbitrary system can be used to good effect, but, as with the norms method, it relies on a relatively long period of data accumulation.

A specific version of using arbitrary standards in evaluation is the pass–fail assignment. The student's performance is compared with a specific criterion-referenced standard that represents a minimum level of acceptable performance or ability. The pass–fail criterion-referenced standard is used extensively in the health-related fitness evaluation of youth. Such a standard is a test score representing a minimum performance value that is associated with a reduced risk of disease or acceptable functional capacity.

Mastery Item 9.11

Examine table 9.12 and determine which test or tests were "passed" by a 10-year-old boy who ran the mile in 9 min 45 s (9:45).

Table 9.12   1 Mile Run Standards of Youth Fitness Batteries (min:sec)

Group/age	Prudential FITNESSGRAM	President's National	President's Presidential	President's Health
Boys				
6	15:00	12:36	10:15	13:00
7	14:00	11:40	9:22	12:00
8	13:00	11:05	8:48	11:00
9	12:00	10:30	8:31	10:00
10	11:00	9:48	7:57	9:30
11	11:00	9:20	7:32	9:00
Girls				
6	16:00	13:12	11:20	13:00
7	15:00	12:56	10:36	12:00
8	14:00	12:30	10:02	11:00
9	13:00	11:52	9:30	10:00
10	12:00	11:22	9:19	10:00
11	12:00	11:17	9:02	10:00

Mastery Item 9.12

The data from table 9.1 are available on the WWW site.

1. Download table 9.1 modified. Total raw scores and total *T* scores have been calculated for each student.
2. Use the **Sort Cases** . . . command under the **Data** menu. Choose the descending sort order. First sort the students by their raw score total. Note the ranking for each of the 25 students.
3. Now use the **Sort Cases** . . . command again and rank the students by *T* score total. Note the ranking for each of the 25 students.
4. Compare each of the 25 students on the two different ranking methods and note how their rankings change. The rankings change because the variabilities of the raw scores are different for each test and the variabilities for the *T* scores are the same for each test.
5. Run descriptive analyses to examine the variabilities of each raw score and *T* score.

Measurement and Evaluation Challenge

After thinking through what objectives he wanted his students to achieve, Tom Jones wanted to weight two cognitive tests and two psychomotor skills tests to make up certain percentages of the final basketball unit grade for each student. To accomplish this he had to convert the scores on all the tests to a standard score format (e.g., *T* scores) and adjust these converted scores in accordance with the percentages he had chosen. However, he had to make a further adjustment to the scores on the dribbling test because lower scores are better on this test than are higher scores (timed test). The four adjusted standard scores were then added together to arrive at a total composite score representing each student's class achievement. Tom then selected one of the methods described in the last section of this chapter to convert the composite scores to final grades.

SUMMARY

Determining and issuing grades to students is difficult. However, because grades are vital to a great number of important decisions, you must take them seriously and determine them fairly and in a meaningful way. This chapter identifies many of the issues involved in the process of grading and suggests things for you to consider when faced with this task. Determine your own answers to some of the philosophical issues of assessing the performance of others, and use the mechanical tools provided here for measuring, weighting, and combining various scores, as well as determining the final grade.

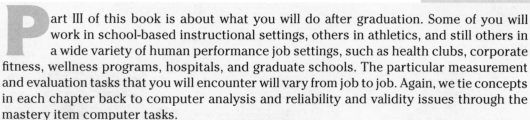

Applications of Measurement and Evaluation

Part III of this book is about what you will do after graduation. Some of you will work in school-based instructional settings, others in athletics, and still others in a wide variety of human performance job settings, such as health clubs, corporate fitness, wellness programs, hospitals, and graduate schools. The particular measurement and evaluation tasks that you will encounter will vary from job to job. Again, we tie concepts in each chapter back to computer analysis and reliability and validity issues through the mastery item computer tasks.

We introduce part III with a brief description of the domains that will be illustrated throughout the remainder of the book. The domains reflect the cognitive, psychomotor, and affective learning that you will be evaluating after graduation. Each of these domains is reflected in a taxonomy. Each level of the taxonomy is built on the levels below it. In the cognitive domain, you must first demonstrate knowledge before you can exhibit comprehension. Likewise, you would not expect young children to be able to achieve well on a difficult (or higher-order) assignment and you would expect that all college athletes would score very high on a simple motor task. The measurement tasks that you conduct must reflect the appropriate level of learning or performance expected of the people with whom you are working. Thus, measurement protocols for each domain must be carefully considered. A key concept in measurement is to be able to design and use measurement protocols that discriminate among people who are actually at different levels of achievement.

Chapter 10 presents information on how to construct and use knowledge tests and questionnaires that are useful for school-based educators, researchers, and other human performance professionals. Assessment in chapter 10 reflects the cognitive domain.

The medical and scientific literature on the relationships among physical fitness and physical activity and the prevention of disease increases yearly. Therefore, exercise scientists must fully understand the issues of reliability and validity as they relate to human performance testing. Reliability and validity in the measurement and evaluation of adults are presented in chapter 11, whereas chapter 12 deals with measurement and evaluation of youth. Chapter 13 presents techniques for reliable and valid measurement of ability and skills assessment for sport and human performance. Chapters 11, 12, and 13 reflect measurement in the psychomotor domain.

Chapter 14 deals with the growing interest in sport psychology and provides specific examples of psychological inventories, as well as issues related to the appropriate use of these batteries in sport and human performance. Measurement issues in chapter 14 reflect the affective domain.

chapter 10

Measuring Cognitive Objectives With Written Tests

Outline

Measurement and Evaluation Challenge

Kate Murphy, an educational researcher, is conducting an experiment on the effectiveness of using computers to teach basic statistical concepts. She randomly assigns students taking the basic statistics course to one of three sections of the class. One group will be taught by the traditional lecture method. A second group will conduct all assignments on the computer using a newly developed multimedia approach. This group will not attend the general lectures. The third group will attend the general lectures but will also conduct enrichment activities using the multimedia presentation. What decisions must Kate consider and what steps does she need to follow to measure how well the students in each group learn basic statistical concepts?

Objectives

After studying this chapter, you will be able to
- plan written tests;
- construct written tests;
- score written tests;
- administer written tests;
- analyze written tests; and
- understand concerns associated with planning, constructing, and enhancing the return of questionnaires.

Often, a primary objective of a research project or a physical education curriculum is to increase participant knowledge and understanding of various aspects of physical activity. To determine if this objective is being met, it is necessary to make measurements in the cognitive domain. *The written test is used for measuring the level of achievement of cognitive objectives. Also, a common objective of research is to assess individuals' attitudes, opinions, or thoughts about a particular topic. Most often this is accomplished through the use of a questionnaire. The construction and administration of a questionnaire are more complex than you might think.*

There are many sources for written tests. In some disciplines nationally normed, standardized tests are available. Textbook publishers (e.g., the one publishing this book) often provide tests or banks of test questions from which you can build your own tests. Some state agencies provide written tests for statewide testing programs to make comparisons among schools or districts. In human performance, however, outside sources of written tests are rare. In physical education, the lack of standardized tests is partly due to the great variety of activities embedded in physical education curricula and the fact that there are fewer textbooks available in physical education than in such classroom subjects as English and math. In our discipline, the most common source of written tests and questionnaires is undoubtedly the researcher or teacher interested in measuring cognitive objectives. This is not all bad, because the person making the assessment should be able to construct the most valid measuring instrument (one that measures what it is intended to measure). However, knowing what to measure and knowing how to measure it are two different things.

There are five requirements for constructing effective written tests. First, you must be knowledgeable in the proper techniques for constructing written tests. Second, you must have a thorough knowledge of the subject area to be tested. Without this knowledge, it is difficult to construct meaningful test questions. Third, you must be skilled at written expression. Test questions devised by people lacking good writing skills are often ambiguous. This ambiguity reduces the validity and reliability of a written test because there is no way of distinguishing whether an incorrect answer is due to a lack of respondent knowledge or to an error in interpretation of the question. Fourth, you must have an awareness of the level and range of understanding in the group to be tested so that you can construct questions of appropriate difficulty. This (as will be explained later) can affect the efficiency of the test. Fifth, the prospective test constructor must be willing to spend a considerable amount of time on the task. Effective written tests are not put together overnight. The following list summarizes the five requirements:

- ▶ Knowledge of proper test construction techniques
- ▶ Knowledge of subject area to be tested
- ▶ Skill at written expression
- ▶ Awareness of level of examinees' understanding
- ▶ Willingness to spend time

If you examine these five requirements carefully, you will notice that the last four are also qualities of a careful researcher or a dedicated teacher. However, we will limit ourselves in this chapter (and book) to presenting information about the first requirement, proper techniques for constructing a written test. Properly constructed tests can result in reliable and valid decisions about the ability being assessed. The following information can be used by Kate as she considers how to construct a test to measure knowledge of basic statistical techniques.

PLANNING THE TEST

First, consider the differences between mastery tests (criterion-referenced tests) and achievement tests (norm-referenced tests). A **mastery test** is used to determine whether a student has achieved enough knowledge to meet some minimum requirement set by the tester. It is not used to determine the relative ranking of students' cognitive abilities but rather to determine each student's compliance, or lack of compliance, with some previously determined standard or criterion. A familiar example of a mastery test is the type of spelling test in which the expected score is perfect or nearly perfect—all words are spelled correctly. Another example is the written portion of the test for obtaining a driver's license, on which you must get a certain minimum number correct to pass.

The purpose of an **achievement test,** on the other hand, is to discriminate among different levels of accomplishment. Because it is usually not reasonable to expect every student to achieve 100% of every cognitive objective put forth, identifying each student's progress toward meeting the objectives is of great interest.

In human performance, both types of tests have important uses. For example, in a potentially dangerous activity such as gymnastics or swimming, using a mastery test of safety rules might be prudent. For the most part, however, this chapter deals with the various phases of constructing and using achievement tests, which are more commonly used for human performance assessments than mastery tests.

Mastery Item 10.1

Describe situations in a research setting in which a written mastery test would be appropriate. Do the same for written achievement tests.

There are two important decisions to make when planning your written test. The first and more important of these involves determining what is to be measured. One technique for ensuring that a written test measures the desired objectives and that the correct emphasis is given to each objective is to develop a table of specifications. The second fundamental decision in planning a written test involves answering several mechanical questions concerning how to measure, including questions about the frequency and timing of testing, the number and type of questions, and the format and scoring procedures.

What to Measure

The question of what the test will measure should be answered before instruction begins. *The objectives of a course of study, the experiences used to meet these objectives, and the implementation and sequence of these experiences must all be determined in advance if an instructional unit is to be effective.* You can alter these elements as instruction progresses, but radical changes should not be necessary. In any event, testing will allow you to measure the degree to which the objectives of the course are being achieved and to evaluate where problems may exist. When the objectives to be assessed lie in the cognitive domain, the initial step in designing a test is the development of a table of specifications.

A **table of specifications** is to a test what a blueprint is to a building. The table identifies the relative importance of each item on the test by assigning it a percentage value. It is a two-way table, with the **content objectives** of the instructional unit along one axis and **educational objectives** along the other. Content objectives are the specific goals the instructor determines, and educational objectives are generic goals suggested by various experts. A table of specifications helps to ensure a test's content validity (the extent to which the items on a test adequately sample the subject matter and abilities that the test is designed to measure).

Let us look at an example that demonstrates the process of formulating a table of specifications for a 60-item test to be used with an instructional unit for badminton. The content objectives of the instructional unit and the tester's decision about their relative importance might be as follows:

History	5%
Values	5%
Equipment	10%
Etiquette	10%
Safety	10%
Rules	20%
Strategy	15%
Techniques of play	25%
Total	**100%**

The educational objectives (see also chapter 1, p. 9) and the tester's weighting of each might be as follows:

Knowledge	30%
Comprehension	10%
Application	30%
Analysis	20%
Synthesis	0%
Evaluation	10%
Total	**100%**

Once you have determined the objectives and their relative importance, you can construct your table of specifications.

For this example, the result is shown in table 10.1. The content objectives and their relative weights are located on the vertical axis, and the educational objectives and their weights are on the horizontal axis. The weight associated with a single cell of the table is found by determining the product of the intersection of the appropriate column and row. For example, the weight for knowledge of history is determined by multiplying 5% (the weight of history) by 30% (the weight of knowledge) to get 0.015, or 1.5%. This product for any cell is an expression of the approximate percentage of the test that should be made up of items combining the two types of objectives intersecting at that cell. The actual number of questions of each combination is found by multiplying the percentage by the proposed length of the test. In this case, we wanted a 60-item test, so for knowledge of history we would multiply 0.015 by 60 to get 0.9. In table 10.1, each cell is divided into two halves; the upper number represents the percentage of the test made up of items combining the appropriate combination of objectives, and the bottom number represents the number of questions of this type based on a total test length of 60 items.

Obviously, it is not possible to include on the test 0.9 of a question dealing with knowledge of the history of badminton; the numbers in the table of specifications are to be used as guides, and usually some rounding and adjusting are required. If the table of specifications is followed closely, the resulting test will contain questions in proportion to the percentages of weighting for each category.

Mastery Item 10.2

Based on the table of specifications in table 10.1, how many questions involving the analysis of techniques of play would be included on a 100-item test?

Table 10.1   **Table of Specifications for a 60-Item Written Test on Badminton**

			Educational objectives						
			Knowledge	Compre-hension	Applica-tion	Analysis	Synthesis	Evaluation	Totals for content objectives
		Weight	30%	10%	30%	20%	0%	10%	100%
Content objectives	History	5%	1.5%	0.5%	1.5%	1.0%	0%	0.5%	
			0.9	0.3	0.9	0.6	0	0.3	3
	Values	5%	1.5%	0.5%	1.5%	1.0%	0%	0.5%	
			0.9	0.3	0.9	0.6	0	0.3	3
	Equipment	10%	3.0%	1.0%	3.0%	2.0%	0%	1.0%	
			1.8	0.6	1.8	1.2	0	0.6	6
	Etiquette	10%	3.0%	1.0%	3.0%	2.0%	0%	1.0%	
			1.8	0.6	1.8	1.2	0	0.6	6
	Safety	10%	3.0%	1.0%	3.0%	2.0%	0%	1.0%	
			1.8	0.6	1.8	1.2	0	0.6	6
	Rules	20%	6.0%	2.0%	6.0%	4.0%	0%	2.0%	
			3.6	1.2	3.6	2.4	0	1.2	12
	Strategy	15%	4.5%	1.5%	4.5%	3.0%	0%	1.5%	
			2.7	0.9	2.7	1.8	0	0.9	9
	Techniques of play	25%	7.5%	2.5%	7.5%	5.0%	0%	2.5%	
			4.5	1.5	4.5	3.0	0	1.5	15
	Totals for educational objectives	100%	18	6	18	12	0	6	Test total = 60

Note: The top number i
the bottom number is the actual number of questions (of the 60 in total) that the percentage represents.

Various educators and test construction experts have identified educational objectives that may be used in tables of specifications. The educational objectives in table 10.1 come from a list published in *Taxonomy of Educational Objectives* (Bloom 1956). In Bloom's book, six educational objectives are defined and divided into several categories, and many examples of test items are given to illustrate the type of cognitive behavior associated with each type of objective. Briefly, knowledge is defined as remembering and being able to recall facts; comprehension as the lowest level of understanding; application as the use of abstractions in real situations; analysis as the division of material into its parts to make clear the relationship of the parts and the way they are organized; synthesis as putting together elements and parts to form a whole; and evaluation as judgments about the value of ideas, works, solutions, methods, and materials.

The following list of questions or tasks gives you an idea of how Bloom's taxonomy would apply to a written test for basketball.

- ▶ **Knowledge:** What is the height of a regulation basketball hoop?
- ▶ **Comprehension:** What area of the court is the forwards' responsibility in a zone defense?
- ▶ **Application:** What defense should be used when the opposing team is much faster than your team?
- ▶ **Analysis:** Prioritize the following basketball skills for each player position: blocking shots, dribbling, passing, shooting.
- ▶ **Synthesis:** Design a practice schedule for the first 3 weeks of the season for a boys high school team having four baskets in the gym, 5 days per week to practice, 90 min for each practice, and 35 boys out for the team.
- ▶ **Evaluation:** Present arguments for and against the following statement: Rather than have separate boys' and girls' basketball teams at the junior high school level, the school should have coed teams.

Another list of educational objectives includes the categories of terminology, factual information, generalization, explanation, calculation, prediction, and recommended actions (Ebel 1965). Bloom's and Ebel's examples indicate some of the educational objectives that can be used in constructing the table of specifications. You can also devise your own lists.

Mastery Item 10.3

Construct a table of specifications for a 50-item test covering a physical fitness unit.

How to Measure

As mentioned previously, determining how to measure usually involves answering several mechanical questions. The answers are often partially resolved by deadlines or by practical considerations, but frequently the answers require understanding the outcomes of various testing procedures.

When to Test

For testing that occurs in a school system, institutional policies may dictate the times testing is done. The type and frequency of grade reporting, a requirement to set aside certain class periods for testing, and various class-scheduling practices may influence the decision of when to test. Most frequently, tests are administered during a regularly scheduled class period at or near the end of each unit of study, and the lengths of the units are designed to coincide with the school's grading periods. These practices are justifiable for the achievement type of test discussed in this chapter. However, there may be valid reasons for administering tests at other times during instruction.

Deadlines, too, usually determine the appropriate time to administer a written test related to a research project a student is completing. Depending on the hypothesis being tested in the student's research, an instructor may plan cognitive assessments prior to, at the conclusion of, or at both ends of an instructional unit embedded in a research project.

Test frequently enough to ensure that you obtain reliable results and yet not so frequently that you needlessly use valuable instructional or subject time. *For obvious reasons, there is no set amount of time that you should reserve for measurement purposes, but it is likely that more errors are made by providing too little time for testing than by incorporating too much.*

A mastery test to ensure knowledge of safety rules near the beginning of a unit or a rules test near the middle of a unit would be appropriate.

How Many Questions

Generally, the reliability of an achievement test increases as its length increases. Reliability increases with length because the more often an assessment of achievement is made, the less overall effect chance occurrences have on the results. Flipping a coin twice and obtaining two heads is meager evidence to support the contention that the coin has two head sides. However, if the coin is tested 50 times and heads occur 50 times, the contention becomes tenable because the chance occurrence of such an event with a normal coin is extremely remote.

The length of a test is a function of other factors in addition to the desire for reliable results. Three other important factors determining the number of questions on a test are

1. the time available for testing,
2. the type of questions used, and
3. the attention span of the students.

In most school situations, the length of the class period is the limiting factor on the length of an achievement test. Often, only the typical 45 to 60 min of class time are available. The number of questions that can be answered in this time largely depends on the type of questions used, such as essay, true–false, or multiple-choice. The time required may vary considerably not only by question type but also within one type. For example, very few essay questions requiring extensive answers can be completed within one class period, but many more essay questions requiring a one- or two-sentence answer can be included. A test composed mainly of factual multiple-choice items can realistically include more questions than one made up of multiple-choice items that require examinees to apply knowledge to novel situations because the factual questions involve mostly recall, whereas the application items require additional thinking and reflection. Finally, differences in the attention spans of the examinees influence the decision of how many questions to include on a test. Schools often account for differences in attention spans by adjusting the length of class periods according to the grade level of the students. The researcher has more flexibility than the schoolteacher in varying test length, so the factor of attention span becomes the most important limiting factor for the researcher.

Another aspect to consider when determining the number of questions to include on a test is that not all students work at the same rate. What percentage of students should be able to complete the test? In most situations, all or nearly all of the individuals being tested should be able to finish the test. With a few exceptions—such as a sport officiating course or an emergency room diagnosis unit in which an objective is to acquire the ability to make rapid and correct decisions—it is generally true that a measurement of the ability to answer questions correctly is of more value than a measurement of the speed with which correct answers can be produced. Furthermore, to construct a test containing more questions than can be completed by most or all of the examinees is an inefficient use of your time, because the questions near the end of the test are seldom used.

The numerous combinations of the factors of available time, question type, attention span, and work rate make it inevitable that a certain amount of trial and error will occur in determining the number of questions on a test. However, we suggest some general guidelines that you may adjust to meet your particular situation. Most students of high school age and older should be able to complete three true–false questions, three matching items, one or two completion questions, two recognition-type multiple-choice items, or one application-type multiple-choice item in 1 min. For younger individuals, these estimates should be reduced appropriately. Few guidelines can be given regarding the number of essay questions; however, allow enough time for a student to organize an answer. Also, in general, many short essay questions measure achievement more effectively than a few lengthy ones.

Mastery Item 10.4

How much time should be allotted for a college-aged person to complete a written test containing a combination of 25 true–false questions, 25 recognition-type multiple-choice questions, and 25 application-type multiple-choice questions?

What Type of Test Format

Achievement tests are most commonly presented orally, by projection onto a screen, or in printed form. Expense, convenience, minimizing the opportunities for cheating, and concern for vision- or hearing-impaired examinees all affect the decision of what format to use. *Most important, the format should maximize the opportunity for every individual to understand and complete the required task or tasks.*

Oral presentation of test questions is, in general, an unsatisfactory procedure for most types of items, with the possible exception of true–false questions. Although the expense and your preparation time for this format are minimal, all examinees are forced to work at the same set pace, and there is little or no opportunity for examinees to check over answers. Projecting the test by means of slides, filmstrips, or transparencies on an overhead projector has basically the same disadvantages as oral presentation. In addition, this format introduces some expense and time-consuming preparation. *Probably the most common and efficient method of presenting achievement tests is in written form, with each examinee receiving a copy of the test questions.* Although this method requires advance preparation (in this case, typing, proofreading, duplicating, and possibly compiling), it maximizes convenience to the examinees. Each can work at his or her own rate, the answers can be checked if time permits, and the questions can be answered in any order. You are free to monitor the test.

Paying attention to the way you lay out your test can help you cut costs and test preparation time as well as enhance the accuracy of responses. When an examinee actually knows the correct answer to a question but makes an incorrect response because of an illegible test copy, the reliability and validity of the test are reduced. Also, carefully proofreading your test before administering it can eliminate the need to orally correct errors in test items, which wastes valuable testing time.

Here are some additional tips to consider:

▷ If various types of questions are used in one test, group together questions of the same type to reduce fluctuation among the types of mental processes required of the examinees.

▷ Group together questions of similar content (i.e., subject area) on achievement tests.

▷ Although ordering test questions from easiest to hardest is not generally recommended, including a relatively simple question or two at the beginning of a test may benefit students by reducing anxiety about the test.

Two interesting variations of the typical written test are the open-book–open-notes test and the take-home test. Each has advantages and disadvantages and under certain conditions can be used effectively. The greatest benefit of both is the reduction of anxiety on the part of students. In addition, an open-book test can allow you to ask fewer trivial questions and more application questions; it forces you to invent novel situations rather than present questions based entirely on circumstances presented in the textbook or lectures. Furthermore, open-book tests reduce the possibilities for cheating because students are allowed to use books, notes, and other materials.

One possible disadvantage of an open-book test is that it may reduce a student's incentive to overlearn and reduce the time a student spends preparing for the test. The examinees tend to rely on being able to obtain answers from their notes and books during the test and thus spend less time studying. Because examinees can look up answers, you will need to set time limits on open-book tests, or some examinees (usually unprepared and in effect studying while taking the test) will take an inordinate amount of time to finish. If an open-book test is well constructed, most examinees will find that the textbook and notes are of little value except for looking up formulas and tables.

Take-home tests can be used in situations in which more time is required to complete a test than is available in a controlled setting. The major problem with them lies in the impossibility of ensuring that each person does his or her own work. Thus, take-home tests should not be used to measure achievement but might be used for illustrating what the individuals should study and as homework assignments.

Mastery Item 10.5

| How can the method of presentation of a written test affect the resulting scores?

What Type of Questions

Questions can be classified into three general categories: semiobjective, objective, and essay. Semiobjective *questions* have characteristics of both of the other categories. There are three types of questions in this category: short-answer, completion, and mathematical questions. For these questions, the examinee must compose the correct answer; the answer is so short that little or no organization of it is necessary. Some subjectivity may be involved in scoring (e.g., awarding partial credit for the correct procedures but a wrong answer for a mathematical problem, or the incorrect spelling of the correct answer). Scoring procedures are generally similar to those used for objective questions: The response is checked to see whether it matches the previously determined correct answer.

Characteristically, the task of an examinee responding to an *objective question* is to select the correct (or best) answer from a list of two or more possibilities provided. This type of question is considered objective because scoring consists of matching the examinee's response to a previously determined correct answer; scoring is relatively free of any subjective or judgmental decision. Types of questions classified as objective include true–false, matching, multiple-choice, and classification items.

When responding to an *essay question,* one's task is to compose the correct answer. Usually the question provides some direction by including such terms as *compare* or *explain.* Or the question may constrain the answer by including such phrases as *Limit your discussion to . . .* or *Restrict your answer to the year. . . .* Essay questions are considered subjective because scoring usually involves judgmental decisions.

Several differences among the categories—other than objective versus subjective and selecting answers versus composing them—have consequences for either the instructor or the examinee. For examinees, much of the time available for testing is consumed in writing (essay questions), reading (objective or semiobjective questions), or calculating the answer (mathematical problems). Hence, because reading is less time consuming than writing or calculating, usually a greater number of objective questions can be included on a test than questions from the other two categories. Also, examinees who are weak in one of these areas (writing, reading, or calculating) may be at a disadvantage in taking tests composed mainly of questions requiring the skill in which they are weak. A poor reader, for example, may do worse on an objective test than on an essay test over the same material.

From the test constructor's point of view, essay and semiobjective questions are easier to prepare than objective questions but harder to score. In addition, the quality of an objective test depends almost entirely on your ability as a test constructor rather than as a scorer, whereas the situation is reversed for an essay or semiobjective test. Thus, your decision about what type of test to construct might be influenced, in part, by the time you have available to construct and score it, or whether your abilities lie in constructing or scoring tests.

It is plausible that individuals study differently for different types of tests (though evidence for this is inconclusive); for example, some believe that objective tests promote the study of factual and general concepts. However, this rests mainly on the mistaken assumption that objective questions cannot measure depth of achievement. Although it is often more difficult to construct, a test composed of objective questions can measure the achievement of almost any objective as well as a test made up of essay questions. *In short, the type of studying promoted by a test is more a function of the quality of the questions than the type of questions.*

It is true, though, that one type of question is more efficient than another in a particular situation. It would be difficult, for example, to conceive how the quality of one's handwriting might be measured efficiently with an objective test, or how the ability to solve mathematical problems might be measured any more validly than by a test composed of mathematical problem questions. However, the fact that it may be more efficient to use objective questions to measure factual knowledge and essay questions to measure the organization and integration of knowledge has stereotyped the way certain questions are used. Also, other testing factors may preclude the use of what appears to be the most efficient type of question. For example, it is often impractical to correct an essay test given to a large number of individuals. Thus, an objective test may be used even though the measurement involves more than just factual information. The almost exclusive use of objective questions on nationally standardized tests is an example of this situation.

Despite the names of the three categories of questions, remember that subjectivity is a part of every test constructed. Subjective decisions are required in the scoring of essay questions and, to a lesser degree, semiobjective questions. Subjectivity is present in the construction of all types of questions: Your decisions in determining both what questions to ask and how to phrase them are subjective in nature. *To increase the reliability of written tests, reduce the amount of subjectivity involved in their construction and scoring as much as possible.* Practices such as formulating a table of specifications (see p. 181) and consulting with colleagues can ensure that you accomplish this.

Mastery Item 10.6

Explain the basis for a type of question being classified as objective, semiobjective, or subjective.

Mastery Item 10.7

Review the term *objectivity* presented in chapter 6. How does the concept of objectivity apply to the administration of a written test? Can you list some of the procedures used in the administration of the ACT or SAT that are meant to increase objectivity?

Regardless of the type or types of questions used on a test, the usefulness of the resulting score depends on its stability (i.e., reliability). A test is designed and constructed to measure the achievement of certain objectives, and the scores resulting from the administration and correction of the test, are supposed to express the degree of achievement. If a different construction, administration, or correction of the test by you or a different person were to result in a different set of scores and a corresponding different ordering of the examinees, the stability and thus the usefulness of the score would be reduced. The type of questions included on a test affects the stability of the scores in various ways.

For example, if two individuals were told to construct a test over the same unit of instruction, it is more likely that the two tests would contain similar questions if the individuals were told to construct an essay test rather than an objective or semiobjective test. On the other hand, if two people each corrected an objective test, a semiobjective test, and an essay test, concurrence is much more probable for the objective test than for the semiobjective or essay test.

Understanding the similarities and differences among the types of questions, and being aware of the advantages and disadvantages of each type of question (see the following section), are necessary in selecting the most efficient types of questions for a particular situation. This knowledge, in addition to proficiency in the general requirements of test construction, will allow you to develop valid and reliable written achievement tests.

Kate, from the measurement and evaluation challenge, has decided to develop a table of specifications to ensure the proper emphasis and weighting of the concepts her test will assess. She also will probably choose to have as lengthy a test as possible and to use either multiple-choice questions or mathematical problems.

CONSTRUCTING AND SCORING THE TEST

Most of the work of the teacher or researcher will be either in the construction or in the scoring of written test items. As we've discussed, essay questions are relatively easy to construct and time consuming to score, whereas multiple-choice questions are the opposite. There are many ways to construct and score the various types of questions to increase their efficiency.

Semiobjective Questions

The three types of semiobjective questions are *short-answer questions, completion questions,* and *mathematical problems*. The short-answer question and the completion question differ only in format: The completion item is presented as an incomplete statement (a fill-in-the-blank), whereas the short-answer item is presented as a question. The task required to answer a mathematical problem is specified by symbols or by words, as in a story problem. We describe the uses, advantages, and limitations and provide construction and scoring suggestions for all three types of questions simultaneously because of their similarities.

Uses and Advantages

Semiobjective questions are especially useful for measuring relatively factual material such as vocabulary words, dates, names, identification of concepts, and mathematical principles. They are also suitable for assessing recall rather than recognition, because the examinee supplies the answer. *The advantages of semiobjective questions include relatively simple construction, almost total reduction of the possibility of guessing on the part of the examinee, and simple and rapid scoring.*

Limitations

Because of the limited amount of information that can be given in one question or incomplete statement, it is often necessary to include additional material to prevent semiobjective questions from being ambiguous. Even when a situation is explained in fair detail, the danger of ambiguity is not completely removed, especially for completion items. Occasionally, a blank left in a sentence can be filled by a word or phrase that can be defined as being correct even though it is not precisely what the test constructor desired. For example, consider the following completion item: "Basketball was invented by _____." The name *James Naismith,* the phrase *a male,* and the date *1900* are three possibilities that correctly complete the sentence. When this situation occurs, an instructor must decide whether to award credit. With mathematical questions, instructors may have to decide whether to award no credit, partial credit, or total credit if a student followed correct procedures but gave the wrong answer. These situations introduce some subjectivity and thus the possibility of inconsistency in the scoring procedure. Specific construction techniques can help reduce (but seldom completely eliminate) this problem.

Recommendations for Construction

Of the three types of semiobjective questions, ambiguity is most likely to occur with completion questions. Rephrasing the incomplete sentence into a question—that is, converting it into a short-answer item—often resolves several problems. However, if you prefer a completion question, these suggestions may reduce some ambiguities.

▷ Avoid or modify indefinite statements for which several answers may be correct and sensible. Do this, in part, by specifying in the incomplete statement what type of answer is required. For example, "Basketball was invented by _____" can be reworded as "The name of the person who invented basketball is _____." A similar method for eliminating ambiguity is to present the item this way: "Basketball was invented by _____(person's name)."

▷ Construct the incomplete sentences, when possible, so that the blank occurs near the end of the statement. This technique better identifies the specific type of answer required than when the blank space occurs early in the statement. For example, in the item "The _____ system of team play in doubles badminton is recommended for beginners," the desired correct answer is *side-by-side,* but the blank could logically be filled in with the phrase *least complex* because it is not clear that the name of the system is desired. Rewording the statement so that the blank occurs near the end solves this problem: "The type of team play recommended for beginners in doubles badminton is called the _____ system."

▷ Do not leave so many blanks in one statement that the item becomes indefinite. Consider this extreme example: "The name of the _____ who invented _____ is _____." As the example demonstrates, the more blanks in the statement, the less information given; answering the question becomes a guessing game. Give additional information by either explaining what is required or making several items from the one.

▷ Do not give inadvertent clues. Occasionally the phrasing of the statement or the use of a particular article (*a* vs. *an*) or verb reduces the number of possible words or phrases that might complete a statement. Use the following format for the indefinite article: "Basketball was invented by a(n) _____(nationality)." If more than one blank occurs in a statement, each blank should be the same length to avoid giving the student information about the length of the correct response.

▷ If a numerical answer is required, indicate the units and degree of accuracy desired. Specifying this information simplifies the scorer's task and eliminates one source of confusion for the examinee.

▷ Use short-answer questions where possible to reduce ambiguity. For example, using a short-answer question such as, "An athlete from what country won the gold medal in the pentathlon in the 2000 Olympics?" rather than the completion item, "The gold medal in the pentathlon in the 2000 Olympics was won by _____" increases the probability that the country will be identified rather than other possible information. Scoring consistency is enhanced because the examinee's task is typically more clearly identified than with completion items. You should phrase short-answer items in such a way that the limits on the length of the response are obvious.

Recommendations for Scoring

If semiobjective questions are well constructed and you encounter no problems (as when two or more answers are plausible for one item), the scoring process is simple, objective, and reliable. The answers can be scored easily by persons other than the test maker.

If the test consists of completion items, you can prepare an answer key by cutting out from a copy of the test a rectangular area where each blank occurs. Write the correct answer immediately below or adjacent to the rectangular area. When the answer key is superimposed on a completed test, each response can be quickly matched with the keyed answer.

Using separate answer sheets for short-answer items speeds the scoring process. Because only one-word or short-phrase answers are expected, you can distribute, along with the test itself, an answer sheet previously prepared with a numbered blank space corresponding to each test item. Usually, you can place two columns of answers on one side of a standard-sized piece of paper. To score short answers efficiently, construct an answer key by recording the correct responses on a copy of the answer sheet and place this alongside each answer sheet. This procedure eliminates the need to search through the pages of all the test booklets to locate the answers.

Mastery Item 10.8

| What are the advantages and disadvantages of semiobjective questions?

Objective Questions

Questions requiring the selection of one of two or more given responses can be scored with minimal subjective judgment and are thus categorized as *objective questions*. Although there are many similarities among types of objective questions, we give separate consideration to true–false, matching, and multiple-choice questions because of their differences.

True–False Questions

Perhaps unfortunately, *true–false questions* have been widely used by teachers and others, probably because these questions are relatively easy to construct and score. Although there are advantages to true–false questions and situations in which their use is justifiable, they are the least adequate type of objective question because of several weaknesses.

Uses and Advantages

Like the various semiobjective questions, true–false items are particularly suited for measuring relatively factual material such as names, dates, and vocabulary words. The advantages of using true–false items include the ease of construction, administration, and scoring and the fact that more true–false items can be answered than any other type of question within a given time.

Limitations

Many of the major weaknesses of true–false questions stem from the fact that an unprepared examinee might answer half of the items correctly by chance alone. This makes it difficult to assess a test taker's level of achievement. A correct answer could be an indication of complete understanding of the concept, a correct blind guess, or any shade of understanding between these two extremes. *In addition, the inordinately excessive influence of chance lowers the amount of differentiation among good and poor examinees and consequently the reliability of the test.*

To be fair and avoid ambiguity, a true–false item should be absolutely true or absolutely false. It is difficult to meet this requirement except when factual knowledge is involved. *True–false questions are not well suited for measuring complex mental processes.* Because of this, ill-composed true–false tests can include trivial questions and reward sheer memory rather than understanding.

Recommendations for Construction   Generally, writing good true–false questions involves avoiding ambiguity. Here are some specific suggestions. Examples of good and poor true–false questions are given at the end of the section.

▶ Avoid using an item whose truth or falsity hinges on one insignificant word or phrase. To do so results in measuring alertness rather than knowledge.

▶ Beware of using indefinite words or phrases. A question whose response depends on the interpretation of such words or phrases as *frequently, many,* or *in most cases* is usually a poor item.

▶ Include only one main idea in each true–false question. Combining two or more ideas in one statement often leads to ambiguity. If the combination introduces the slightest amount of falsity in an otherwise true statement, the examinee must decide whether to mark true or false on the basis of the amount of truth rather than on the basis of absolute truth.

▶ Avoid taking statements directly out of textbooks or lecture notes. Out of context, the meaning of the resulting item can be confusing. Very few statements made in a text or lecture can stand alone meaningfully. In addition, using textbook sentences as true–false items results in rewarding memorization.

▶ Use negative statements sparingly and avoid double negatives completely. Inserting the word *not* to make a true statement false borders on trickery and may result in a measurement of vigilance rather than knowledge. Statements containing double negatives, especially if false, are often needlessly confusing and complex.

▶ Beware of giving clues to the correct choice through specific determiners or statement length. Specific determiners are words or phrases that inadvertently provide an indication of the truth or falsity of a statement. For example, true–false items containing words such as *absolutely, all, always, entirely, every, impossible, inevitable, never,* or *none* are more likely to be false than true because an exception can usually be found to any such sweeping generalization. Conversely, such qualifying words as *generally, often, sometimes,* or *usually* are more common in true statements than in false statements. Because it often takes several qualifications to make a statement absolutely true, take care to avoid a pattern of long statements being true and false statements being short.

▶ Include approximately the same number of true statements and false statements on a test. Having too many of one or the other can bias responses. There is some evidence that false statements are slightly more discriminating, perhaps because an unprepared examinee is more inclined to mark true. For this reason it may be advantageous to include a slightly higher percentage of false statements.

▶ Do not arrange a particular pattern of correct responses. Regulate the placement of true and false statements by chance to avoid the possibility that the examinees will detect a pattern of responses.

▶ Arrange for a colleague to review the true–false questions before administering them. This may help you remove ambiguity from questions.

Modifications Test makers have attempted to modify true–false questions with the intent of reducing excessive blind guessing. One method is to require the examinee to identify the portion of a false statement that makes it false. A further modification requires the correction of the inaccurate portion. Although these two modifications partially eliminate the effect of chance on the final score, they simultaneously introduce other problems. Ambiguity may result, as in the following: "James Naismith invented the game of volleyball." The statement is false but may be corrected by replacing the name *James Naismith* with the name *William Morgan,* or by replacing the word *volleyball* with the word *basketball.* These kinds of true–false questions can introduce some subjectivity into the scoring. Furthermore, the advantage of quick scoring is lost.

Another way to modify true–false questions involves changing the answering and scoring procedure to reflect the degree of confidence examinees have in their responses. The intent is to discriminate between those who get an answer wrong because they do not know the correct response and those who know something but not enough to prevent a "bad luck" choice. Several scoring systems have been devised to accomplish such *confidence weighting* of the response to a true–false item. In the system presented in table 10.2, if the examinee marks A, for example, and the correct answer is "true," the examinee is awarded two points, but if the correct answer is "false," two points are deducted from the examinee's score.

Table 10.2   System for Confidence Weighting Answers to True–False Questions: Scoring Procedure

		Points awarded or subtracted	
Response	Mark	Correct	Incorrect
Definitely true	A	2.0	–2.0
Likely true	B	1.0	0.0
Omit or don't know	C	0.5	0.5
Likely false	D	1.0	0.0
Definitely false	E	2.0	–2.0

This modification, although increasing the discriminatory power of a true–false test, may introduce some undesirable variables. For example, differences in personality traits among examinees (some more willing to gamble than others) and the importance of knowledge in the subject being tested as well as an awareness of the nature of one's knowledge become factors influencing the final tests results. Thus, these modifications may well increase the reliability and discriminatory power of a true–false test but simultaneously reduce its validity.

Recommendations for Scoring   As is true of most semiobjective and objective questions, using a separate answer sheet facilitates the scoring procedure. Because of the similarity between the letters *T* and *F,* it is not a good idea to have the examinees write the letters *T* and *F* on a sheet of paper when completing a true–false test. A previously prepared answer sheet on which examinees block out, circle, or underline the correct response eliminates such scoring problems. Special answer sheets that can be scored by machine are available for most objective questions, including true–false questions. You (or someone not even familiar with the subject matter) can efficiently score the test by hand by matching each response on an answer sheet with a previously prepared correct answer sheet.

Mastery Item 10.9

What are the advantages and disadvantages of true–false questions?

Examples of True–False Questions for Basketball

Good Questions

1. Kicking the ball is a team foul. (False)
2. It is generally better to dribble than to pass. (False) (Earlier, *generally* was identified as a specific determiner, and its was use discouraged. However, notice that in this question it is used in a false statement rather than the normally expected true statement.)
3. A jump shot is best executed from a dribble. (True)
4. A double violation occurs when a player commits two fouls at the same time. (False)

Poor Questions

1. Basketball was first introduced in 1901. (False) (Very trivial)
2. The overhead pass should always be used by short players. (False) (Use of the specific determiner *always*)

(continued)

3. The shovel, underhand, and hook passes are made while holding the ball with both hands. (False) (Part of the statement is true and part is false.)

4. In most cases, teams play one-on-one or zone defense. (True) (Use of indefinite phrase *in most cases*)

5. A time-out shouldn't be wasted when the team isn't in trouble. (True) (Double negative)

Matching Questions

Matching questions generally involve a list of questions and a list of possible answers. The examinee's task is to match the correct answer to the proper question. At times, instead of involving a question-and-answer format, this type of question involves matching an item in one list with the item most closely associated with it in the second list.

Uses and Advantages As with true–false questions, matching questions are most efficient for measuring relatively superficial types of knowledge. Measurements of vocabulary, dates, events, and simple relationships, such as authors to books, can be effectively obtained with matching questions. *Basically, matching questions are used to measure who, what, where, and when rather than how or why.* Among the advantages of matching questions are relative ease of construction and the rapidity, accuracy, and objectivity of scoring. Matching questions require developing a cluster of similar questions and similar answers. The most discriminating matching questions often are those used in conjunction with a graph, chart, map, diagram, or similar device for which labels on the illustration are matched with functions, names, or similar categories of answers.

Limitations It is difficult, although not impossible, to construct matching questions that require the examinee to use higher-order mental processes. However, the most limiting aspect of matching questions is that they require similarity within each of the two lists that make up the item. As compliance to this requirement lessens, the discriminating power of the matching item also usually diminishes.

Recommendations for Construction Because it is easier to write questions that measure relatively superficial knowledge than it is to write questions that measure higher-order cognitive processes such as application, analysis, and evaluation, refer often to the table of specifications developed for a test when constructing matching questions. This ensures that you will achieve the desired balance among the areas measured. Unless you exercise caution, a test composed mainly of matching items may concentrate more heavily on factual material than warranted by the table of specifications. Here are some additional suggestions for constructing matching questions. Examples of good and poor matching questions are given at the end of the section.

▶ Present clear and complete directions. In general, include three details in the instructions:

    ▶ the basis for matching the item in the two lists,

    ▶ the method to record the answers, and

    ▶ whether a response in the second column may be used more than once.

An instruction such as, "Match the statements in column 1 with those in column 2" does not include any of the three points; contrast it with the following complete instruction: "For each type of physical activity listed in column 1, select the physical benefit from column 2 that is most likely to be derived from it. Record your choice on the line preceding the question number. An item in column 2 may be used once, more than once, or not at all."

▶ Avoid providing clues. Every word or phrase in one column must be a logically and grammatically acceptable answer to every question in the other. Use the same verb tense, either singular or plural words, and the same articles if possible in all questions.

An independent test reviewer can help find and correct ambiguities, errors, and clues.

▶ Avoid including too many questions in one matching item. To be effective, the list of questions and the list of answers in a matching item must be somewhat homogeneous. As the length of the list of questions or answers increases, meeting the requirement of homogeneity becomes increasingly difficult. In most cases five or six questions is the practical limit for each matching item.

▶ Make sure all questions and answers appear on the same page of the test.

▶ Include a greater number of answers than questions or allow the repeated use of some answers. This removes the possibility of using the process of elimination to obtain the answer to one question of a matching item.

▶ Keep the parts of the matching questions as short as possible without sacrificing clarity. Examinees must completely reread the list of possible answers as they respond to each item. Needlessly lengthy answers consume valuable testing time.

▶ Arrange the two lists of questions and answers in a random fashion. There should not be any particular pattern to the sequence of correct responses.

▶ Place the answers in a logical order (e.g., alphabetical, chronological) if one exists. This allows an examinee who knows the answer to locate it quickly.

Recommendations for Scoring   Because matching questions are generally answered on the test itself rather than on a separate answer sheet, arrange the items on the test so that a key can be placed next to the margin for quick scoring. Scoring a matching item can be done by someone not familiar with the subject matter tested.

Examples of Matching Questions

Good Question

For each person listed in column 1, select the sport from column 2 for which he or she is most noted. Record your choice on the line preceding the question number. A sport in column 2 may be used once, more than once, or not at all.

___1. Aaron, Hank
___2. Brown, Larry
___3. Williams, Serena
___4. Mickelson, Phil
___5. Karolyi, Bela
___6. Ruth, Babe
___7. Hamm, Mia
___8. Armstrong, Lance

a. Baseball
b. Basketball
c. Cycling
d. Football
e. Golf
f. Gymnastics
g. Soccer
h. Swimming
i. Tennis
j. Track

(continued)

Poor Question

Match column 1 and column 2.

___1.	Sit-and-reach	a.	Muscle fibers
___2.	50 yard dash	b.	Golf
___3.	Pull-up	c.	Tennis
___4.	Shuttle run	d.	$\dot{V}O_2max$
___5.	Balke treadmill	e.	Agility
___6.	Dyer backboard volley	f.	Arm strength
___7.	Disch putting	g.	Speed
___8.	Biopsy	h.	Flexibility

This is a poor matching question because

- the instructions do not indicate basis for matching, how to record answers, or how many times items in column 2 may be used;
- the items in each column are too heterogeneous, making the answers too obvious; and
- both columns contain the same numbers of items, so the last item could be answered by elimination.

Mastery Item 10.10

What are some topics of interest to you for which matching questions would be particularly appropriate?

Multiple-Choice Questions

A multiple-choice question includes two parts: the *stem,* which may be in the form of a question or an incomplete statement, and at least two responses, one of which best answers the question or best completes the statement. The task is to select the correct or best response to the question presented in the stem.

Uses and Advantages Multiple-choice questions are used on almost all nationally standardized written tests for several reasons:

▷ Questions can be scored and analyzed efficiently, quickly, and reliably.

▷ There is less ambiguity than with other types of questions.

▷ Questions with more than two responses are not as susceptible to chance errors caused by blind guessing.

▷ Questions can be used to measure the higher-order cognitive processes, such as application and analysis.

▷ Questions can measure almost any educational objective.

Because multiple-choice questions are capable of measuring all levels of cognitive behavior, are applicable to nearly any subject or grade level, and can be used to measure virtually any educational objective, they can be used in almost any situation. If you are testing a large group of examinees or are reusing a test, multiple-choice tests are most efficient in terms of the time it takes to construct, administer, score, and analyze them. In the event that fairly rapid feedback is important, multiple-choice tests, because of their quick and accurate scoring characteristics, should be used. Generally, you can include a fairly large number of multiple-choice questions on a test because the time required to answer each item is short. Because of this, and because multiple-choice questions can be constructed to measure most educational objectives, it is less difficult to construct a test fitting the table of specifications by using multiple-choice questions than any other type of question. Finally, scoring is fast and can be done by someone not familiar with the subject area.

Limitations   Multiple-choice questions, because of their versatility, do not have many intrinsic weaknesses. However, the required investment of time makes multiple choice inefficient for small groups or one-time use. A few objectives are not as efficiently measured by multiple-choice questions as by other types of questions. For example, organization, grammatical construction of sentences, and other writing characteristics are probably best measured by essay questions (although appropriate multiple-choice tests could probably be devised).

Recommendations for Construction   Writing good multiple-choice questions requires paying careful attention to many aspects, such as constructing the stem and the responses and avoiding clues. Examples of good and poor multiple-choice questions are given at the end of the section.

General Considerations

▶ As you write the initial draft, realize that each question will probably require revision.

▶ Set up a computer file to allow for revision and the addition of information. Record with each question the course objectives and educational objectives it measures so you quickly can determine its place in the table of specifications. Also record the location of the source of the idea around which the question is built, since this information is often lost with the passage of time.

▶ Base each question on an important, significant, and useful concept. Usually the most successful multiple-choice questions are those based on generalizations and principles rather than on facts and details. For example, a question requiring knowledge of the general organization of Bloom's Taxonomy of Educational Objectives is more valuable than a question requiring the examinee to know that the third category of the taxonomy is "application."

▶ Use novel situations when possible. Generally, effective questions result when you avoid using the specific illustrative materials used in the textbook or lecture and use novel situations requiring the application of knowledge instead.

▶ Phrase each question such that one response can be defended as being the best of the alternatives. It is not always necessary that the response keyed as being correct is the best of all possible answers to the question, but it must be the best of the choices listed. Also in this regard, avoid asking a question that requests an opinion, because this results in a "no-best-answer" situation. For example, consider the following stem: "What do you consider to be the best defense against the fast break in basketball?" Because this asks for an opinion, any choice marked must be regarded as correct, whether or not it agrees with the opinions of basketball authorities.

▶ Phrase each question clearly and concisely. Ideally the stem should contain enough information that examinees understand what is being asked and yet should be brief enough that no testing time is wasted reading unnecessary material. Occasionally it is necessary to include a sentence or two to clarify a situation and avoid ambiguity. However, avoid the practice of "teaching on the test," including unnecessary information (called "window dressing" by some test-construction experts) or flowery and imaginative language. Flowery language can increase possible interpretations of questions, which then can lead to ambiguity.

▶ Avoid negatively stated questions. When you do use them, capitalize or underline the negative words. The purpose of asking a question is to determine whether the examinee knows the answer, not to see who reads carelessly and who can work through the confusion that sometimes arises with negatively stated questions.

▶ Do not include a question that all examinees will answer correctly or incorrectly unless you determine that the question must be included to increase the validity of the test.

A question that every examinee answers correctly (or incorrectly) is of little value on an achievement test because no discrimination results. In fact, it can be shown mathematically that maximum discrimination can occur only when a question is of middle difficulty—that is, when approximately half the examinees answer the question correctly and half incorrectly. Although it is difficult to estimate the proportion of examinees who will answer a question correctly the first time the question is used, you should attempt to structure multiple-choice questions so they will be of middle difficulty. (Recall that one of the requirements for writing good test questions is to be aware of the level and range of understanding of the group being tested.) The difficulty of a multiple-choice question is most effectively altered by changing the homogeneity of the responses; the more homogeneous the responses, the more difficult the question. A method for obtaining an index describing the difficulty of a multiple-choice question is presented on page 213.

▶ Arrange to have the questions reviewed by someone knowledgeable in the same subject. Often an independent reviewer can locate ambiguities, grammatical mistakes, idiosyncrasies, and clues—all of which can affect a test negatively. If it is not possible to arrange for another person to review the questions, reread them yourself a few days after you have written them. (An implication of this suggestion is that you should not write the questions the night before the test is to be administered. One of the requirements for writing good test questions is the willingness to spend a considerable amount of time on them.)

▶ Consider layout issues when formatting and printing out the test. List each response starting on a new line rather than immediately following one after another. Also, unless each response is long (an unlikely event), print the items in two columns instead of across the page. Use letters instead of numbers to identify the responses (this avoids confusion between questions and answers). Keep all the responses to a question on the same page as the question stem. Separate groups of related questions from other questions by a space or dotted line.

Writing the Stem

If a multiple-choice question is to be meaningful and important, keep in mind a definite concept around which the question is built. In expressing this concept, the most important part of the multiple-choice question is the stem, and it is the first part constructed.

The stem can take two forms: a direct question or an incomplete sentence. It is usually wise (especially for novice question writers) to use questions rather than incomplete stems so that the examinee's task is clearly defined. *No matter which form is used, it is important that when the examinee finishes reading the stem, a definite problem has been identified so that the search for the correct response can begin.* A stem such as "Badminton experts agree that . . ." does not provide a specific question or task, because badminton experts agree on many things. The examinee is forced to read through all the responses to determine what exactly is being asked. This stem would not be improved greatly by changing it to the question, "On what do badminton experts agree?" If the stem is revised to "On what do badminton experts agree regarding the learning of the rotation strategy by beginning badminton players?" the examinee can begin reading the responses to locate the correct one rather than to determine what is being asked. Using incomplete stems is more likely to result in incomplete specification of tasks than is using direct questions. The suggestions provided previously under General Considerations are especially germane to writing the stem of a multiple-choice question.

Writing the Response

Following the stem of a multiple-choice question are usually four or five words, phrases, or sentences known as responses. One of the responses is predetermined to be the correct response (usually called the **keyed response**). The remaining responses are labeled *foils* or **distractors.** When constructing a multiple-choice question, write the keyed response immediately after you write the stem. Following this procedure helps ensure that the question is based on an important concept. On the test, of course, the position of the keyed response among the responses should be determined by some random procedure.

There is no reason that a multiple-choice question must contain any set number of responses or that all the multiple-choice questions on a test have to have the same number of responses. Four or five responses are commonly used because this represents a compromise between the problem of finding several adequate, plausible possibilities and including enough responses so that, as happens with true–false questions, chance does not become an important factor.

The distractors, the last part of a multiple-choice question developed, should not be constructed for the purpose of tricking the knowledgeable examinee into selecting one of them. However, one should make the distractors "attractive" to the unprepared examinee. *All the responses should be plausible answers to the question.* Often, using statements that are true but do not answer the question or using statements that include stereotypical words or phrases as distractors are effective methods for making them attractive to unprepared examinees. Using a ridiculous distractor unlikely to be chosen by any examinee is a waste of testing time.

Take care not to word the keyed response more precisely than the distractors. Recall that the keyed response needs only to be the best of the listed choices, not unequivocally correct under any circumstances. Keep all responses as similar as possible in appearance, length, and grammatical structure to avoid the selection of any response for reasons other than its correctness. As with the stem, keep the responses simple, clear, and concise to avoid ambiguity and to keep reading time to a minimum. If a natural order exists among the responses (such as with dates), list them in that order to remove one possible source of confusion.

Essentially, you want the distractors to appear equally correct to the examinee who is not familiar with the content of the item. However, examinees who fully understand the concept being tested should be able to ascertain the correct response. In other words, you want an item to appear ambiguous to the ill-prepared student (i.e., to have **extrinsic ambiguity**). If an item appears ambiguous to well-prepared examinees, it has **intrinsic ambiguity.** Extrinsic ambiguity is desirable; intrinsic ambiguity is not. Figure 10.1 depicts the difference between these types of ambiguity.

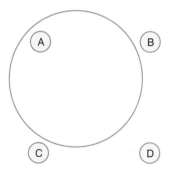

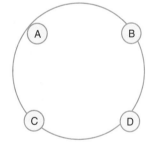

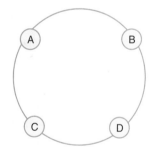

Item that is too easy	Extrinsic ambiguity	Intrinsic ambiguity
Response A (the correct response) is clearly shown to be correct, being inside the realm of the larger circle, and the other responses are clearly incorrect, falling outside of the larger circle. This item will be answered correctly by nearly every examinee and thus will not discriminate among them.	Response A is the best response, but the other responses are reasonable (have some degree of acceptability), and the unprepared student will find it difficult to choose among the possibilities. The prepared student will most likely select A, perceiving it as being better than the other responses.	All the responses could be considered to be correct, although response A is depicted as being slightly better than the other responses. This type of item will be ambiguous to well and ill-prepared students alike and will likely not discriminate among them.

Figure 10.1   The difference between extrinsic and intrinsic ambiguity. "A" is the correct response in each example.

Often, when sufficient plausible distractors are difficult to invent, it is tempting to use "None of these" as the final response. To avoid confusion, however, do not use this unless the keyed response is absolutely correct (as in a mathematical problem) and not merely the best response. When all the responses are partially correct (even if one of them is more correct than the others), the response "None of these" might be defended as being correct because none of the partially correct answers is absolutely correct. Without the "None of these" response in this situation, the most correct response is defensible as the best and thus the correct answer. A similar problem exists with the response "All of these." When there is not an absolutely correct answer and all responses contain some element of correctness, the response "All of these" could be considered the keyed response, but the examinee is put in a difficult position if one of the responses is a little more correct than the others. If you use these types of responses, make sure they are the keyed response occasionally (especially at the beginning of the test), so that examinees realize that they are to be considered seriously as possible correct answers.

Clues

Ideally, an examinee will answer a multiple-choice question correctly only if he or she knows the answer and incorrectly if he or she does not. Two factors, however, can adversely affect this situation. An examinee may blindly guess the correct answer to a question—there is no way to determine whether a correct response indicates knowledge or luck. However, in the long run, everyone has an equal chance to be lucky, and the effects of chance can be mathematically accounted for. The second and more serious factor is that of clues included within multiple-choice questions or tests. Because all examinees are not equally adept at spotting clues, the effects are not as predictable as those resulting from chance. The only way to eliminate the problem is to eliminate the clues.

Some clues are rather obvious; others are subtle. For example, it is usually easy to spot the use of a key word in both the stem and the correct response, or a keyed response that is the only one that grammatically agrees with the stem (e.g., stem calls for a plural answer and all but one of the responses are singular). Clang associations, words that sound as if they belong together, such as bats and balls, shoes and socks, up and down, are often relatively difficult for the test constructor to spot but provide immediate clues to test takers. We've suggested using stereotypical words or phrases as a method of securing attractive distractors. However, do not use these in the correct response, because an unprepared student may select the keyed response because it sounds good rather than because he or she knows it to be the correct answer.

In the process of asking one question, test constructors may inadvertently give information that answers another item on the test. Such interlocking questions provide clues for the test-wise examinee. This is especially likely to happen if you construct a test by selecting several questions from a file of possible questions, or if you add new questions or revise old questions on a subsequent test. To prevent interlocking items, read the test in its entirety once you have assembled it.

Variations Several variations of multiple-choice questions have been devised to meet the needs of particular situations. For example, the classification question is an efficient form of the multiple-choice format if the same set of responses is applicable to many questions. An example of a classification item follows:

> For questions 89 through 92, determine the type of test best described by each statement or phrase. For each item, blacken answer space
>
> A. if an essay test is described.
>
> B. if a true–false test is described.
>
> C. if a matching test is described.
>
> D. if a classification test is described.

E. if a multiple-choice test is described.

89. Test limited by difficulty in securing sufficiently similar stimulus words or phrases. (C)

90. Responses generally cover all possible categories. (D)

91. Quality determined by skill of reader of answers. (A)

92. Simplest to prepare. (B)

Another variation of the multiple-choice question involves using pictures or diagrams. This is illustrated in figure 10.2.

If the shaded circle represents a top view of a tennis player making a crosscourt forehand stroke, in what location should the ball be when contacted by the racket: A, B, C, or D?

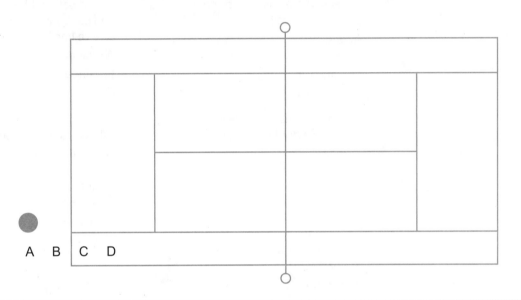

Figure 10.2   Sample of a diagram used on a written test.

You can create other variations to serve particular functions as long as the examinees are able to understand their task in answering. Most of the suggestions presented previously will apply to these diverse variations.

Recommendations for Scoring   Typically, examinees record the answers to multiple-choice questions on the test itself or on a separate answer sheet. Having students mark directly on the test slightly reduces the chances of mismarking an answer and is convenient when discussing a test after it has been administered. If this procedure is used, you can facilitate the scoring process by arranging the questions so that the answers are recorded along the margins of the test and by using an answer key overlay spaced to match each page.

Although not as convenient to the examinee, recording answers on a separate answer sheet has many advantages to the scorer. You can score the answer sheets quickly and accurately by constructing a key from one of the answer sheets. Punch holes corresponding to the positions of the keyed responses on the answer sheet. When the key is superimposed on an examinee's answer sheet, you can count the number of correct responses. You can also use machine-scorable answer sheets that allow data to be scored and analyzed by the machine and a computer program.

Good Questions

Notice that these questions are printed in two columns with each item completed in the column or on the page on which it starts. Notice also that responses are stacked and identified by letters. (Asterisk denotes correct answer for good questions and poor questions.)

1. What is called if a served ball hits the server's partner, who is standing in the correct area?
   A. Short
   B. Fault
   C. Handout
   *D. Dead ball

2. How many outs are there in the first inning of a doubles game?
   A. One
   B. Two
   *C. Three
   D. Four

3. Which of the following shots is used to get your opponent to move into the backcourt?
   A. Kill shot
   B. Passing shot
   *C. Clear shot
   D. Front wall angle shot

4. What is called if the server stops a served ball that has hit the front wall and has bounced twice in front of the service line?
   A. Handout
   B. Hinder
   C. Fault
   *D. Short

5. What is called if a served ball hits the server on the rebound from the front wall?
   A. Short
   B. Fault
   *C. Handout
   D. Dead ball

Poor Questions

1. In handball

   A. One short and one hinder put the server out. *B. The fist may be used to hit the ball. C. Receivers can make points. D. The game may be played by only two or four persons.

   (The stem does not ask a question and thus the examinee must read everything to determine what is being asked. Moreover, the responses have been run together and printed across the page.)

2. What is called if a served ball that hits the front wall, side wall, floor, back wall, and the other side wall is not returned by the receiver?

1. Lucky *2. Point 3. 911 4. Strikeout

(All distractors are not plausible answers. Also the question has been printed across the page, numerals have been used to identify responses, and the responses have been run together and printed across the page.)

3. What is called if a player gets in the way of his or her opponent?

A. Point

B. Short

C. Kill

*D. Hinder

("Gets in the way of" and "Hinder" represent a clang association. Most examinees would answer correctly with little or no knowledge of the game.)

4. What is called if, in receiving a serve, one receiver causes the ball to hit his or her partner before it touches the front wall or floor?

A. Hinder

B. Fault

C. Handout

*D. Point, because hitting your partner is your own fault

(Wording the keyed response more precisely than the distractors to ensure its correctness will cause examinees to select it even if they are not sure of the answer.)

5. What is the best shot to use in racquetball?

A. Passing shot

*B. Ceiling shot

C. Kill shot

D. None of the above

("None of the above" could be defended as being as correct as the keyed response because there is no absolutely "best" shot in every situation.)

Mastery Item 10.11

Using the suggestions just presented, write five multiple-choice questions on a topic of your choice. Critique your classmates' questions.

Essay Questions

To complete an essay question, the examinee must read the question, conceive a response, and write the response. The essay question has many uses, such as requiring the examinee to give definitions, provide interpretations, make evaluations or comparisons, and demonstrate knowledge of relationships. For the response to an essay question to be scored accurately, the evaluator must be knowledgeable in the topic being assessed.

Uses and Advantages

Although almost any type of question can effectively measure the ability to organize, analyze, synthesize, and evaluate information, essay questions are easier to construct for this purpose than any other type. The contention that essay questions promote the study of generalizations rather than facts seems reasonable but has not been and probably never will be conclusively substantiated. Essay questions can effectively measure opinions and attitudes; however, we are seldom interested in measuring these attributes in an instructional unit. Questionnaires are often used to measure opinions and attitudes as well. Essay questions on questionnaires are referred to as open-ended questions. Information about questionnaires as measuring instruments is presented at the end of this chapter. In some situations, using essay questions is more efficient or convenient, regardless of the mental processes or subject matter involved. For example, the total time required to construct and correct an essay test is often less than for other types of questions.

You should also consider your personal preferences. If you are confident in your ability to construct and score essay questions but lack confidence in using other types of questions, you should probably use essay tests. However, be aware of the limitations of essay questions and the ways you can eliminate or minimize those limitations. Finally, when scheduling circumstances dictate little time for test construction but ample time for test correction, use essay tests.

Limitations

Even with careful preparation and scoring methods, at least three problems can arise when essay questions are used to measure achievement.

Inability to Obtain a Wide Sample of Achievement   Because of the time required to organize and write answers, it is not always possible to include enough essay questions on a test to measure the achievement of each content and educational objective. Consequently, there is some lack of content validity. You can alleviate this problem by constructing a table of specifications, using several essay questions requiring relatively short answers rather than a few questions requiring extended answers, and testing frequently to reduce the amount of material measured by each.

Inconsistencies in Scoring Procedures   *The most serious problem associated with essay questions is the unreliability of the scoring procedures.* Not only does it take a substantial amount of time to correct an essay question properly, but several factors cause inconsistencies in the scores obtained. Because of the freedom the examinee has in constructing the essay answer, it is often necessary for you to decide, sometimes subjectively, whether an examinee has achieved an objective. You can reduce (although not completely eliminate) the subjectivity if you are very knowledgeable in the subject matter tested and if you make it clear what task is required of the examinee for each question.

Another problem is the "halo effect," or generalization—the part of an examinee's score that reflects your overall opinion of him or her. Giving the benefit of the doubt on one question to an examinee who has done well on most of the other questions on a test or to an examinee who has impressed you favorably in the past is an example of this phenomenon. By devising a coding system so that examinees' names do not appear on the answer sheets and by correcting the test question by question rather than paper by paper, you can diminish the consequences of this problem.

Handwriting, spelling, and grammar, for example, can positively or negatively affect the correction of an essay answer. Unless these are specific objectives of the test, the score should not reflect these elements but should be influenced only by the achievement in the area being measured.

Difficulties in Analyzing Test Effectiveness   After you have constructed, administered, and corrected a test, you will want to analyze how well the test measured what it was intended to measure, especially if you will use the test again. Analyzing a test generally includes obtaining indications of the overall reliability, validity, and objectivity of the test and the strengths and weaknesses of the test's individual items. Although some of these characteristics can and should be investigated for essay tests, essay questions do not lend themselves to this scrutiny as conveniently as do objective questions.

Mastery Item 10.12

Describe conditions under which an essay test would be most appropriate and least appropriate.

Recommendations for Construction

The following six suggestions on the construction of essay questions will help you overcome some of the weaknesses and problems associated with their scoring.

▶ Phrase the question such that the mental processes required are clearly evident. The objective of a question might be to determine whether mastery of factual material has occurred (e.g., "What are the outside dimensions of a regulation tennis court?"), to ascertain the degree to which a student can apply learned material to novel situations (e.g., "If the rules were changed to allow the shot-put circle to be raised by 2 ft [61.5 cm], would this increase or decrease the distance the shot would travel if all other factors were kept equal? Why?"), or to evaluate the ability to organize an answer in a logical manner (e.g., "Trace the development of public school physical fitness tests from the Kraus and Weber Low Back Test to the FITNESSGRAM."). The examinee should be able to recognize the type of answer required by the manner in which a question is stated.

▶ Use several essay questions requiring relatively short answers rather than a few questions requiring extended answers. This practice usually leads to two positive results: a wider sampling of knowledge and a test made up of relatively specific questions, the answers to which can normally be scored more reliably.

▶ Phrase the question so that the task of the examinee is specifically identified. Avoid asking for opinions when measuring educational achievement. Begin essay questions with such words or phrases as *Explain how, Compare, Contrast,* and *Present arguments for and against.* Do not start essay questions with such words or phrases as *Discuss, What do you think about,* or *Write all you know about.* Also, unless the purpose of a question is to measure the mastery of relatively factual material, do not begin an essay question with such words as *List, Who, Where,* or *When.*

▶ Set guidelines to indicate the scope of the answer required. Build limiting factors into the question, such as: "Explain how, during the years 2000-02, . . ." or "Limiting your answer to team games only, compare . . ." Other methods to indicate how involved the answer should be are to specify the amount of time to be spent on the answer, the number of words needed to provide a "best answer" or the number of points the answer is worth in the margin beside the question, or the amount of space in which the answer is to be completed; however, spacing each question differently to provide "sufficient" room for each may penalize those examinees with large handwriting.

▶ Prepare for yourself an ideal answer to the question. Because this requires that you identify exactly what the question is intended to measure, ambiguities often become apparent. This practice also increases the reliability of the scoring process.

▶ Avoid giving a choice of essay questions to be answered. If an essay test is designed to measure achievement of the objectives in a group of students all exposed to the same instruction, each student should be required to answer the same questions. The common base of measurement is lost if a choice of questions is given. Optional questions add another variable and increase the possibility of inaccurate assessment.

Recommendations for Scoring

Certain practices reduce some of the unreliability inherent in the process of scoring an essay answer. Several of these procedures are related to or follow from the previous suggestions for construction.

▸ Decide in advance what the essay question is intended to measure. If an essay question is designed to measure application of facts, evaluate the answers to the question on that basis and not on the basis of organization, spelling, grammar, neatness, or some other standard. Ignore elements not dealing with the question's objective.

▸ Use the ideal answer previously prepared as a frame of reference for scoring. This is especially important if you secure an independent rating of the answers (see p. 207 for more on this).

▸ Determine the method of scoring. Use one of three systems:

▸ **Analytic scoring** involves identifying specific facts, points, or ideas in the answer and awarding credit for each one. An answer receiving a perfect score would necessarily include all the specific items occurring in the ideal answer. This type of scoring is especially effective when the question's objective is to measure whether the student acquired the factual material.

▸ **Global scoring** consists of reading the answer and converting the general impression obtained into a score. In theory, the general impression is a function of the completeness of the answer in comparison to the ideal answer. Of the three grading methods, this is the most subjective and the one most likely to be affected by extraneous factors.

▸ **Relative scoring** consists of reading all the students' answers to one question and arranging the papers in order according to their adequacy. You may accomplish this by setting up several categories (such as good, fair, and poor; or excellent, above average, average, and below average) and assigning each answer to one of the categories. Second, third, and possibly more readings may be necessary to arrange the papers within each category, and you may occasionally shift one paper to another category. The end result is an ordering of all the papers in respect to the correctness of the answers to the one question evaluated. After sorting, scores may be assigned to each answer. There is no reason the top paper must be assigned an A or the bottom paper an F; your evaluations should also be influenced by the comparison of each answer to the ideal answer. This ordering of the answers enhances consistency in the scoring procedure and is especially effective when the objective of a question is to measure relatively complex mental processes. Repeat the procedure for each of the remaining questions.

▸ Develop a system so that you don't know whose paper is being scored. Examinees could sign their names on a piece of paper next to a number corresponding to the number on their test booklet, or they could mark their test copy with a unique design or pattern that only they will recognize. Having each answer recorded on a separate sheet of paper also eliminates the bias caused by noticing the scores given to a previous answer. If several answers do occur on one answer sheet (as would be the case if short answers are required), record points awarded for each answer on a separate sheet of paper, thus helping to eliminate the halo effect. This procedure is also useful if tests are rescored to check reliability. The second reader, who may or may not be you, will not be influenced by the score previously awarded.

▸ Score everyone's answers to one question rather than score one complete paper at a time. This process is required if you use global or relative scoring. Although not required for analytic scoring, the process usually leads to more consistent scoring because it is easier to compare all the answers to one question if answers to other questions do not intervene.

> ▶ Arrange for a second scoring of the question. Ensuring the reliability and objectivity of essay test scoring requires that each answer be scored twice and the two scores compared. Ideally, these two scores should be awarded by two different scorers to ensure that they are independently obtained. If it is possible to arrange for another person knowledgeable in the area covered by the test to score the paper, supply that scorer with the ideal answers to the questions so that the two scores thus obtained have a common basis. However, if it is not feasible to obtain an independent scorer, score the answers yourself on two different occasions, perhaps separated by a week, in an effort to secure some evidence about the consistency of the scoring procedures used.

As should be obvious by this point, the process of constructing and scoring a reliable essay test can be very tedious and time consuming. However, to be fair to the examinees, the procedures explained here should be followed if an essay test is used to measure cognitive objectives.

Mastery Item 10.13

What problems are introduced when you grade a set of essay exams by scoring each paper in its entirety before moving on to the next one?

ADMINISTERING THE TEST

As we have noted, there are problems involved in testing. Before and during a testing session, the anxiety level of some examinees can increase beyond desirable levels; during a testing session cheating can occur; and afterward, when examinees learn of their scores, feelings of humiliation or haughtiness may be experienced. However, these undesirable circumstances do not *have* to occur. The suggestions presented here should help eliminate or reduce many of the objectionable occurrences that are often associated with test administration. *Although the written test itself and the scoring procedures used have some influence on these occurrences, the administration of the test itself probably has the greatest impact on whether problems will arise before, during, and after the test.*

Before the Test

▶ **Prepare the examinees for the test.** Generally, less anxiety is associated with a test announced well in advance than with surprise tests, and discussing the content of an upcoming test with students can help reduce their apprehension. It is not logical (or ethical) to include on an achievement test questions on topics that have not been covered or assigned. Items such as the general areas to be measured, the approximate amount of testing time devoted to each area, the types of questions that will be asked (essay, multiple-choice), and the length of the test represent legitimate concerns of the examinee. In the final analysis, a written test, if properly constructed, should be a precise expression of the objectives of the instructional unit. It is difficult to imagine a situation in which knowledge of these objectives should be withheld from the examinees.

▶ **Eliminate the test-wise advantage for some examinees.** Use proper test construction techniques outlined previously in this chapter (avoiding grammatical clues, specific determiners, interlocking items, and the like) and provide examinees with test-taking suggestions. For example, the following recommendations might be made to examinees:

- ▸ Realize that all the material measured by a good test cannot be learned the night before the test. Spend this time reviewing, not learning.
- ▸ Read the instructions to the test before beginning to answer the questions. Know how the test will be scored. Be aware of (a) whether all questions have the same value; (b) whether neatness, grammar, and organization will be accounted for in the score; and (c) whether a correction-for-guessing formula will be applied.

▸ Pace yourself.

▸ Plan an essay answer before starting to write it down.

▸ Check often to see that you are writing answers in the correct place on the answer sheet.

▸ Check over your answers if time permits.

▸ **See the comprehensive list of test-taking skills on pages 209-210 as well as on the Web site for this book.**

▸ **Give any unusual or lengthy instructions before test administration time.** This will save time on the day the test is given and, more important, will enable the examinees to begin the test as soon as possible. This reduces the time available for anxiety to build, especially for those who feel pressured by time.

▸ **Proofread the test before it is reproduced.** Proofreading helps to ensure that each examinee will receive a legible copy free of typographical, spelling, and other errors. It also eliminates or reduces the time spent during the test clearing up these errors.

▸ **Give a practice test to reduce the examinees' anxiety.**

During the Test

▸ **Organize an efficient method for distributing and collecting the tests.** With a small group, this is seldom a concern; however, with 60 or so examinees spread out in a large room, an efficient collection procedure is vital to keeping the test secure.

▸ **Help the examinees pace themselves.** This can be accomplished by quietly marking on a blackboard the time remaining as well as a rough estimate of the portion of the test on which the examinees should be working.

▸ **Answer individual questions carefully and privately.** To avoid disturbing others, answer an individual question at the examinee's or the proctor's desk. However, take care that your response does not give any examinee an advantage over others.

▸ **Control cheating.** Obviously, cheating negates the validity of a set of test scores. Of more serious concern, though, are the negative attitudes generated toward those who cheat, toward the proctor who does not control cheating, and toward testing in general.

▸ **Control the environment.** In the final analysis, any factor that prevents an examinee from doing his or her best on a written test lowers the reliability, validity, and usability of the resulting set of scores. Some of these factors—examinee motivation and reading habits—are not under the direct control of the tester, although they can be influenced. You can, however, provide adequate lighting, eliminate noise distractions, maintain a comfortable temperature, and provide adequate space in which to work.

After the Test

▸ **Correct the tests and report the scores as quickly as possible.** The rapidity of this operation depends, of course, on the type and length of the test administered. However, examinees generally appreciate prompt results.

▸ **Report test scores anonymously.** Let the examinee decide whether to make his or her score known to others. Use a confidential identification number system if you post scores.

▸ **Avoid misusing and misinterpreting test scores.** By following the suggestions in this section, you will be able to improve the reliability and validity of your written tests. However, remember that no test is perfectly reliable and valid. Because of this, do not base crucial decisions on the results of one written test. For example, do not interpret a 1-point variation between two examinees' scores on a written test as showing a significant difference between the examinees. Such an interpretation is a misuse of test scores. Along with other forms of measurement, consider written test results when evaluating individuals, but allow these results to influence the evaluations only to the degree their accuracy permits.

Some Test-Taking Skills

Preparing for the test

- Schedule your time ahead—plan when good study times are available.
- Know when, where, and how you will be tested. Ask the instructor.
- Go to the instructor when you encounter a difficult or problematic area while studying.
- Be in the best physical and mental shape possible.
- Be motivated and positive in your attitude toward the test.
- Be mechanically prepared, with pencils, space, text, tables, and notes.
- Practice, practice, practice by finding practice tests to take. Generally people who are more familiar with taking tests perform better. Practice efforts are generally better on timed tests. The less the interval between practice and testing, the better the practice effect.
- Carefully read the summaries of each chapter. Look at highlighted text, figures, and tables.
- Study with classmates.
- Avoid cramming.
- Be test-wise, even though it may not help with well-developed tests. It will help on poorly constructed tests.
- If studying for an essay exam, make up and answer practice questions in advance.
- If studying for a completion test, find out if spelling will count.
- If studying for a matching test, find out if you can use an answer in more than one place.
- If preparing for an open-book test, mark particularly important pages or sections with tags to make them easy to find during the test.
- If studying for a take-home test, find out what sources you will be permitted to use for help.
- Look over all materials related to the course. (But don't stay up all night.)
- Get a good night's rest.
- Do not use stimulants or tranquilizers.
- Do not drink a lot of fluids or eat a big meal just before the test.
- Get to the testing area early and familiarize yourself with the area.
- Avoid last-minute panic questions. Panic is contagious. Do not talk with friends immediately before the examination.
- Relax.

Getting started and taking the test

- Sit where you feel comfortable—whether that means near a window, near an outlet, or where you generally sit in class. Don't sit near annoying people.
- Read and listen to instructions carefully. There is important information in the instructions including oral directions or corrections.
- Find out how the test will be scored, whether some items are worth more than others, and whether neatness counts.
- Know how much time you have to complete the examination, and be aware of time remaining during the examination.
- Look through the test before starting so that you can plan ahead and gauge your time.

(continued)

- Check to see that you have all the pages and items before beginning.
- Pace yourself, budget your time, and don't spend too much time on any one item.
- Concentrate on the test; do not pay attention to others in the room.
- Think positively.
- Stay calm if you don't know the answer; make an educated guess.
- Ask the instructor if you do not understand something.
- If you are stuck, move on to the next question and come back to a difficult one later. Activity reduces anxiety.
- Be aware of when time is almost up so that you can review and check the test.
- Do not worry about other students (e.g., if they leave earlier than you, or the questions they ask).
- If you are using a separate answer sheet, check often that you are in the correct column and row. Check that you answered all the items. If there is time, reread questions and answers.
- Just before you submit the answer sheet, count the number of answers you have blackened. Make sure that the number of answers blackened equals the number of test items.

After taking the test

- Write down what you remember about the test.
- If you feel you did poorly, go to the instructor's office and review the test with the instructor.
- Appeal when you feel that you are correct about an "incorrect" answer.

Mastery Item 10.14

Explain how careful preparation, administration, and reporting of results can eliminate many students' testing concerns.

ANALYZING THE TEST

To determine the amount of confidence that can be placed in the set of scores resulting from a test administration, examine the reliability and the validity of the test. This is based on how closely (validity) and consistently (reliability) the test actually measures what is intended. Evidence for the reliability and validity of a test is both global (overall test performance) and specific (quality of individual questions).

Reliability

If a test were perfectly reliable, each examinee's observed score would be an exact representation of his or her level of achievement of whatever the test measures. Each observed score would be a true score, uncontaminated by error. In actuality, of course, an observed score consists of two parts: the true score and the error score. The error score may be positive or negative, increasing or decreasing the observed score. As the error portion for the observed score increases, reliability decreases. Unfortunately, there are several sources of error in written tests:

▶ **Inadequate sampling.** The questions that appear on a test represent only a sample of the infinite population of possible questions that could have been selected. Error is introduced if the sample selected does not adequately represent the desired population of possible questions. An examinee's failure to be credited with understanding, or being penalized for not comprehending, a particular notion because no question was included on the test to measure that comprehension is an example of how sampling error might reduce test reliability (and validity).

▶ **An examinee's mental and physical condition.** Illness, severe anxiety, overconfidence, or fatigue can alter one's score and thus lower the reliability of a test.

▶ **Environmental conditions.** Poor lighting, poor temperature control, excessive noise, or any other similar variable that negatively affects concentration can cause observed scores to misrepresent true scores.

▶ **Guessing.** Because each examinee has, in theory at least, the same chance for good luck (and bad luck) when blindly guessing during an objective test, it would seem that in the long run the total effect would balance out and there would be no error introduced. However, one administration of a test does not represent the long run, and test reliability might be reduced because some examinees could, on one administration of a test, be luckier guessers than their peers.

▶ **Changes in the field.** Sometimes error is introduced not by the measuring instrument but by the fact that the variable being measured is changeable. Lack of a consistent definition (e.g., disagreement by authorities on the definition of "physical fitness") and fluctuations in the amount of the attribute being measured (e.g., attitude toward physical activity can change from time to time) make constructing a reliable test difficult in some areas.

Thus, many factors, some of which are at least partially under your control, can introduce error and consequently reduce the reliability of a written test. As indicated in chapter 6, there are several methods of calculating a coefficient to express the reliability of a test; each of these methods reflects one or more of the sources of error. If test questions are scored correct (1) or incorrect (0), the alpha coefficient (identical to the Kuder–Richardson formula 20, or KR_{20}) can be used to estimate the test's reliability. The KR_{20} is actually the average of all possible split-half reliability coefficients and as such is a relatively conservative estimate of a test's reliability. Obtaining a satisfactory reliability coefficient when using a conservative procedure is good because using other less conservative procedures would only result in higher estimates. KR_{20} is defined as:

$$KR_{20} = \frac{K}{K-1}\left[1 - \frac{\Sigma pq}{s^2_{total}}\right] \qquad (10.1)$$

where K is the number of test items, s^2_{total} is the variance of the test scores and Σpq is the sum of the difficulty (p) times q where q is defined as ($1 - p$). You will learn more about p (Difficulty or Diff.) on pages 213-214.

Another method of estimating a written test's reliability when it is reasonable to assume that all items on the test are equally difficult is the KR_{21}. The formula follows:

$$KR_{21} = \frac{K}{K-1}\left[1 - \frac{M\left(1 - \bar{p}\right)}{s^2_{total}}\right] \qquad (10.2)$$

where K is the number of questions on the test, s^2_{total} is the variance of the test scores, M is the mean test score, and $\bar{p}$ is the average difficulty defined as M/K. Note the similarity between KR_{20}, KR_{21}, and the alpha coefficient (see equation 6.3, p. 90). The alpha coefficient is actually equivalent to KR_{20}. The KR_{21} reliability estimate is relatively easy to calculate, but the assumption of equally difficult items is rarely true. Violation of this assumption results in this formula's underestimating the test's reliability; therefore, the formula is a conservative estimate of the test reliability. Thus KR_{20} will always be greater than or equal to KR_{21}. Obtaining a satisfactory reliability coefficient when using a conservative procedure is a good idea because using other less conservative procedures would only result in higher estimates.

Mastery Item 10.15

Use the KR_{21} formula to estimate the reliability of a 60-item test having a mean of 45 and a standard deviation of 6.

Validity

If a written test does not measure what it is designed to measure (even though it may measure something consistently), the resulting test scores are of little value. As noted in chapter 6 (pp. 94-100), there are various types of validity and several methods of assessing them.

For a written test, one of the most important types of validity is content validity. This is generally determined subjectively by the extent to which the individual test items represent a sufficient sample of the educational and content objectives included in a course of instruction. In other words, by examining a copy of a test, you determine the degree of content validity the test has for the particular situation. Following the proper procedures for constructing a written test, especially using a table of specifications, helps to ensure that your test will have content validity.

ITEM ANALYSIS

Analyzing the responses to the test items is important for several reasons but especially to continually improve the items and consequently the test. The difficulty level and the discriminating power (ability of the question to separate strong from weak examinees) of each item are the keys to item improvement. **Item analysis** can also improve your instruction by identifying weaknesses in the examinees as a group, in instructional methods, or in the curriculum. It can also improve your skill in constructing written tests. Most of the illustrations and examples presented involve multiple-choice questions because there are efficient methods for analyzing them. However, you can modify most of the following steps of item analysis for other types of objective items, and you can apply the principles involved to most types of questions. The procedures for item analysis follow.

- ▶ **Step 1**—Score the tests.
- ▶ **Step 2**—Arrange the answer sheets in order from high to low score.
- ▶ **Step 3**—Separate the answer sheets into three subgroups: (a) the upper group, which consists of the upper 27% (approximately) of the answer sheets; (b) the middle group, which consists of the middle 46% (approximately); and (c) the lower group, which consists of the same number of answer sheets as placed in the upper group. You will use only the answer sheets of the two extreme groups—the upper and the lower—in the item analysis. Test authorities suggest that to include as many responses as possible and maximize the differences between the types of responses, the upper and lower groups should each be composed of 27% of the answer sheets. Generally, as long as there is an equal number in each of these groups, use the most convenient number of answer sheets between 25% and 33%. For example, if 60 answer sheets were available for analysis, the highest and lowest 15 to 20 could be used.
- ▶ **Step 4**—Count and record for each item the frequency of selection of each possible response by the upper group.
- ▶ **Step 5**—Count and record for each item the frequency of selection of each possible response by the lower group.

Steps 4 and 5 are the most time-consuming portion of the item analysis. Several procedures can reduce the tedium of this task:

▶ Use previously prepared "scorecards" for each item.

▶ Use a computer to speed the process of recording responses (e.g., assign five adjacent keys to each five possible responses, so responses can be tabulated quickly).

▶ Cooperate with another scorer, with one person reading and the other recording, or use an optical scanner to have a computer accomplish these steps.

An example of a possible organization of the resulting data is shown in figure 10.3. (These data were obtained from a question included on a nationally standardized test of physical fitness knowledge administered to college senior physical education majors.)

At the completion of step 5, the necessary data are available to calculate an **index of difficulty** and an **index of discrimination** for each item. The data in figure 10.3 illustrate the calculation of these two indexes and how change suggested by the response pattern can improve an item. In this example, the left side of the figure contains the initial draft of the question and the data (as described previously) resulting from its administration to approximately 185 examinees. The right side contains the revised question and the data resulting from its administration to more than 1000 examinees.

▶ **Step 6**—Calculate and record the index of difficulty for each item; this is the estimated percentage of examinees who answered the item correctly. The formula is as follows:

$$\text{Diff} = \frac{U_c + L_c}{U_n + L_n} \times 100 \tag{10.3}$$

where Diff is the index of difficulty, U_c is the number of examinees in the upper group answering the question correctly, L_c is the number of examinees in the lower group answering the question correctly, U_n is the number of examinees in the upper group, and L_n is the number of examinees in the lower group (recall that $U_n = L_n$).

Inspection of this formula reveals that the index of difficulty is the percentage of examinees answering the question correctly; thus, the higher the index, the easier the question. The following examples illustrate the use of the index of difficulty formula (see figure 10.3).

Source: *Handbook of Physical Fitness*									Topic: Physical fitness								
First draft: In the opinion of most authorities, three of the following factors have contributed to a lowering of the national level of physical fitness. Which has NOT had this effect? A. An increase in life span B. A decrease in the physical effort required for daily living C. An increase in the number of occupations involving sedentary activity *D. An increase in school consolidation									*Revision:* In the opinion of most authorities, three of the following have contributed to a lowering of the national level of physical fitness. Which has NOT had this effect? A. An increase in the number of senior citizens B. A decrease in the physical effort required for daily living C. An increase in the number of occupations involving sedentary activity *D. An increase in school consolidation								
Item 5	Test: Form D trial			Date: 6/68			*n* = 185		Item 25	Test: Final form A			Date: 9/00			*n* = 1112	
Responses	A	B	C	D*	E	Omit	Diff.	Net D	Responses	A	B	C	D*	E	Omit	Diff.	Net D
Upper 27% = 50	28	2	1	19		0			Upper 27% = 300	69	10	5	216		0		
Lower 27% = 50	24	8	1	17		0	36%	4%	Lower 27% = 300	89	52	54	104		1	53%	37%

Figure 10.3 One way to organize data for item analysis.

First-draft results: n = 185; therefore U_n = L_n = 185 × 0.27 = 50.

$$\text{Diff} = \frac{19 + 17}{50 + 50} \times 100$$

Revision results: n = 1112; therefore U_n = L_n = 1112 × 0.27 = 300.

$$\text{Diff} = \frac{216 + 104}{300 + 300} \times 100\% = 53\%$$

The maximum amount of discrimination can occur only when an item has an index of difficulty of exactly 50%. If this criterion were met by every question on a test, the mean score of the test would be equal to one half the number of items on the test. For example, the mean score of a test containing 80 items would be 40. This ideal, however, assumes that no element of chance is involved. On an 80-item multiple-choice test on which each item has 4 possible responses, random marking of the answer sheet should produce approximately 20 correct responses (i.e., 1/4 × 80 = 20). Considering chance, the mean score of the test just described should be 50. This value is obtained by determining what score lies halfway between the chance score and the highest possible score (80 items – 20 correct by chance = 60 items; if each examinee answers 50% of these 60 items correctly, he or she would have 30 items correct, plus the 20 by chance, resulting in a score of 50). If the index of difficulty of each of the 80 items was 62.5%, the mean score for the test would be 50 (80 × 0.625 = 50).

Obviously, it is not possible, especially on the first draft, to produce an item having exactly some predetermined difficulty index. *The point is, to maximize an item's discrimination power, an attempt should be made to write each item in such a manner that half or slightly more than half of the examinees will answer it correctly.* One further point should be noted. Maximum discrimination can occur only for an item of middle difficulty, but meeting this condition does not necessarily guarantee that it will occur. Figure 10.4 describes the relationship between discrimination and difficulty and shows that as difficulty increases from 0 to 0.50, the potential discrimination increases. However, as difficulty continues to increase from 0.50 to 1.0, *potential* discrimination decreases.

▷ **Step 7**—Calculate and record the index of discrimination for each item; this is an estimate of how well an item discriminates among examinees who have been categorized by some criterion.

$$\text{Net D} = \frac{U_c - L_c}{U_n} \times 100 \qquad (10.4)$$

where **Net D** is the index of discrimination. (Note that either U_n or L_n may be used in the denominator.) The index of discrimination presented here, known as the Net D, is only one of nearly 100 discrimination indexes that have been devised. The most commonly cited discrimination indexes are correlational techniques for quantifying the relationship between the score on a particular item and a criterion score (usually the total test score). Flanagan's r, Davis' index, biserial correlation, and tetrachoric correlation are used extensively. However, we use the Net D because it is relatively simple to calculate, uses the same data as required to determine the difficulty index, and is fairly simple to interpret. The following examples, again using the data presented in figure 10.3, illustrate the use of the Net D formula.

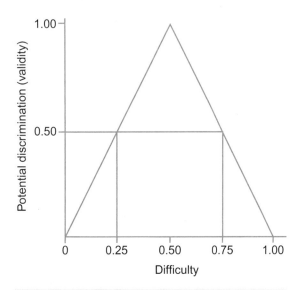

Figure 10.4 The relationship between discrimination and difficulty.

First-draft results: $n = 185$; therefore $U_n = L_n = 50$.

$$\text{Net D} = \frac{19 - 17}{50} \times 100 = 4\%$$

Revision results: $n = 1112$; therefore $U_n = L_n = 300$.

$$\text{Net D} = \frac{216 - 104}{300} \times 100 = 37\%$$

The criteria usually used to examine the discriminating power of an item are the scores on the entire test on which the item appears. In general, if the examinees who performed well on the entire test did well on the item and the examinees who did poorly on the entire test did poorly on the item, the item is considered a good discriminator. If approximately the same number of "good" and "poor" examinees answer an item correctly, it is considered to have little or no discriminatory power. If the item is answered correctly by more of the "poor" examinees than by the "good" examinees, it is considered a negative discriminator. *Discrimination is the most important characteristic of an item. A test cannot be reliable or valid unless the individual items discriminate among the examinees.*

Note that the higher the value of Net D, the higher the discriminating power of the item, and that the Net D formula could produce a negative number indicating an item that discriminates negatively. In fact, the value obtained is actually the net percentage of "good," or positive, discriminations achieved by an item, thus the name, Net D. Figure 10.5 illustrates this concept.

No discrimination occurred between Bill, Kelly, Pete, Alicia, Judy, and Gregg, because all answered the item correctly. Similarly, no discrimination occurred between Fred, Michelle, Dave, and Stephanie, because all answered the item incorrectly. The discrimination that occurred between Bill (or Kelly, Pete, or Alicia) and Michelle (or Dave or Stephanie) is considered a good, or positive, discrimination because of the groups in which these examinees have been placed based on their total test scores. Altogether a total of 12 (4 × 3) positive discriminations occurred. Conversely, the discrimination that occurred between Fred and Judy (or Gregg) is considered a bad, or negative, discrimination because

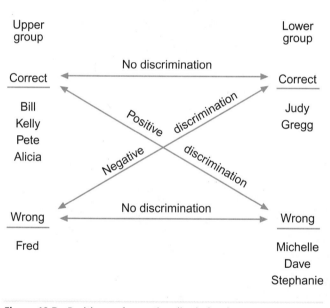

Figure 10.5 Positive and negative discrimination.

Fred is in the upper group and Judy and Gregg are in the lower group. A total of two (2 × 1) negative discriminations occurred. The maximum number of discriminations possible with five examinees in each group is 25 (5 × 5). Of these 25, 12 were positive, 2 were negative, and 11 did not occur. Subtracting the 2 negative discriminations from the 12 positive discriminations results in a net of 10 positive discriminations. The ratio of net positive discriminations to the total possible (10/25) is 40%. Using equation 10.4 for Net D results in this same value:

$$\text{Net D} = \frac{4 - 2}{5} \times 100 = 40\%$$

Attempt to keep the index of discrimination of an item on an achievement test as high as possible. Most test construction authorities agree that an item with a discrimination index of 40% or higher is a very good item. Items with an index of discrimination below 20%, and especially those with negative discrimination indexes, are poor and should probably be discarded from future tests. Discrimination indexes between 20% and 40% are acceptable but may indicate the need for revision, especially as the value approaches 20%.

> **Step 8**—Examine the pattern of responses to determine how an item might be improved.

According to the previous suggestions for retaining and discarding questions based on their discrimination index, the initial draft of the question in figure 10.3 probably should have been discarded. However, the response pattern of the examinees revealed a possible solution. Although it is often difficult to understand why certain responses are selected or ignored and even more difficult to determine possible alterations of the responses or stem that will improve an item, examining the response patterns often suggests possibilities. For example, response A to the first draft of the item displayed in figure 10.3 was chosen by more than 50% of the combined upper and lower groups of examinees, even though it is incorrect. Rewording this distractor in the revision resulted in the keyed response becoming more attractive than the first response, especially to the examinees in the upper group. The positive changes in the difficulty and discrimination indexes indicate that the alteration of this one response improved the item considerably.

Mastery Item 10.16

How many answer sheets should be used in an item analysis for a written test taken by 250 examinees?

Mastery Item 10.17

Calculate the index of difficulty and the Net D discrimination index for a multiple-choice question answered correctly by 40 of the 60 examinees in the upper group and by 10 of the 60 examinees in the lower group.

Mastery Item 10.18

To demonstrate the relationship between the difficulty and potential discrimination of an item, calculate the indexes of difficulty and discrimination for the following five items:

Item no.	Upper group n = 10	Lower group n = 10
1.	2 correct	0 correct
2.	5 correct	5 correct
3.	10 correct	5 correct
4.	10 correct	0 correct
5.	5 correct	10 correct

SOURCES OF WRITTEN TESTS

When a written test is given in the field of physical education, the chances are great that the examination was constructed locally. Generally, the number of sources for written tests in our discipline is relatively limited. With a few exceptions, nationally standardized written tests are not available.

When you are constructing a written test, it is often helpful to examine similar tests to obtain ideas for questions. Some possible sources for similar tests are professionally constructed tests, textbooks, periodicals, theses, and dissertations.

QUESTIONNAIRES

The questionnaire is a close relative of the written test. Both of these data collection instruments require careful construction and thoughtful analysis of the data they produce. However, the written test is primarily designed to assess the amount of knowledge a subject has and to discriminate among subjects on the basis of their cognitive behavior, whereas questionnaires are typically used to measure affective domain concerns such as attitudes and opinions.

Questionnaire responses provide the independent and dependent variables for survey research. Cox (1997) provided an extensive presentation of questionnaire development. Thomas and Nelson (2001) listed eight steps in conducting survey research:

1. Determining the objectives
2. Delimiting the sample
3. Constructing the questionnaire
4. Conducting the pilot study
5. Writing the cover letter
6. Sending the questionnaire
7. Following up
8. Analyzing the results and preparing the report

Using a mailed questionnaire to collect information has both advantages and disadvantages. On the plus side, the questionnaire can be relatively efficient in terms of money and time. Because a questionnaire can be sent to all respondents simultaneously, data collection can be completed over a period of a few weeks. Respondents can be widely spread geographically, and they can respond at their own convenience. If important, anonymity can be ensured, and each respondent is exposed to exactly the same instrument.

On the negative side, the value of questionnaire data can be reduced by low response rate, inability to clarify a question that the respondent finds ambiguous, unanswered questions, and lack of assurance regarding who actually completes the questionnaire. Some of these concerns can be addressed through careful planning, but they can never be totally eliminated.

Planning the Questionnaire

Time spent planning a questionnaire is invaluable. Before constructing a questionnaire, clarify the purpose of the instrument and study and formulate relevant hypotheses, so it will be possible to determine specifically what data the items on the questionnaire are designed to obtain. Unfortunately, this direct link between the items on the questionnaire and their exact purpose is not always carefully considered, resulting in either the collection of unnecessary information, the inability to respond to some hypotheses, or both. To prevent this, ask yourself exactly how each item on the questionnaire is to be analyzed. In general, if you can't answer this question for a particular item, it should be omitted.

As with a question on a written test, it is difficult to know how an item on a questionnaire will function the first time it is used. This is why it is necessary to do a few pilot studies before finalizing your questionnaire. Perhaps the best advice is to conduct pilot work with the questionnaire as you are developing it. In the first trial run, ask some colleagues to examine potential items for ambiguities, personal idiosyncrasies, and problems in the directions to the respondents. After these are addressed, get feedback on the next draft of the instrument from a small sample of individuals (a focus group) who are in the potential respondent pool. Their task is not only to respond to the questionnaire but also to indicate any problems they encounter. Then address these problems and attempt to input and analyze the resulting data to determine if the correct information to address the hypotheses is being secured and if there are any data entry problems (e.g., multiple responses to items, inappropriate responses).

Constructing the Questionnaire

One of the first decisions in constructing a questionnaire is to decide whether to use open-ended or closed-ended questions. Open-ended questions are those for which response categories are not specified. Essentially, the respondent provides an essay-type answer to the question. An example is, "What benefits do children gain from participation in an organized sports program?" Closed-ended questions are those requiring the respondent to select one or more of the listed alternatives—for example, "How many days a week should elementary school children participate in physical education classes? 1 2 3 4 5." Both types of question have advantages and disadvantages.

Open-Ended Questions

The advantages of open-ended questions are that they

- ▹ allow for creative answers and allow the respondent freedom of expression,
- ▹ allow the respondent to answer in as much detail as desired,
- ▹ can be used when it is difficult to determine all possible answer categories, and
- ▹ are probably more efficient than closed-ended questions when complex issues are involved.

The disadvantages of open-ended questions are that they

- ▹ do not result in standardized information from each respondent, making analysis of the data more difficult,
- ▹ require more time of the respondent, which ultimately could reduce the return rate of the questionnaire,
- ▹ are sometimes ambiguous because they attempt to solicit a general response, which can result in the respondent not being certain what is being asked, and
- ▹ can provide data that are not relevant.

Closed-Ended Questions

The advantages of closed-ended questions are that they

- ▹ are easy to code for computer analysis,
- ▹ result in standard responses, which are easiest to analyze and to use in making comparisons among respondents, and
- ▹ are usually less ambiguous for the respondent, and the ease of the respondent's task increases the return rate of the questionnaire.

The disadvantages of the closed-ended questions are that they

- ▹ may frustrate a respondent if an appropriate category is omitted,
- ▹ can result in a respondent selecting a category even if he or she doesn't know the answer or have an opinion,
- ▹ may require too many categories to cover all possible responses, and
- ▹ are subject to possible recording errors (e.g., the respondent checks B but meant to check C).

Deciding which type of question to use depends on several factors, such as the complexity of the issues involved, the length of the questionnaire, the sensitivity of the information sought, and the time available for construction and analysis of the questionnaire. In general, closed-ended questions work best when the responses are discrete, nominal in nature, and few in number. Descriptive information such as gender, years of education, and marital status most often are measured with closed-ended questions. Closed-ended questions may have some advantages when seeking sensitive information. For example, a respondent may be willing to provide information about annual income if it is done by checking the appropriate range (e.g., between $80,000 and $100,000) rather than by listing the salary specifically. Closed-ended questions should be used if you have more time to construct the questionnaire than to analyze it. Open-ended questions are relatively simple to construct, but the coding and interpreting necessary once the responses are returned are anything but simple.

Other Item Concerns

Questionnaire items need to be simple and avoid containing more than one element. Do not include two questions in one. For example, how would you answer yes or no to this question: "Do you think physical education or music should be kept or dropped from the curriculum?" Avoid ambiguous questions. Consider the responses to this item: "Do you think the punishment was appropriate? Yes or No." If a person answers no, you do not know if they thought the punishment was too light or too harsh. To avoid misinterpretation on the part of the respondent, avoid vague terms, slang, colloquial expressions, and lengthy questions. Be certain that the level of wording is appropriate for the subjects who will be responding. Avoid the use of leading questions. For example, how would you likely respond to this item? "Most experts believe that regular moderate exercise produces health benefits. Do you agree?"

Factors Affecting the Questionnaire Response

Besides the instrument itself, many ancillary materials and techniques affect the success of collecting data with a questionnaire. Computer responses are now being used to obtain survey data. Some problems with these techniques are that the sample choosing to respond may not represent the population to which you want to generalize, a person might respond more than once, and although the availability of computers is becoming widespread, not everyone has access to them or feels comfortable answering personal questions on the computer because of perceived security and privacy issues.

Cover Letter

Next to the questionnaire itself, the most important item sent to the respondent is the cover letter. This brief document has the important tasks of describing the nature and objectives of the questionnaire and soliciting the cooperation of the respondent. Personalize cover letters (address them to the respondent rather than "Dear Sir or Madam"), use some slight form of flattery (e.g., "Because of your extensive background . . ."), and include the endorsement of someone known to the respondent. In addition, be sure that the letter's appearance is neat and attractive. These strategies enhance the probability that the questionnaire will be returned.

Ease of Return

Providing clear instructions as to how and when the questionnaire is to be returned and enclosing a self-addressed postage-paid envelope have been shown to produce increased response rates. If the number of questionnaires to be returned is large, it may be advantageous to set up a postage-due arrangement with the post office. Under this arrangement, you pay only for the questionnaires that are returned rather than putting postage on return envelopes that may never be returned. The ease of mailing for the respondent is the same in either case.

The respondents must be representative of the population that was sampled. It is important that you have data to provide evidence that the sample reflects the population and is not systematically biased in some way. For example, do the ethnic, age, or gender characteristics of your respondents look like the group to whom the original questionnaire was sent?

Neatness and Length

It is logical that if you spend time making the questionnaire and cover letter neat, easy to read, free of grammatical errors, and uncluttered, the respondent may be more inclined to spend time responding. The shorter you can make the questionnaire, the more likely it will be returned.

Inducements

The inclusion of a pencil or a pen ("so you won't have to locate one"), a penny ("for your thoughts"), a dollar ("for a cup of coffee while you complete the questionnaire"), or a lottery ticket ("winner to be chosen from returned questionnaires") are examples of inducements people have used to encourage respondents to return the questionnaires. At the University of Colorado, a $2 bill was enclosed with a student satisfaction survey done each year on a sample of students. The hope is to instill a sense of obligation. For some individuals, it may be difficult to put the money in their wallet and the survey in the wastebasket.

Timing and Deadlines

It is best not to have the questionnaire arrive just before a major holiday or other significant event (such as the beginning or end of the school year). The return rate of the student satisfaction survey mentioned previously would likely be low if it arrived to students during finals week. The inclusion of reasonable deadlines should enhance return rates. Receiving the questionnaire the day before (or even after) it is due back gives the respondent an easy excuse for not completing the questionnaire. Allowing too long a time before it is to be returned may cause the instrument to be put aside and never surface again.

Follow-Up

It is generally believed that at least one follow-up procedure can enhance the return rate of questionnaires. After one or two reminders, the effectiveness of follow-up procedures generally decreases dramatically. A typical procedure is to send the original questionnaire and cover letter, wait until the responses trickle down to a few, and then send out a reminder letter. The reminder letter can potentially increase the response rate. If further follow-up seems necessary, it is common to next send a duplicate questionnaire and cover letter; if still no response is received, a telephone reminder is the next possibility. Follow-up procedures beyond this are mostly without success.

Questionnaire Reliability

The procedures presented in chapters 6 and 7 are most often used to validate and estimate the reliability of questionnaire responses. To estimate the reliability of a single item, you must ask the specific item on at least two occasions. However, affective and cognitive domain subscales completed on a questionnaire can have their reliability estimated with the alpha coefficient. An important issue is the stability reliability of the responses. To estimate stability reliability, you must administer the questionnaire to the same people on two or more occasions. The typical time between testings to determine stability reliability is 2 to 4 weeks. A period longer than this could actually result in changes in respondents' opinions. If there are changes in responses that reflect true changes of opinion, the estimated reliability will be reduced. The specific type of reliability estimate to use depends on the nature of the questions asked. For example, are the items nominally (e.g., "What is your gender?") or intervally scaled (e.g., a series of "attitude" questions)? See chapters 6 and 7 for the specific methods to estimate reliability for the obtained responses.

Questionnaire Validity

The most important issue of a questionnaire, as with any measuring instrument, is the validity of the responses. It is important that the respondents truthfully respond to the items and are not responding based on what they believe the socially acceptable response would be. Developing quality questionnaire items, having items reviewed by experts, conducting pilot testing, and ensuring confidentially or anonymity are ways to increase the validity of responses. Most questionnaires are validated with content-related procedures (presented in chapter 6). However, there are ways to cross-check the responses with additional data to determine if the respondent is answering truthfully. For example, if a respondent says that he or she votes for a certain candidate in an election, there is no way to determine the actual vote. However, you can verify through public records whether the respondent is actually registered to vote and voted in the specific election. Finally, whether the respondent sample is representative of the population to which one desires to generalize is an important validation issue to consider.

Booth, Okely, Chey, and Bauman (2002) provided an example of estimating the reliability and validity of a questionnaire in their examination of the Adolescent Physical Activity Recall Questionnaire.

Mastery Item 10.19

The data for MI 10.19 are available on the WWW site.

1. Download MI 10.19. These data represent test scores for 10 people on a 15-item test. The items are scored 1 if correct and 0 if incorrect.
2. Use SPSS reliability to confirm that the reliability of the 15-item test is .568.
3. Use the Spearman–Brown prophecy formula (equation 6.2 on page 87) to estimate the reliability if you increased the test length to 30 items.
4. Confirm that the estimated reliability for the 30-item test is .724.

Measurement and Evaluation Challenge

To measure how well three groups of students learned basic statistical concepts after each group had been exposed to different teaching methods, Kate constructed a 60-item multiple-choice test based on a table of specifications she developed before her research began. The table of specifications reflected the content and importance of the material. She initially developed a 100-item test and then conducted a pilot test followed by an item analysis to evaluate the individual items and the overall test. Using the item analysis, she was able to select 60 items that each demonstrated acceptable difficulty and discrimination indexes and, as a total test, an acceptable reliability. She also asked two experts who have taught statistical concepts for nearly 20 years to evaluate the items. These experts' suggestions helped her ensure that the items had acceptable content validity.

The test was then administered to students in each of the three teaching-method groups. The mean score on the test was used as the dependent variable in an analysis of variance (see chapter 5) to see if the groups differed in their knowledge of statistics.

SUMMARY

When a research project or a physical education curriculum requires the assessment of objectives in the cognitive domain, the instrument of choice is usually a written test. When the assessment of attitudes or opinions is desired, a questionnaire is typically used. The procedures for planning, constructing, scoring, administering, and analyzing the results from application of such instruments have been presented in this chapter. All of the procedures described in this chapter focus on making the written test or the questionnaire as objective, reliable, and valid as possible.

11

Physical Fitness and Activity Assessment in Adults

Measurement and Evaluation Challenge

Jim is a new graduate with a major in kinesiology. He was seeking to become a certified teacher but then became interested in working in the health and fitness industry. He interviewed with a new branch of the YMCA for a position as director of physical fitness. The executive director of the YMCA is interested in having good physical activity and fitness programs but also wants to have a good fitness assessment program available to all YMCA members. Jim believes that his interview went very well. His grades at school were good and he has worked part-time in fitness facilities for the last 2 years. The director asked him if he was certified in fitness instruction by any professional organizations. Jim had to respond that he was not currently certified but would certainly be happy to seek an appropriate certification. The director asked Jim to investigate the certification he would seek and outline a suggested adult fitness assessment program. They are going to meet again in a week, at which time Jim is to report on these two issues. The director concluded, "Jim, if the meeting goes well, I think that we will ask you to join our team." Jim is excited but also nervous. Because his career may depend on his responses to the director, Jim has to do some research and thinking.

Objectives

After studying this chapter, you will be able to

- identify and define the components of health-related physical fitness;
- identify and define an apparently healthy adult;
- use reliable and valid methods of measurement of aerobic capacity, body composition, and muscular fitness;
- identify and use test items specifically developed for the older adult population; and
- understand the issues associated with reliable and valid measurement of physical activity in all populations.

No more important objective exists in the exercise and sport sciences than the attainment of physical fitness. Physical fitness is a multifaceted objective, with different meanings to different people—a cardiologist might define physical fitness very differently than a gymnastics coach. Whatever the definition or understanding of physical fitness, its importance to you in your professional career is related to two primary factors:

1. The citizens and state and federal governments of many industrialized countries have taken the position that the general public should have sufficient levels of physical activity and fitness as this will improve health and enable citizens to deal with the physical challenges that they may confront. The U.S. government established public health objectives for improved levels of physical activity and fitness largely through its initiative called *Healthy People 2000.*

Physical Activity and Health: A Report of the Surgeon General, released in 1996 by the U.S. Department of Health and Human Services (USDHHS 1996), was a landmark scientific presentation of the health benefits of physical activity and fitness. The report summarized the physiological and psychosocial benefits that individuals of all ages gain from a physically active lifestyle.

Continuing the efforts of Healthy People 2000, the Centers for Disease Control and Prevention (CDC) and the U.S. Public Health Service included physical activity and fitness as a component of *Healthy People 2010.* Additionally, the World Health Organization's position on physical activity and health has led other countries, governments, and agencies to promote physical activity and fitness.

2. The basic justification for professions in exercise and sport sciences is the improvement and maintenance of physical activity and fitness as an important step in developing a healthy lifestyle.

Professionals, like you, need to know and understand these factors and the effect they will have on your career. There are many excellent sources of information on physical fitness testing. Indeed, entire books are devoted to physical fitness training and assessment (e.g., Golding, Myers, and Sinning 1989). This chapter provides examples of adult physical fitness tests, focusing on reliability and validity.

Mastery Item 11.1

What is your definition of physical fitness? Write it out. Compare it with the definition developed in the following paragraphs.

Because physical fitness is multifaceted, an effective definition must be broad and encompassing. Two factors, the purposes of the tests and the defined population, provide a framework for defining physical fitness for any person. As you see from table 11.1, we might have different objectives (different fitness tests) for different groups of people. The purposes of fitness assessment are related to the specific population to be tested. Thus, we can define physical fitness based on who and what are to be measured. In this chapter we primarily address normal, healthy adults. *Therefore, we define health-related physical fitness as the attainment or maintenance of physical capacities that are related to good or improved health and are necessary for performing daily activities and confronting expected or unexpected physical challenges.* This definition is consistent with the health-related physical fitness definition presented by Pate (1988) and supported by the American College of Sports Medicine (ACSM 2000). In this chapter we examine tests of health-related physical fitness and basic functional capacity.

Mastery Item 11.2

From table 11.1, what would be some different reasons for a 30-year-old mom and a 17-year-old cross country runner to take a test to determine their cardiovascular endurance?

Table 11.1   **Populations and Purposes of Physical Fitness Testing**

Population	Health-related	Motor	Diagnosis	Military preparation	Functional capacity
Youth	*	*		*	*
Adults	*				*
The aged	*				*
Special					
Mentally impaired	*		*		*
Physically impaired	*		*		*
Athletes		*	*		
The ill or injured			*		

HEALTH-RELATED PHYSICAL FITNESS

The ACSM has identified three fitness factors that are health-related; these are listed in table 11.2 and defined in subsequent sections of this chapter. The evidence to support these factors as related to health has come from the branch of medicine called **epidemiology,** which examines the incidence, prevalence, and distribution of disease. For example, a large majority of epidemiologic studies have indicated that physically active groups have lower relative risks of developing fatal cardiovascular disease (CVD) than sedentary groups (Caspersen 1989). **Relative risk** refers to the risk of mortality (death) or morbidity (disease) associated with one group compared with another. Physically active groups, logically, should have higher levels of **cardiovascular endurance,** which is the body's ability to extract and use oxygen in a manner that permits continuous exercise, physical work, or physical activities. Studies have shown an inverse relationship between death rates and cardiovascular endurance (Blair, Kohl, et al. 1989; Blair et al. 1996; Ekelund et al. 1988). Figure 11.1 demonstrates the findings of Ekelund and colleagues.

The poorest cardiovascular endurance quartile death rate was 8.5 times higher than the most fit quartile. People who suffer from obesity have higher rates of CVD, cancer, and diabetes. Thus, body composition is included in a health-related fitness battery to determine percent body fat and the presence of obesity (ACSM 2000). The factors of muscular strength, muscular endurance, and flexibility do not have the same level of research evidence to support their relationship to good health. However, a minimum level of muscular fitness is essential for accomplishing daily activities and being prepared to deal with expected or unexpected physical challenges (see table 11.2).

Table 11.2   **Health-Related Fitness Factors and Benefits**

Factor	Benefits
Cardiovascular endurance	Reduction in risk of cardiovascular disease
Body composition	Reduction in risk of cardiovascular disease, adult-onset diabetes, and cancer
Muscular strength, muscular endurance, and flexibility	Reduction in risk of low-back pain Improved posture and functional capacity Ability to conduct daily activities

Adapted from ACSM (1995).

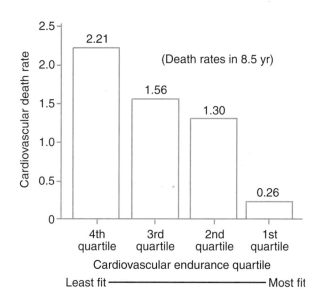

Figure 11.1   Relationship between cardiovascular endurance and cardiovascular death rate.

APPARENTLY HEALTHY PEOPLE

In adult fitness assessment, one of the critical issues is establishing criteria for testing that do not require medical clearance or physician supervision. The criteria established by the ACSM involve describing the adult as *apparently healthy* (ACSM 2000). The apparently healthy adult is a male less than 45 years of age or a female less than 55 years of age who has none or only one of the major CVD risk factors listed in table 11.3 (ACSM 2000). One can determine apparent healthy status by administering the Physical Activity Readiness Questionnaire (PAR-Q) of the British Columbia Ministry of Health (ACSM 2000; table 11.4). Table 11.5 provides the recommendations for physician supervision of exercise testing in the adult population (ACSM 2000).

Table 11.3   Major Risk Factors and Classifications for Cardiovascular Disease

Family history	Father or brother: CVD <55 years of age Mother or sister: CVD <65 years of age
Cigarette smoking	Current or recently quit smoking
Hypertension	Systolic blood pressure ≥140 or diastolic blood pressure ≥90
Hypercholesterolemia	Total cholesterol >200 mg/dl or HDL <35 mg/dl
Impaired fasting glucose	Fasting blood glucose ≥110 mg/dl
Obesity	BMI ≥30 or waist girth >100 cm
Sedentary lifestyle	Not meeting minimal physical activity recommendations of the U.S. Surgeon General's report
Risk classification	
Low risk	Men <45 or women <55 who are asymptomatic and meet no more than one risk factor threshold
Moderate risk	Men ≥45 and women ≥55 or individuals who meet two or more of the risk factors
High risk	Symptomatic or individuals with known cardiovascular, pulmonary, or metabolic disease

Note: HDL = high-density lipoprotein.
Adapted from ACSM 1995.

Table 11.4 Physical Activity Readiness Questionnaire

These are the types of questions that you will find on the PAR-Q & YOU form, developed by the Canadian Society for Exercise Physiology and Health Canada

Yes	No	Question
		Have you ever felt pain in your chest when exercising?
		Have you ever felt pain in your chest when inactive?
		Are you currently taking medication for a cardiovascular condition?
		Have you been advised by your physician that you should restrict your activities?

Table 11.5 Conditions Requiring Physician's Supervision of Exercise Test

	Low risk	Moderate risk	High risk
Submaximal test	Not necessary	Not necessary	Recommended
Maximal test	Not necessary	Recommended	Recommended

Mastery Item 11.3

Consider the questions in table 11.4 and complete the online PAR-Q & YOU available at www.phac-aspc.gc.ca/sth-evs/english/parq.htm. Are you apparently healthy?

In the following sections of this chapter, we examine some of the tests and protocols available for assessing health and fitness of adults. It is impossible to cover every test, but we emphasize some of the more important test protocols and measurement issues related to them.

MEASURING AEROBIC CAPACITY

As mentioned earlier, physical activity and cardiovascular endurance are related to the risk of CVD. The exercise physiologist's concept of cardiovascular endurance is an individual's aerobic capacity, or **aerobic power,** which is the ability to supply oxygen to the working muscles during physical activity.

Laboratory Methods

Fitness assessment in laboratories and clinical settings involves expensive and sophisticated equipment and exacting test protocols. From a measurement perspective, these tests are often criterion referenced.

Measuring Maximal Oxygen Consumption

The single most reliable ($r_{xx'} > .80$) and valid measure of aerobic capacity is the **maximal oxygen consumption**, or **$\dot{V}O_2$max** (ACSM 2000; Safrit et al. 1988). $\dot{V}O_2$max is a measure of the maximal amount of oxygen that can be used by a person during exhaustive exercise. In laboratory testing of $\dot{V}O_2$max, the subject performs a **maximal exercise test** on an

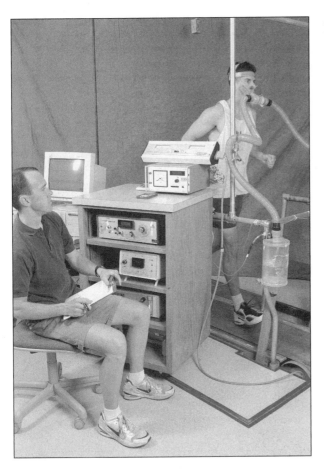

Figure 11.2 Gas exchange analysis during a maximal exercise test.

ergometer, such as a treadmill, stationary cycle, step-bench, swimming flume, or arm crank device (figure 11.2). The subject performs the exercise under a specific protocol until he or she reaches exhaustion. While the subject is exercising, expired gases are monitored with a gas analysis system. Most modern exercise physiology laboratories use an automated and computerized metabolic system.

A variety of exercise protocols are available in the literature for determining $\dot{V}O_2$max; all focus on increments in work rate until the subject reaches exhaustive levels of physical exertion (ACSM 2001). *$\dot{V}O_2$max is achieved when the work rate is increased, but the oxygen consumption ($\dot{V}O_2$max) does not increase or has reached a plateau.* Other indications of $\dot{V}O_2$max are a respiratory exchange ratio greater than 1.1 and heart rates near age-predicted maximal levels. When these physiological criteria are not clearly achieved, then the maximal oxygen consumption measured during the test is called $\dot{V}O_2$peak. $\dot{V}O_2$max and $\dot{V}O_2$peak are highly correlated and represent a valid measure of a subject's aerobic capacity. Blair, Kohl, and colleagues (1989) estimated that $\dot{V}O_2$max values of 31.5 ml · kg⁻¹ · min⁻¹ for females and 35 ml · kg⁻¹ · min⁻¹ for males represent the minimal levels of aerobic capacity associated with a reduced risk of disease and death. Figure 11.3 demonstrates the Balke treadmill protocol for determining $\dot{V}O_2$max. Table 11.6 provides evaluative norms for $\dot{V}O_2$max.

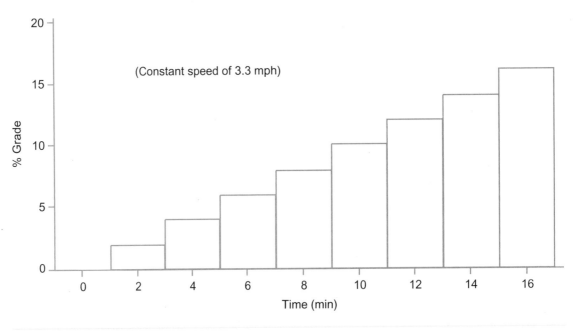

Figure 11.3 The Balke treadmill protocol.

Table 11.6 Male and Female Norms for $\dot{V}O_2$max (ml · kg^{-1} · min^{-1})

	Age (years)					
Male rating	18-25	26-35	36-45	46-55	56-65	66+
Excellent	80-63	70-58	77-53	60-47	58-43	50-38
Good	59-53	54-50	49-44	43-40	39-37	36-33
Above average	51-47	47-44	42-40	38-35	35-33	32-29
Average	46-43	42-40	38-35	35-32	31-30	28-25
Below average	41-38	39-35	34-32	31-29	29-26	25-22
Poor	35-31	34-31	30-27	28-26	25-22	21-20
Very poor	29-20	28-20	25-19	23-18	21-16	18-15
Female rating						
Excellent	71-58	69-54	66-46	64-42	57-38	51-33
Good	54-48	51-46	44-39	39-35	36-32	31-28
Above average	46-42	43-40	37-34	33-31	31-28	27-25
Average	41-39	38-35	33-31	30-28	27-25	24-22
Below average	37-34	34-31	30-28	27-25	24-22	22-20
Poor	32-29	30-26	26-23	24-21	21-19	18-17
Very poor	26-18	25-20	21-18	19-16	17-14	16-14

Adapted from Golding, Myers, and Sinning (1989).

Mastery Item 11.4

What level of $\dot{V}O_2$max would you like to have based on the norms in table 11.6?

Mastery Item 11.5

Based on the values of Blair, Kohl, and colleagues (1989), what level of $\dot{V}O_2$max would you like to have?

Mastery Item 11.6

What types of evaluation standards do the values in table 11.6 and those of Blair, Kohl, and colleagues represent?

Estimating $\dot{V}O_2$max

Although $\dot{V}O_2$max is the criterion measure of aerobic capacity, it is a difficult measure to determine because it requires expensive metabolic equipment, exhaustive exercise performance, and a lot of time. Consequently, researchers in exercise science have developed techniques for estimating, or predicting, $\dot{V}O_2$max reliably and validly. The estimations are calculated from measurements of maximal or submaximal exercise performance or submaximal heart rate; the same or similar exercise protocols and ergometers discussed previously are used.

Maximal Exercise Performance $\dot{V}O_2$max can be accurately estimated from the maximum exercise time of a maximal treadmill exercise test (Pollock et al. 1976). Although this procedure requires exhaustive exercise, it does not require the metabolic measurement of expired gases. Thus, the test is greatly simplified, and expensive metabolic equipment is not necessary. Published correlations (concurrent validities) between $\dot{V}O_2$max and maximal exercise time exceed .90. Baumgartner, Jackson, Mahar, and Rowe (2003) provide a figure of $\dot{V}O_2$max estimates for maximal treadmill times for several treadmill protocols.

Submaximal Exercise Testing Submaximal estimates of $\dot{V}O_2$max are based on the linear relationship among heart rate, workload, and $\dot{V}O_2$max. Such estimates are based on **submaximal exercise tests,** which require less than maximal effort. As figure 11.4 indicates, a subject with good aerobic capacity has a higher $\dot{V}O_2$max than one with poor aerobic capacity (both have a maximal heart rate of 200 beats/min). The slopes of the lines representing the linear relationship between heart rate and $\dot{V}O_2$max are different for each subject. In figure 11.5, we see the difference in workloads each subject can achieve for a fixed submaximal heart rate, 160 beats/min.

Several exercise test protocols are available for estimations of $\dot{V}O_2$max (ACSM 2001). These estimations are based on the linear relationships of workload, heart rate, and oxygen consumption. One of the classic procedures is referred to as the Åstrand–Rhyming nomogram (Åstrand and Rhyming 1954). This was originally established as a cycle ergometer test that coordinated workload and heart rate responses into a prediction of $\dot{V}O_2$max. Baumgartner and colleagues (2003) converted the nomogram into an equation that can be used to produce the same predictions of $\dot{V}O_2$max from cycling or treadmill tests (Jackson et al. 1990), allowing computer calculations of predicted aerobic capacity. The ACSM (2000, 2001) and the YMCA (Golding, Myers, and Sinning 1989) provide descriptions of specific treadmill and cycle test protocols for estimating $\dot{V}O_2$max.

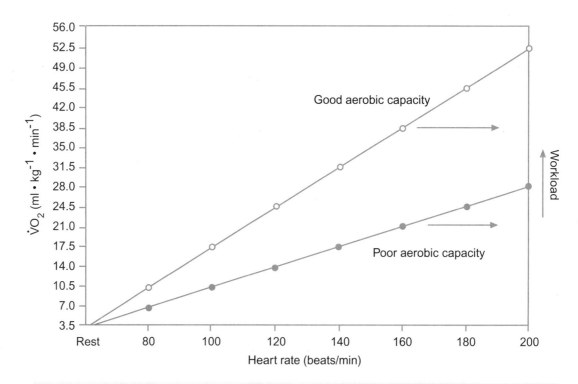

Figure 11.4 Linear relationship among oxygen consumption, heart rate, and workload.

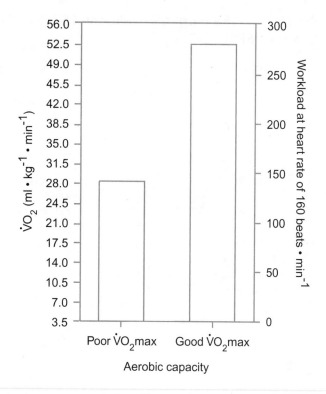

Figure 11.5 How maximal oxygen consumption affects a submaximal workload.

Perceptual Effort During Exercise Testing

Borg (1962) pioneered the measurement of perceptual effort, or **perceived exertion,** during exercise testing. Perceived exertion is the mental perception of the intensity of physical work. The measurement of perceived physical effort or stress has been named *rating of perceived exertion (RPE)*. Borg (1998) presented RPE scales for assessment of perceived effort during exercise testing (table 11.7). RPE scale values correlate with exercise variables such as heart rate, ventilation, lactic acid production, percent $\dot{V}O_2$max, and workload (ACSM 2000). The subject simply gives a verbal or visual score from the scale during the exercise test as the workload increases or as time progresses. RPE is typically monitored during exercise tests and is used in exercise prescription to control exercise intensity.

Mastery Item 11.7

In figure 11.4, which subject would achieve the higher heart rate and RPE for any submaximal workload?

Laboratory tests for assessment of aerobic capacity tend to be reliable and valid. However, sources of measurement error are present in these laboratory situations as in any other testing situation. The subject, the test and protocol, and the test administrator can all be sources of measurement error. The following list provides some important facts you should know concerning laboratory assessment of aerobic capacity.

▷ Equipment, treadmills, cycles, and gas analysis systems should be calibrated and checked regularly.

▷ Test administrators should be trained and qualified.

▷ Practice test administrations should be required for subject and administrator to become familiar with test protocols and equipment.

Table 11.7   The Borg RPE Scale, the 15-Grade Scale for Ratings of Perceived Exertion (RPE)

6	No exertion at all		0	Nothing at all	"No P"
7			0.3		
8	Extremely light		0.5	Extremely weak	Just noticeable
9	Very light		1	Very weak	
10			1.5		
11	Light		2	Weak	Light
12			2.5		
13	Somewhat hard		3	Moderate	
14			4		
15	Hard (heavy)		5	Strong	Heavy
16			6		
17	Very hard		7	Very strong	
18			8		
19	Extremely hard		9		
20	Maximal exertion		10	Extremely strong	"Max P"
			11		
			✶		
	Borg RPE scale		●	Absolute maximum	Highest possible
	© Gunnar Borg, 1970, 1985, 1994, 1998				

Borg CR10 scale
© Gunnar Borg, 1981, 1982, 1998

▶ Standardized test procedures should be established and followed; this creates a focused test environment.

▶ Treadmill $\dot{V}O_2$max values will be greater than values from cycle ergometer tests for most subjects.

▶ Many Americans seldom ride bicycles, so cycle exercise tests can produce artificially low values of $\dot{V}O_2$max attributable to test cessation from localized fatigue in the legs.

▶ Submaximal estimates of $\dot{V}O_2$max have a standard error of estimate greater than $5.0 \text{ ml} \cdot \text{kg}^{-1} \cdot \text{min}^{-1}$.

Field Methods

Field methods include ways to assess aerobic capacity and are feasible for mass testing. Generally, field methods require little equipment and are less expensive in time and costs than laboratory methods.

Distance Runs

Distance runs to achieve the fastest possible time or greatest distance covered in a fixed period of time are some of the more popular field tests of aerobic capacity. In adults, distances of 1 mi or greater are used. Safrit and colleagues (1988) indicated that distance runs tend to be reliable ($r_{xx'} > .78$) and have a general concurrent validity coefficient of .741 ± .14. The 12 min run for distance developed by Cooper (1968) is an example of a distance-run test. AAHPERD has published norms for the 1 mi run for college students (AAHPERD 1985; see table 11.8).

Distance runs are useful for educational situations in which testing entire classes is required in a short amount of time. However, to ensure reliability, validity, and safety (i.e., to correct pacing and provide proper physical conditioning), subjects should receive aerobic training and practice trials on the test. It is important that distances, timers, and recording procedures be used in a standardized manner. **Older adults** or people with poor aerobic capacity should undergo one of the other field tests or procedures discussed next.

Table 11.8   **Percentile Norms for the 1 Mile Run for College Students (min:sec)**

Percentile	Males	Females
90	5:44	7:26
75	6:12	8:15
50	6:69	9:22
25	7:32	10:41
10	8:30	12:00

Reprinted from AAHPERD (1995).

Step Tests

Several step-test protocols are available for estimating aerobic capacity. These tests are based on the linear relationships among workload, heart rate, and $\dot{V}O_2$max discussed previously. Generally, the subject steps up and down to an up, up, down, down cadence until a specific workload, heart rate, or time is achieved. The aerobic capacity is then estimated from the heart rate response or the recovery heart rate. Subjects with higher aerobic capacity will have a faster return to lower heart rates.

The YMCA 3-Minute Step Test is one of the simplest step tests to administer and is useful for initial testing of unfit subjects.

YMCA 3-Minute Step Test

Purpose

To assess aerobic fitness in mass testing situations with adults.

Objective

To step up and down to a set cadence for 3 min and take the resulting heart rate.

Equipment

12 in. (30.5 cm) high bench
Metronome set at 96 beats/min
Watch or timer
Stethoscope (carotid pulse can be used)

Instructions

The subject listens to the metronome to become familiar with the cadence and begins when ready and the time starts. The subject steps up, up, down, down to the 96 beat/min cadence, which allows 24 steps/min. This continues for 3 min. After the final step down, the subject sits down and the heart rate is counted for 1 min.

Scoring

The 1 min recovery heart rate is the score for the test. Table 11.9 provides evaluative norms for test results.

Table 11.9 **Male and Female Norms for Recovery Heart Rate Following the 3-Minute Step Test (beats/min)**

	Age (years)					
Male rating	18-25	26-35	36-45	46-55	56-65	66+
Excellent	70-78	73-79	72-81	78-84	72-82	72-86
Good	82-88	83-88	86-94	89-96	89-97	89-95
Above average	91-97	91-97	98-102	99-103	98-101	97-102
Average	101-104	101-106	105-111	109-115	105-111	104-113
Below average	107-114	109-116	113-118	118-121	113-118	114-119
Poor	118-126	119-126	120-128	124-130	122-128	122-128
Very poor	131-164	130-164	132-168	135-158	131-150	133-152
Female rating						
Excellent	72-83	72-86	74-87	76-93	74-92	73-86
Good	88-97	91-97	93-101	96-102	97-103	93-100
Above average	100-106	103-110	104-109	106-113	106-111	104-114
Average	110-116	112-118	111-117	117-120	113-117	117-121
Below average	118-124	121-127	120-127	121-126	119-127	123-127
Poor	128-137	129-135	130-138	127-133	129-136	129-134
Very poor	142-155	141-154	143-152	143-152	142-151	135-151

Adapted from Golding, Myers, and Sinning (1989).

Rockport 1-Mile Walk Test

Kline and colleagues (1987) presented a field method for estimating $\dot{V}O_2$max that has been called the Rockport 1-Mile Walk Test. The procedure involves using the time of a 1 mi walk, gender, age, body weight, and ending heart rate to estimate $\dot{V}O_2$max. The 1 mi walk requires the subjects to walk as fast as possible; their heart rates are taken immediately at the end of the walk. The 1 mi walk test was shown to be reliable ($r_{xx'}$ = .98; Kline et al. 1987). The prediction equation (equation 11.1) produced a concurrent validity coefficient of .88 with a standard error of estimate of 5.0 ml · kg⁻¹ · min⁻¹:

$$\dot{V}O_2 \max = 132.853 - (.0769) \times wt - (.3877) \times age + (6.315) \times gv - (3.2469) \times 1 \text{ mi walk time} - (.1565) \times \text{heart rate}$$

(11.1)

with wt as weight in pounds, age in years, gv (gender values) of 0 for females and 1 for males, 1 mi walk time in minutes (to the hundredths of a minute), heart rate in beats per minute at end of walk, and $\dot{V}O_2$max in ml · kg⁻¹ · min⁻¹.

The original study used a sample with an age range of 30 to 69 years. Further research supported the validity (r_{xy} = .79; s_e = 5.68 ml · kg⁻¹ · min⁻¹) of the equation for adults aged 20 to 29 years (Coleman et al. 1987). As with any physical performance test, the 1 mi walk can have improved reliability and validity if a practice trial is administered (Jackson, Solomon, and Stusek 1992). Evaluative norms for people aged 30 to 69 years for the 1 mi walk test and percentile norms for people aged 18 to 30 years are provided in tables 11.10 and 11.11, respectively.

Table 11.10   **Norms for the 1-Mile Walk Test (Subjects Aged 30-69 Years; min:sec)**

Rating	Males (n = 151)	Females (n = 150)
Excellent	< 10:12	< 11:40
Good	10:13-11:42	11:41-13:08
High average	11:43-13:13	13:09-14:36
Low average	13:14-14:44	14:37-16:04
Fair	14:45-16:23	16:05-17:31
Poor	> 16:24	> 17:32

Table 11.11   **Norms for the 1 Mile Walk Test (Subjects Aged 18-30 Years; min:sec)**

Percentile	Males (n = 400)	Females (n = 426)
90	11:08	11:45
75	11:42	12:49
50	12:38	13:15
25	13:38	14:12
10	14:37	15:03

Predicting $\dot{V}O_2$max Without Exercise

Jackson and colleagues (1990) developed an equation for estimating $\dot{V}O_2$max without an exercise test of any kind. The equations had reasonable validity coefficients ($r_{xy} > .79$) and standard errors of estimation ($s_e < 5.7$ ml $\cdot$ kg^{-1} $\cdot$ min^{-1}); the latter is comparable to submaximal exercise test and field test standard errors of estimation. This technique allows for accurate estimation of aerobic capacity in situations in which large numbers of subjects need to be evaluated, such as in epidemiologic research. The equation follows:

$$\dot{V}O_2 \text{ max} = 50.513 + 1.589 \times \text{self-reported physical activity} - .0289 \times \text{age in years} - 0.522 \times \text{percent body fat} + 5.863 \times \text{gender (female} = 0, \text{male} = 1) \tag{11.2}$$

Mastery Item 11.8

If you are apparently healthy, estimate your aerobic capacity. Perform the YMCA 3-Minute Step Test and the Rockport 1-Mile Walk Test. Are the results consistent in estimating your aerobic capacity?

MEASURING BODY COMPOSITION

Obesity is a risk factor in the development of CVD, cancer, and adult-onset diabetes. As a consequence—and because the United States has a large percentage (>25%) of adults who are obese—measuring obesity in an accurate manner is an important measurement goal.

The term *obesity* refers specifically to overfatness, not overweight. A well-muscled athlete who is extremely fit may be considered overweight on a height and weight table but may actually be quite lean. In health-related fitness, the measurement of **body composition** involves estimating a person's percent body fat, which requires that his or her body density be determined. A good method for conceptualizing body composition is to divide the body into two compartments: *lean,* which includes muscle, bone, and organs and is of high density, and *fat,* which is of low density. For a fixed body weight, a leaner person with a lower percent body fat will have a higher body density than a fatter person of the same weight. In body density estimation, lean tissue is assumed to have an average density of 1.10 g/cm^3, whereas fat tissue is assumed to have an average density of .90 g/cm^3. This assumption leads to one of the errors in body composition measurement: Lean tissue and fat tissue do not have the same density, and different types of lean tissue (e.g., bone vs. muscle) have different densities. This variable source of measurement error is present in the methods discussed next.

There are approximately a dozen methods of body density and percent fat measurement, including the following:

- Chemical analysis of cadavers
- Hydrostatic weighing
- Volumetry (body volume)
- Helium dilution
- Radiographic (X-ray) analysis
- K^{40} counting (radiation emissions)
- Total body water
- Ultrasound
- Anthropometry (skinfolds and girths)
- Bioelectrical impedance
- Total body electrical conductivity (TBEC)
- Infrared interactance

Nieman (1995) provided a summary of the advantages and disadvantages of 13 techniques of measurement of body composition. These methods, their reliability, and their validity are results of the development of relatively new technologies.

Laboratory Methods

The main laboratory procedure for assessing body composition involves hydrostatic weighing.

Hydrostatic Weighing

Hydrostatic weighing (underwater weighing), which is based on Archimedes' principle, is the most popular method of laboratory assessment of body density (figure 11.6). This method has provided the criterion measurements for validating such field methods as skinfold and girth measurements. The basic comparison in this method is one's dry land body weight and underwater body weight. For two people of the same weight but different percent body fat, the leaner person, who has a higher body density, will have a higher underwater weight than the fatter person. Equation 11.3 can help you calculate body density from hydrostatic weighing:

$$BD = \frac{Wt_d}{\dfrac{(Wt_d - Wt_w)}{D_w} - (RV + 100ml)} \qquad (11.3)$$

where BD = body density in g/cm³, Wt$_d$ = subject's dry weight in kilograms, Wt$_w$ = his or her weight under water in kilograms, D$_w$ = density of water in g/cm³ at the temperature of measurement, and RV = residual volume in liters.

    To ensure reliability and validity of the body density measured by hydrostatic weighing, underwater weighing should be repeated as many as 10 times or until a consistent weight is determined, and the **residual volume,** which is the air left in the lungs after maximal forced expiration of air, should be measured, not predicted or estimated. Measuring residual volume is a complex laboratory process that requires sophisticated equipment; however, if you use estimated residual volumes, your measurement error of body density will be quite large (Morrow, Jackson, Bradley, and Hartung 1986). Nieman (1995) provides an excellent step-by-step description of the hydrostatic weighing technique.

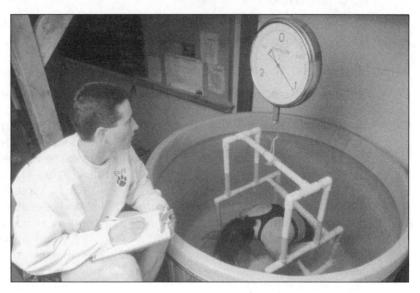

Figure 11.6 Hydrostatic weighing.

Using Body Density

Once the body density is determined, the percent body fat can be estimated from Siri's (1956) equation:

$$\%\text{fat} = (495 \div BD) - 450 \tag{11.4}$$

Estimating percent body fat allows the calculation of other useful measures of body weights: fat weight, lean weight, and **target weight.** The target weight is the weight a person should achieve to reach a target percent body fat. This establishes an easily measured goal for a weight-reduction program. The following equations are used for determining these weights:

$$\text{fat weight} = (\%\text{fat} \div 100) \times \text{body weight} \tag{11.5}$$

$$\text{lean weight} = \text{body weight} - \text{fat weight} \tag{11.6}$$

$$\text{target weight} = \text{lean weight} \div \left[1 - \left(\text{target }\%\text{fat}\right)\right] \tag{11.7}$$

Table 11.12 provides an example of these calculations.

Mastery Item 11.9

A male weighs 200 lb and has 30% body fat. His target percent is 15%. What is his target weight?

Table 11.12   **Example Calculation of Fat, Lean, and Target Weights**

Example: Male = 150 lb; Percent body fat = 30%; Target percent body fat = 25%

Component	Calculation	Result
Fat weight	150 × (30/100)	45 lb
Lean weight	150 – 45	105 lb
Target weight	105/(1 – 25/100)	140 lb

Field Methods

Field methods for body composition assessment include skinfold measurements, using the body mass index (BMI), and the waist–hip ratio.

Skinfolds

Determining body composition with hydrostatic weighing is necessary in research studies, but it is not a feasible method for body composition measurement in the field. One of the most feasible, reliable, valid, and popular methods of field estimation of body composition is the skinfold technique, which is the measurement of skinfold (actually "fatfold") thicknesses at specific body sites. The measurements are performed with skinfold calipers, such as those manufactured by Lange and Harpenden (figure 11.7).

Two research studies (Jackson and Pollock 1978; Jackson, Pollock, and Ward 1980) developed valid generalized equations for predicting body density from skinfold measurements for males and females with an age range of 18 to 61 years. The equations were adapted for the YMCA (Golding, Myers, and Sinning 1989) and provided predictions of percent body fat. The seven skinfolds used were the chest, axilla, triceps, subscapular, abdomen, suprailium, and thigh (figure 11.8). The skinfolds were each highly correlated ($r > .76$) with hydrostatically determined body density. During their analysis, the researchers found that the skinfolds had a nonlinear, quadratic relationship to body density; age was also a useful predictor. Table 11.13 provides the relevant equations.

Skinfold measurements predict body density and percent body fat in a valid manner. *However, to ensure reliability of your skinfold measures, you should have plenty of practice.* Properly trained testers should be able to produce measurements with high reliability ($r_{xx'} > .90$) and standard errors of skinfold measurement of <1.3% (Baumgartner, Jackson, Mahar, and Rowe 2003).

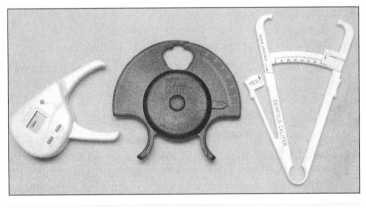

Figure 11.7   Skinfold calipers.

The recommended steps for taking skinfold measures are as follows:

1. Lift skinfolds two or three times before placing the skinfold caliper and taking a measurement.

2. Place the calipers below the thumb and fingers and perpendicular to the fold so that the dial can be easily read; release the caliper grip completely; and read the dial 1 to 2 s later.

3. Repeat the process at least three times; the measures should not vary by more than 1 mm. The median value should be used as the measure. An interval of at least 15 s should occur between each measurement to allow the site to return to normal. If you get inconsistent values, you should go to another site and then return to the difficult one. Baumgartner and colleagues (2003) caution that as many as 50 to 100

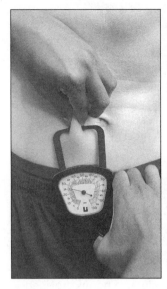
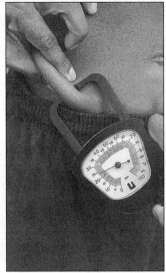
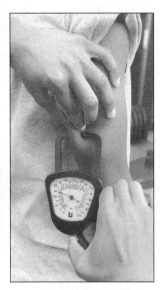
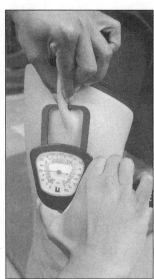

Figure 11.8   Where to measure skinfolds: abdomen *(a)*, ilium or hip *(b)*, triceps *(c)*, and thigh *(d)*.

Table 11.13   YMCA Equations for Estimation of Percent Body Fat

Four sites: Abdomen, suprailium, triceps, and thigh

Males

%fat = .29288 × (sum of 4) - .0005 × (sum of 4)^2 + .15845 × (age) – 5.76377

r = .90 s_e = 3.49%

Females

%fat = .29669 × (sum of 4) - .00043 × (sum of 4)^2 + .02963 × (age) + 1.4072

r = .846 s_e = 3.89%

Three sites: abdomen, suprailium, and triceps

Males

%fat = .39287 × (sum of 3) - .00105 × (sum of 3)^2 + .15772 × (age) – 5.18845

r = .893 s_e = 3.63%

Females

%fat = .41563 × (sum of 3) – .00112 × (sum of 3)^2 + .03661 × (age) + 4.03653

r = .825 s_e = 3.98%

Adapted from Golding, Myers, and Sinning (1989).

practice sessions on different subjects may be needed to develop reliable skinfold techniques; with proper preparation, you can achieve reliable measures (Morrow, Fridye, and Monaghen 1986). Nieman (1995) provided specific rules for skinfold measurements.

Properly taken skinfold measures are useful field estimates of body composition. Keep in mind that, as the data in table 11.13 show, there is a standard error of estimate of up to 3.98% fat. When you report a subject's percent body fat, it is a good idea to inform him or her that it is an estimate and to state the potential error of that estimate—for example, "Your percent body fat is 15 with a potential range of 11 to 19." In dealing with very obese subjects, you may not be able to take skinfolds. You may need to use another technique.

Purpose

To estimate a person's percent body fat.

Objective

To provide a field method of accurately estimating body composition characteristics.

Equipment

Skinfold calipers

Instructions

Take skinfolds at the abdomen, ilium, triceps, and thigh sites with the procedures described previously.

Scoring

Convert the skinfold measures to percent body fat using the equations in table 11.13. Compare the values to the recommended percent body fat levels and the evaluative norms provided in tables 11.14 and 11.15.

Table 11.14   **ASCM-Recommended Levels of Percent Body Fat**

Gender	Age (years)				
	20-29	30-39	40-49	50-59	60+
Male	14	15	17	18	19
Female	20	21	22	23	24

Adapted from ASCM (1988).

Table 11.15   **Norms for Percent Body Fat in Males and Females**

Male rating	Age (years)					
	18-25	26-35	36-45	46-55	56-65	66+
Very lean	4-7	8-12	10-14	12-16	15-18	15-18
Lean	8-10	13-15	16-18	18-20	19-21	19-21
Leaner than average	11-13	16-18	19-21	21-23	22-24	22-23
Average	14-16	19-21	22-24	24-25	24-26	24-25
Fatter than average	18-20	22-24	25-26	26-28	26-28	25-27
Fat	22-26	25-28	27-29	29-31	29-31	28-30
Overfat	28-37	30-37	30-38	32-38	32-38	31-38
Female rating	18-25	26-35	36-45	46-55	56-65	66+
Very lean	13-17	13-18	15-19	18-22	18-23	16-18
Lean	18-20	19-21	20-23	23-25	24-26	22-25
Leaner than average	21-23	22-23	24-26	26-28	28-30	27-29
Average	24-25	24-26	27-29	29-31	31-33	30-32
Fatter than average	26-28	27-30	30-32	32-34	34-36	33-35
Fat	29-31	31-35	33-36	36-38	36-38	36-38
Overfat	33-43	36-48	39-48	40-49	39-46	39-40

Adapted from Golding, Myers, and Sinning (1989).

Work with a classmate and measure skinfolds needed for the equations in table 11.13. Use the equations to calculate your percent body fat.

Body Mass Index

The body mass index (BMI) is a simple measure expressing the relationship of weight to height that is correlated with fatness. It is used in epidemiological research and has a moderately high correlation (r_{xy} = .69) with body density. It is easily calculated from the following formula:

$$\text{BMI} = \frac{\text{Weight}}{\text{Height}^2} \tag{11.8}$$

where weight is measured in kilograms and height in meters.

The following ratings have been applied to the BMI by the National Heart, Lung, and Blood Institute of the National Institutes of Health:

Underweight <18.5

Normal 18.5-24.9

Overweight 25.0-29.9

Obesity class I 30.0-34.9

Obesity class II 35.0- 39.9

Extreme obesity 40.0+

In the field, the BMI can serve as an acceptable substitute for skinfold measurements on very obese subjects; however, do not use it on lean or normal individuals, for whom skinfolds are more accurate. Table 11.16 provides the BMI for a given height and weight.

Distribution of Body Fat

It is well known that excessive body fat is a health risk, but another factor is the distribution of this body fat. People with excessive body fat on the trunk (android obesity) as compared with the lower body (gynoid obesity) have a higher risk of coronary heart disease (CHD; ACSM 2000). CHD is a component (along with stroke) of cardiovascular disease (CVD). A simple measure of this body composition risk factor is the **waist–hip girth ratio;** the circumference of the waist is divided by the hip circumference. Ratios greater than 1.0 for males and .80 for females are associated with a significantly increased risk of CHD (AHA 1994). Men with a waist circumference greater than 102 cm (40 in) and women with greater than 88 cm (35 in) are considered to have increased risk for type 2 diabetes, hypertension, and cardiovascular disease by the National Heart, Lung, and Blood Institute.

A set of skinfold data is collected on a group of women. The sum of skinfolds is used to estimate percent fat based on hydrostatic weighing. These data are contained in the MI 11.11 data set on the text Web page.

1. Use SPSS correlation to confirm that the correlation between the sum of skinfolds and percent fat is .840.

2. Next use **Analyze → Regression → Linear** to confirm that the standard error of estimate for predicting percent fat from the sum of skinfolds is 5.45. (Hint: Put percent fat in the Dependent cell and sum of skinfolds in the Independent cell.)

Table 11.16 Body Mass Index

Weight (lb)	Height (in.) 48	49	50	51	52	53	54	55	56	57	58	59	60	61	62	63	Weight (kg)
100	30.6	29.3	28.2	27.1	26.1	25.1	24.2	23.3	22.5	21.7	21.6	20.2	19.6	18.9	18.3	17.8	45.5
105	32.1	30.8	29.6	28.4	27.4	26.3	25.4	24.5	23.6	22.8	22.0	21.3	20.5	19.9	19.2	18.6	47.7
110	33.6	32.3	31.0	29.8	28.7	27.6	26.6	25.6	24.7	23.9	23.0	22.3	21.5	20.8	20.2	19.5	50.0
115	35.2	33.7	32.4	31.2	30.0	28.8	27.8	26.8	25.8	24.9	24.1	23.2	22.5	21.8	21.1	20.4	52.3
120	36.7	35.2	33.8	32.5	31.3	30.1	29.0	27.9	27.0	26.0	25.1	24.3	23.5	22.7	22.0	21.3	54.5
125	38.2	36.7	35.2	33.9	32.6	31.4	30.2	29.1	28.1	27.1	26.2	25.3	24.5	23.7	22.9	22.2	56.8
130	39.8	38.1	36.6	35.2	33.9	32.6	31.4	30.3	29.2	28.2	27.2	26.3	25.4	24.6	23.8	23.1	59.1
135	41.3	39.6	38.0	36.6	35.2	33.9	32.6	31.4	30.3	29.3	28.3	27.3	26.4	25.6	24.7	24.0	61.4
140	42.8	41.1	39.5	37.9	36.5	35.1	33.8	32.6	31.5	30.4	29.3	28.3	27.4	26.5	25.7	24.9	63.6
145	44.3	42.5	40.9	39.3	37.8	36.4	35.0	33.8	32.6	31.4	30.4	29.3	28.4	27.5	26.6	25.7	65.9
150	45.9	44.0	42.3	40.6	39.1	37.6	36.2	34.9	33.7	32.5	31.4	30.4	29.4	28.4	27.5	26.6	68.2
155	47.4	45.5	43.7	42.0	40.4	38.9	37.5	36.1	34.8	33.6	32.4	31.4	30.3	29.3	28.4	27.5	70.5
160	48.9	47.0	45.1	43.3	41.7	40.1	38.7	37.3	35.9	34.7	33.5	32.4	31.3	30.3	29.3	28.4	72.7
165	50.5	48.4	46.5	44.7	43.0	41.4	39.9	38.4	37.1	35.8	34.6	33.4	32.3	31.2	30.2	29.3	75.0
170	52.0	49.9	47.9	46.6	44.3	42.6	41.1	39.6	38.2	36.9	35.6	34.4	33.3	32.3	31.2	30.2	77.3
175	53.5	51.4	49.3	47.4	45.6	43.9	42.3	40.8	39.3	37.9	36.7	35.4	34.2	33.1	32.1	31.1	79.5
180	55.0	52.8	50.7	48.8	46.9	45.1	43.5	41.9	40.4	39.0	37.7	36.4	35.2	34.1	33.0	32.0	81.8
185	56.6	54.3	52.1	50.1	48.2	46.4	44.7	43.1	41.6	40.1	38.7	37.4	36.2	35.0	33.9	32.8	84.1
190	58.1	55.8	53.5	51.5	49.5	47.7	45.9	44.3	42.7	41.2	39.8	38.5	37.2	36.0	34.8	33.7	86.4
195	59.6	57.2	55.0	52.8	50.8	48.9	47.1	45.4	43.8	42.3	40.8	39.5	38.2	36.9	35.7	34.6	88.6
200	61.2	58.7	56.4	54.2	52.1	50.2	48.3	46.6	44.9	43.4	41.9	40.5	39.1	37.9	36.7	35.5	90.9
205	62.7	60.2	57.8	55.5	53.4	51.4	49.5	47.7	46.1	44.5	42.9	41.5	40.1	38.8	37.6	36.4	93.2
210	64.2	61.6	59.2	56.9	54.7	52.7	50.7	48.9	47.2	45.5	44.0	42.5	41.1	39.8	38.5	37.3	95.5
215	65.7	63.1	60.6	58.2	56.0	53.9	51.9	50.1	48.3	46.6	45.0	43.5	42.1	40.7	39.4	38.2	97.7
220	67.3	64.6	62.0	59.6	57.3	55.2	53.2	51.2	49.4	47.7	46.1	44.5	43.1	41.7	40.3	39.1	100.0
225	68.8	66.0	63.4	60.9	58.6	56.4	54.4	52.4	50.5	48.8	47.1	45.5	44.0	42.6	41.2	39.9	102.3
230	70.3	67.5	64.8	62.3	59.9	57.7	55.6	53.6	51.7	49.9	48.2	46.6	45.0	43.5	42.2	40.8	104.5
235	71.9	69.0	66.2	63.7	61.2	58.9	56.8	54.7	52.8	51.0	49.2	47.6	46.0	44.5	43.1	41.7	106.8
240	73.4	70.4	67.6	65.0	62.5	60.2	58.0	55.9	53.9	52.0	50.3	48.6	47.0	45.4	44.0	42.6	109.1
245	74.9	71.9	69.0	66.4	63.8	61.5	59.2	57.1	55.0	53.1	51.3	49.6	47.9	46.4	44.9	43.5	111.4
250	76.4	73.4	70.5	67.7	65.1	62.7	60.4	58.2	56.2	54.2	52.4	50.6	48.9	47.3	45.8	44.4	113.6
Height (m)	1.22	1.24	1.27	1.30	1.32	1.35	1.37	1.40	1.42	1.45	1.47	1.50	1.52	1.55	1.57	1.60	

Weight (lb)	Height (in.)															Weight (kg)
	64	65	66	67	68	69	70	71	72	73	74	75	76	77	78	
100	17.2	16.7	16.2	15.7	15.2	14.8	14.4	14.0	13.6	13.2	12.9	12.5	12.2	11.9	11.6	45.5
105	18.1	17.5	17.0	16.5	16.0	15.5	15.1	14.7	14.3	13.9	13.5	13.2	12.8	12.5	12.2	47.7
110	18.9	18.3	17.8	17.3	16.8	16.3	15.8	15.4	14.9	14.5	14.2	13.8	13.4	13.1	12.7	50.0
115	19.8	19.2	18.6	18.0	17.5	17.0	16.5	16.1	15.6	15.2	14.8	14.4	14.0	13.7	13.3	52.3
120	20.6	20.0	19.4	18.8	18.3	17.8	17.3	16.8	16.3	15.9	15.4	15.0	14.6	14.3	13.9	54.5
125	21.5	20.8	20.2	19.6	19.0	18.5	18.0	17.5	17.0	16.5	16.1	15.7	15.2	14.9	14.5	56.8
130	22.4	21.7	21.0	20.4	19.8	19.2	18.7	18.2	17.7	17.2	16.7	16.3	15.9	15.4	15.1	59.1
135	23.2	22.5	21.8	21.2	20.6	20.0	19.4	18.9	18.3	17.8	17.4	16.9	16.5	16.0	15.6	61.4
140	24.1	23.3	22.6	22.0	21.3	20.7	20.1	19.6	19.0	18.5	18.0	17.5	17.1	16.6	16.2	63.6
145	24.9	24.2	23.5	22.8	22.1	21.5	20.8	20.3	19.7	19.2	18.7	18.2	17.7	17.2	16.8	65.9
150	25.8	25.0	24.3	23.5	22.9	22.2	21.6	21.0	20.4	19.8	19.3	18.8	18.3	17.8	17.4	68.2
155	26.7	25.8	25.1	24.3	23.6	22.9	22.3	21.7	21.1	20.5	19.9	19.4	18.9	18.4	17.9	70.5
160	27.5	26.7	25.9	25.1	24.4	23.7	23.0	22.4	21.7	21.2	20.6	20.0	19.5	19.0	18.5	72.7
165	28.4	27.5	26.7	25.9	25.1	24.4	23.7	23.1	22.4	21.8	21.2	20.7	20.1	19.6	19.1	75.0
170	29.2	28.3	27.5	26.7	25.9	25.2	24.4	23.8	23.1	22.5	21.9	21.3	20.7	20.2	19.7	77.3
175	30.1	29.2	28.3	27.5	26.7	25.9	25.2	24.5	23.8	23.1	22.5	21.9	21.3	20.8	20.3	79.5
180	31.0	30.0	29.1	28.3	27.4	26.6	25.9	25.2	24.5	23.8	23.2	22.5	22.0	21.4	20.8	81.8
185	31.8	30.8	29.9	29.0	28.2	27.4	26.6	25.9	25.1	24.5	23.8	23.2	22.6	22.0	21.4	84.1
190	32.7	31.7	30.7	29.8	28.9	28.1	27.3	26.6	25.8	25.1	24.4	23.8	23.2	22.6	22.0	86.4
195	33.5	32.5	31.5	30.6	29.7	28.9	28.0	27.3	26.5	25.8	25.1	24.4	23.8	23.2	22.6	88.6
200	34.4	33.4	32.3	31.4	30.5	29.6	28.8	28.0	27.2	26.4	25.7	25.1	24.4	23.8	23.2	90.9
205	35.3	34.2	33.2	32.2	31.2	30.3	29.5	28.7	27.9	27.1	26.4	25.7	25.0	24.4	23.7	93.2
210	36.1	35.0	34.0	33.0	32.0	31.1	30.2	29.4	28.5	27.8	27.0	26.3	25.6	25.0	24.3	95.5
215	37.0	35.9	34.8	33.7	32.8	31.8	30.9	30.0	29.2	28.4	27.7	26.9	26.2	25.5	24.9	97.7
220	37.8	36.7	35.6	34.5	33.5	32.6	31.6	30.7	29.9	29.1	28.3	27.6	26.8	26.1	25.5	100.0
225	38.7	37.5	36.4	35.3	34.3	33.3	32.4	31.4	30.6	29.7	28.9	28.2	27.4	26.7	26.1	102.3
230	39.6	38.4	37.2	36.1	35.0	34.0	33.1	32.1	31.3	30.4	29.6	28.8	28.1	27.3	26.6	104.5
235	40.4	39.2	38.0	36.9	35.8	34.8	33.8	32.8	31.9	31.1	30.2	29.4	28.7	27.9	27.2	106.8
240	41.3	40.0	38.8	37.7	36.6	35.5	34.5	33.5	32.6	31.7	30.9	30.1	29.3	28.5	27.8	109.1
245	42.1	40.9	39.6	38.5	37.3	36.3	35.2	34.2	33.3	32.4	31.5	30.7	29.9	29.1	28.4	111.4
250	43.0	41.7	40.4	39.2	38.1	37.0	35.9	34.9	34.0	33.1	32.2	31.3	30.5	29.7	29.0	113.6
	1.63	1.65	1.68	1.7	1.73	1.75	1.78	1.80	1.83	1.85	1.88	1.91	1.93	1.96	1.98	
	Height (m)															

MEASURING MUSCULAR STRENGTH AND ENDURANCE

Many people, even professionals in exercise science, use the terms *strength, force, power, work, torque,* and *endurance* almost interchangeably. However, it is important for you as a measurement specialist to understand that each of the terms has a distinct meaning. **Work** is the result of the physical effort that is performed. It is defined by the following equation:

$$\text{work (W)} = \text{force (F)} \times \text{distance (D)} \tag{11.9}$$

For example, the equation 150 ft lb = 150 lb × 1 ft means that a weight of 150 lb was moved 1 ft in distance. **Power** is the amount of work performed in a fixed amount of time. It is defined by the following equation:

$$\text{power (p)} = (\text{F} \times \text{D}) \div \text{time (T)} = \text{W} \div \text{T} \tag{11.10}$$

For example, a power value of 150 ft lb/s means that a weight of 150 lb was moved 1 ft in distance in 1 s. **Muscular strength** is the force that can be generated by the musculature that is contracting. **Torque** is the effectiveness of a force for producing rotation about an axis. Many of the computerized dynamometers report the torque produced during muscular contractions as well as the force or strength.

Muscular endurance is the physical ability to perform work. The pull-up test is often called a test of arm and shoulder strength, but in reality it is a measure of muscular endurance if you can do more than one. You count the maximum number of pull-ups performed, which is a measure of the amount of work that was completed. Muscular endurance can be categorized as relative endurance and absolute endurance. **Relative endurance** is a measurement of repetitive performance related to maximum strength. **Absolute endurance** is a measurement of repetitive performance at a fixed resistance. For example, measuring a subject's maximum strength on the bench press and then having him or her perform as many repetitions as possible at 75% of maximum strength is a relative endurance test. Performing the maximum number of repetitions at a fixed weight of 100 lb would be a test of absolute endurance. Absolute endurance is highly correlated with maximum strength, but relative endurance has a low correlation with maximum strength.

Mastery Item 11.12

What are you measuring if you administer the 1 min sit-up test, which requires the subject to perform the maximum number of sit-ups possible in 1 min?

Muscular actions are defined and categorized by specific terms, the most common of which are *concentric, eccentric, isometric, isotonic,* and *isokinetic.* The terms are defined as follows:

▸ **Concentric contraction**—The muscle generates force as it shortens.

▸ **Eccentric contraction**—The muscle generates force as it lengthens.

▸ **Isometric contraction**—The muscle generates force but remains static in length and causes no movement.

▸ **Isotonic contraction**—The muscle generates enough force to move a constant load at a variable speed through a full range of motion.

▸ **Isokinetic contraction**—The muscle generates force at a constant speed through a full range of motion.

In laboratory testing, researchers assess muscular performance generally by measuring the force, torque, work, and power generated in concentric, eccentric, isokinetic, and isometric contractions. In field situations, muscular performance is assessed with concentric, isotonic contractions.

Laboratory Methods

Similar to laboratory measurements of aerobic capacity, measuring muscular strength and endurance in the laboratory requires expensive and sophisticated equipment and very precise testing protocols. They are the most valid assessments but are difficult to administer.

Computerized Dynamometers

State-of-the-art muscular fitness measurement techniques involve using *computerized dynamometers,* which integrate mechanical ergometers, electronic sensors, computers, and sophisticated software. Computerized dynamometers allow for detailed measurements of force, work, torque, and power generated in terms not only of maximal values but of values throughout a range of motion. Isokinetic dynamometers allow the speed of movement to be controlled during testing. These devices are used in orthopedic clinics by physicians, physical therapists, and athletic trainers in rehabilitation programs for patients recovering from orthopedic injury or surgery; researchers in exercise science use the devices in strength and endurance studies. Strength is measured in terms of peak force and endurance in terms of fatigue rates in force production during a set of repetitions. There are a variety of these devices, each with its own advantages and disadvantages. Because devices can be expensive, selection of a device requires an analysis of cost–benefit ratios.

The Biodex, a computerized isokinetic dynamometer (Biodex System 2 and Advantage Software 4.0, Biodex Medical Systems, Inc., Shirley, NY) is a device in widespread clinical use. Figure 11.9 illustrates graphic output of forces or torques that the system is able to generate.

The Biodex can produce very reliable ($r_{xx'} > .90$) force measurements. However, as with any physical performance test, practice trials for both testers and test subjects are needed to reduce measurement error. Mayhew and Rothstein (1985) provided a thorough discussion of the measurement issues associated with muscle performance assessed with dynamometers. These devices must be calibrated regularly to ensure reliable and valid measurements. Computerized dynamometry for muscular strength and endurance assessment should be

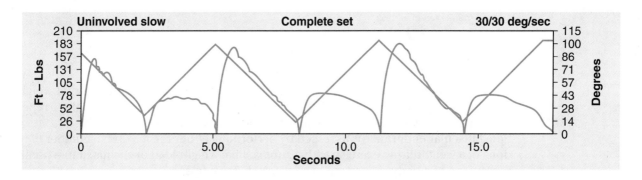

Figure 11.9   Graphic output of forces or torques produced on the Biodex dynamometer.

used to provide criterion measures for concurrent validity research on more feasible field tests of muscular fitness. This application of computerized dynamometry would be valuable, because most field assessments in muscular fitness have only content validity to support their use.

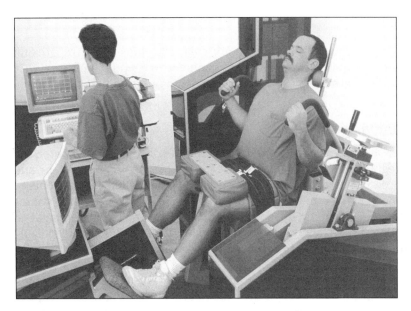

Figure 11.10   MedX Lumbar Extension Machine.

Back Extension Strength Test

Low-back pain is a serious, prevalent health problem in the adult population. Lack of strength and endurance in the extensor muscles of the low-back has been associated with low-back pain (Suzuki and Endo 1983). Graves and colleagues (1990) examined measurement issues associated with isometric strength of the back extensors using the MedX Lumbar Extension Machine (MedX Corp., Ocala, Florida) to measure isometric strength (figure 11.10). Their test protocol has the subject begin at 72° of lumbar flexion and continue at every 12°, down to 0° of lumbar flexion. At each angle of flexion, the subject exerts a maximal isometric contraction of the extensor muscles of the back. The maximal isometric torque is measured by the dynamometer's automated system. This protocol has produced reliable ($r_{xx'} > .78$) torque results. Average strength values for this test are provided in table 11.17.

Table 11.17   Mean Isometric Torque Values for Lumbar Extension (N · m per kg body weight)

Gender	Degrees of lumbar flexion						
	0	12	24	36	48	60	72
Male	3.0	3.8	4.4	4.8	5.2	5.5	6.0
Female	2.2	2.7	3.0	3.1	3.3	3.5	3.9

Adapted from Graves et al. 1990.

Field Methods

Field assessment of muscular strength and endurance involves lifting external weights or the repetitive movement of the body. The measure of maximum strength is the 1-repetition maximum (1RM), which is the maximum amount of weight a person can lift for one repetition. Muscular endurance is assessed by performing either the maximum number of repetitions of a weightlifting exercise with a submaximal weight load or the maximum number of repetitions of a body movement exercise such as sit-ups.

Upper- and Lower-Body Strength and Endurance

The ACSM (1988, 1995) recommends 1RM values of the bench press and the leg press strength tests as valid measures of upper- and lower-body strength. The ACSM further suggests that the 1RM values be divided by the body weight of the subject to present a strength measure that is equitable across weight classes.

The following steps present a method for assessing the 1RM value for any given exercise. These steps were used to produce reliable ($r_{xx'} > .92$) and valid measures of upper- and lower-body strength (Jackson, Watkins, and Patton 1980).

1. Have the subject warm up with stretching and light lifting.
2. Have the subject perform a lift below what you estimate to be his or her maximum. A practice session is extremely useful for novice subjects.
3. To prevent fatigue, have the subject rest at least 2 min between lifts.
4. Increase the weight by a small increment, 5 or 10 lb (2.3 or 4.5 kg), depending on the exercise and weight increments available.
5. Continue the process until the subject fails an attempt.
6. The last weight successfully completed is the 1RM weight.
7. Divide the 1RM by the subject's body weight.

If more than five repetitions are needed to determine the 1RM value, retest the subject after a day's rest with a heavier starting weight.

Tables 11.18 and 11.19 provide norms for the bench press and leg press tests. These norms reflect measurements taken from the Universal Gym Weight Lifting Machine, a rack-mounted weight machine. These standards are not valid if your testing is done with free weights or another type of machine. One of the difficulties of strength measurement is that different strength testing devices—free weights, computerized dynamometers, or rack-mounted devices—produce different results. Use standards appropriate for your testing situation; it may be necessary for you to develop your own local standards.

Table 11.18   **Bench Press Strength (1RM lb/lb body weight)**

Rating	Age (years) 20-29	30-39	40-49	50-59	60+
Men					
Excellent	> 1.26	> 1.08	>0.97	> 0.86	> 0.78
Good	1.17-1.25	1.01-1.07	0.91-0.96	0.81-0.85	0.74-0.77
Average	0.97-1.16	0.86-1.00	0.78-0.90	0.70-0.80	0.64-0.73
Fair	0.88-0.96	0.79-0.85	0.72-0.77	0.65-0.69	0.60-0.63
Poor	< 0.87	< 0.78	< 0.71	< 0.64	< 0.59
Women					
Excellent	> 0.78	> 0.66	> 0.61	> 0.54	> 0.55
Good	0.72-0.77	0.62-0.65	0.57-0.60	0.51-0.53	0.51-0.54
Average	0.59-0.71	0.53-0.61	0.48-0.56	0.43-0.50	0.41-0.50
Fair	0.53-0.58	0.49-0.52	0.44-0.47	0.40-0.42	0.37-0.40
Poor	< 0.52	< 0.48	< 0.43	< 0.39	< 0.36

Table 11.19 **Upper Leg Press Strength (1RM lb/lb body weight)**

	Age (years)				
Rating	20-29	30-39	40-49	50-59	60+
Men					
Excellent	>2.08	>1.88	>1.76	>1.66	>1.56
Good	2.00-2.07	1.80-1.87	1.70-1.75	1.60-1.65	1.50-1.55
Average	1.83-1.99	1.63-1.79	1.56-1.69	1.46-1.59	1.37-1.49
Fair	1.65-1.82	1.55-1.62	1.50-1.55	1.40-1.45	1.31-1.36
Poor	<1.64	<1.54	<1.49	<1.39	<1.30
Women					
Excellent	>1.63	>1.42	>1.32	>1.26	>1.15
Good	1.54-1.62	1.35-1.41	1.26-1.31	1.13-1.25	1.08-1.14
Average	1.35-1.53	1.20-1.34	1.12-1.25	0.99-1.12	0.92-1.07
Fair	1.26-1.34	1.13-1.19	1.06-1.11	0.86-0.98	0.85-0.91
Poor	<1.25	<1.12	<1.05	<0.85	<0.84

Reprinted from The Cooper Institute for Aerobics Research, Dallas, TX.

The YMCA Bench Press Test (Golding, Myers, and Sinning 1989) is used to assess upper-body endurance. The Canadian Standardized Test of Fitness uses a push-up test to exhaustion to measure upper-body endurance. These tests are described in the following highlight boxes.

YMCA Bench Press Test

Purpose

To assess the absolute endurance of the upper body.

Objective

To perform a set of bench press repetitions to exhaustion.

Equipment

Barbells (35 and 80 lb [15.9 and 36.4 kg])
Metronome set at 60 beats/min
Weightlifting bench

Instructions

The weight is 35 lb (15.9 kg) for women and 80 lb (36.4 kg) for men. The consecutive repetitions are performed to a cadence of 60 beats/min, with each sound indicating a movement up or down.

Scoring

The test continues until exhaustion or until the subject can no longer maintain the required cadence. Table 11.20 provides norms for this test. Because this is an absolute endurance test, it is positively correlated with maximal strength and body weight or size.

Table 11.20 Male and Female Norms for the YMCA Bench Press Test (number completed)

	Age (years)					
Male rating	18-25	26-35	36-45	46-55	56-65	66+
Excellent	45-38	43-34	40-30	35-24	32-22	30-18
Good	34-30	30-26	28-24	22-20	20-14	14-10
Above average	28-25	25-22	22-20	17-14	14-10	10-8
Average	22-21	21-18	18-16	13-10	10-8	8-6
Below average	20-16	17-13	14-12	10-8	6-4	4-4
Poor	13-9	12-9	10-8	6-4	4-2	2-2
Very poor	8-0	5-0	5-0	2-0	0	0
Female rating						
Excellent	50-36	48-33	46-28	42-26	34-22	26-18
Good	32-28	29-25	25-21	22-20	20-16	14-12
Above average	25-22	22-20	20-17	17-13	15-12	11-9
Average	21-18	18-16	14-12	12-10	10-8	8-5
Below average	16-13	14-12	11-9	9-6	7-4	4-2
Poor	12-8	9-5	8-4	5-2	3-1	2-0
Very poor	5-1	2-0	2-0	1-0	0	0

Adapted from Golding, Myers, and Sinning (1989).

Canadian Standardized Test of Fitness—Push-Up Test

Purpose

To assess upper-body endurance.

Objective

To perform push-ups to exhaustion.

Equipment

Mat

Instructions

Females perform the test with the knees bent and touching the ground; males perform the test with the toes touching the ground (figure 11.11).

Scoring

The number of correct repetitions is compared to the norms provided in table 11.21.

(continued)

Figure 11.11 Traditional *(a)* and modified *(b)* push-ups.

Table 11.21 **Male and Female Norms for the Push-Up Test (number completed)**

	Age (years)					
Male rating	**15-19**	**20-29**	**30-39**	**40-49**	**50-59**	**60-69**
Excellent	39+	36+	30+	22+	21+	18+
Above average	29-38	29-35	22-29	17-21	13-20	11-17
Average	23-28	22-28	17-21	13-16	10-12	8-10
Below average	18-22	17-21	12-16	10-12	7-9	5-7
Poor	17-	16-	11-	9-	6-	4-
Female rating						
Excellent	33+	30+	27+	24+	21+	17+
Above average	25-32	21-29	20-26	15-23	11-20	12-16
Average	18-24	15-20	13-19	11-14	7-10	5-11
Below average	12-17	10-14	8-12	5-10	2-6	1-4
Poor	11-	9-	7-	4-	1-	1-

Adapted from Nieman (1995).

Johnson and Nelson (1979) indicated that you can achieve a high reliability ($r_{xx'}$ = .93) when these standardized test procedures are followed. *The results of this test will be negatively correlated with the body weight of the subjects tested.*

Trunk Endurance

The most universally used test of abdominal endurance is the 1 min timed sit-up test. The YMCA's test protocol is described here.

YMCA 1-Minute Timed Sit-Up Test

Purpose

To assess abdominal endurance.

Objective

To perform the maximum number of sit-ups in 1 min.

Equipment

Stopwatch
Mat

Instructions

The subject performs the test with bent knees, feet flat about 18 in. (45.7 cm) from the buttocks as shown in figure 11.12. A partner holds the subject's feet as the exercise is performed. The subject touches the elbow to the alternate knee with each sit-up. The subject performs as many sit-ups in 1 min as possible.

Scoring

The number of correct repetitions is compared with the norms provided in table 11.22.

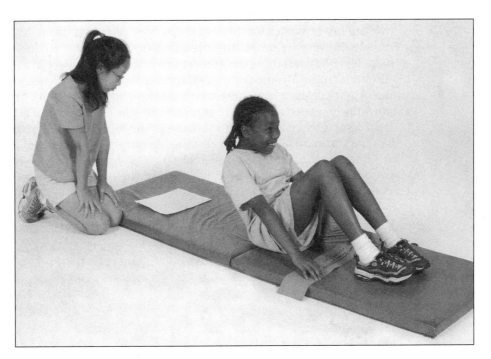

Figure 11.12   The sit-up test.

(continued)

Table 11.22   Male and Female Norms for the YMCA 1-Minute Sit-Up Test (number completed)

Male rating	Age (years)					
	18-25	26-35	36-45	46-55	56-65	66+
Excellent	60-50	55-46	50-42	50-36	42-32	40-29
Good	48-45	45-41	40-36	33-29	29-26	26-22
Above average	42-40	38-36	34-30	28-25	24-21	21-20
Average	38-36	34-32	29-28	24-22	20-17	18-16
Below average	34-32	30-29	26-24	21-18	16-13	14-12
Poor	30-26	28-24	22-18	17-13	12-9	10-8
Very poor	24-12	21-6	16-4	12-4	8-2	6-2
Female rating						
Excellent	55-44	54-40	50-34	42-28	38-25	36-24
Good	41-37	37-33	30-17	25-22	21-18	22-18
Above average	36-33	32-29	26-24	21-18	17-13	16-14
Average	32-29	28-25	22-20	17-14	12-10	13-11
Below average	28-25	24-21	18-16	13-10	9-7	10-6
Poor	24-20	20-16	14-10	9-6	6-4	4-2
Very poor	17-4	12-1	6-1	4-0	2-0	1-0

Adapted from Golding, Myers, and Sinning (1989).

Reliability estimates for this test range from .68 to .94 (Baumgartner, Jackson, Mahar, and Rowe 1999). Acceptable reliability requires the use of standardized procedures and practice trials. The body weight of the subject is negatively correlated to the results of this test.

Robertson and Magnusdottir (1987) presented an alternative field test of abdominal endurance. Their curl-up test requires the subject to lift the head and upper back a distance that allows the extended arms and fingers to move 3 in. (7.6 cm) parallel to the ground rather than sit up completely. This requires less use of the hip flexors and more use of the abdominals than the standard sit-up test. The subject completes as many of these as possible in 60 s. The authors report excellent reliability ($r_{xx'}$ = .93) for the test; however, because the sample sizes were small, more research is necessary before this protocol becomes acceptable for general use.

Mastery Item 11.13

Test yourself for upper-body and abdominal endurance using the push-up and sit-up tests. To ensure a reliable measure, perform practice trials of both tests on day 1 and then do two performance trials of each test on days 2 and 3. Compare your scores with norms for these tests.

MEASURING FLEXIBILITY

The measurement of **flexibility,** or range of motion of a joint or group of joints, is an important aspect of assessment of patients recovering from orthopedic surgery or orthopedic injuries. *Because flexibility is specific to a joint and its surrounding tissues, there are* *no valid tests of general flexibility.* For example, if you have very flexible ankle joints, you may not have very flexible shoulder joints.

Laboratory Methods

Miller (1985) summarized the measurement issues associated with the assessment of joint range of motion. He described clinical measurement techniques, including the following:

- Goniometry—manual, electric, and pendulum goniometers
- Visual estimation
- Radiography
- Photography
- Linear measurements
- Trigonometry

Miller asserted that radiography is the most reliable and valid method but that it has limited feasibility owing to problems with radiation. Goniometry is the most feasible method of clinical assessment of flexibility; it can be very reliable and valid if proper test procedures are followed. Reliability estimates for goniometric measurement of hamstring flexibility have exceeded .90 in past research (Jackson and Baker 1986; Jackson and Langford 1989). Clinical tests of flexibility should serve as criterion measures for validity studies of field tests of flexibility for health-related fitness. Such concurrent validity studies improve the validity of measurement of flexibility in the field.

Field Methods

As reported previously, the ACSM (1995) indicates that low back pain is a prominent health problem in the United States. In theory, lack of flexibility in the low back should be associated with low back pain, but valid research has yet to establish a relationship between the two (ACSM 2000). Flexibility of the low back is measured with tests of trunk flexion and extension. Most adult fitness batteries include a test of trunk flexion that is a version of the sit-and-reach test.

Trunk Flexion

The sit-and-reach test is a universally used test of trunk flexion. Specifically, it was designed to measure the flexibility of the low back and hamstring muscles. The YMCA adult fitness test version of the sit-and-reach test is described next.

YMCA Adult Trunk Flexion (Sit-and-Reach) Test

Purpose

To assess trunk flexibility.

Objective

To reach as far forward as possible.

Equipment

Yardstick
Masking tape

(continued)

Instructions

Place a yardstick on the floor and put an 18 in. (45.7 cm) piece of tape across the 15 in. (38.1 cm) mark on the yardstick. The tape should secure the yardstick to the floor. The subject sits with the 0 end of the yardstick between the legs. The subject's heels should almost touch the tape at the 15 in. mark and be about 12 in. (30.5 cm) apart. With the legs held straight, the subject bends forward slowly and reaches with parallel hands as far as possible and touches the yardstick. The subject should hold this reach long enough for the distance to be recorded (figure 11.13).

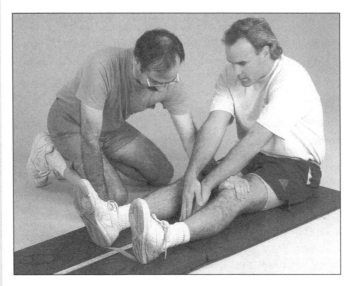

Figure 11.13 The sit-and-reach test.

Scoring

Perform three trials. The best score, recorded to the nearest quarter inch, is compared with the norms supplied in table 11.23.

Table 11.23   Male and Female Norms for the YMCA Trunk Flexion Test (inches)

	Age (years)					
Male rating	18-25	26-35	36-45	46-55	56-65	66+
Excellent	26-20	25-20	24-19	23-19	21-17	20-17
Good	20-18	19-18	19-17	17-16	17-15	15-13
Above average	18-17	17-16	17-15	15-14	13	13-11
Average	16-15	16-15	15-13	13-12	11	11-9
Below average	14-13	14-12	13-11	11-10	9	9-8
Poor	12-10	12-10	11-9	9-7	7-5	7-5
Very poor	9-2	9-2	7-1	6-1	5-1	4-0
Female rating						
Excellent	27-24	26-23	25-22	24-21	23-20	22-20
Good	23-21	22-20	21-19	20-18	19-18	19-18
Above average	21-20	20-19	19-17	18-17	17-16	17-16
Average	19-18	18	17-16	16-15	15	15-14
Below average	18-17	17-16	15-14	15-14	14-13	13-12
Poor	16-14	15-14	13-11	13-11	12-10	11-9
Very poor	13-8	13-8	10-6	10-4	9-3	8-2

Adapted from Golding, Myers, and Sinning (1989).

Keep in mind that flexibility of the low back has not been empirically related to low back pain. Research has shown that the sit-and-reach test is a valid measure of hamstring flexibility and is highly reliable ($r_{xx'} > .90$) but is poorly correlated with a clinical measure of low back flexibility and probably is not a valid field test of low back flexibility (Jackson and Baker 1986; Jackson and Langford 1989). Further research (Jackson et al. 1998) documented no relationship between reported low back pain and performance on the sit-and-reach or sit-up tests. These observations provide little support for including sit-and-reach tests in health-related fitness assessments, but they are often present in both adult and youth fitness batteries.

Mastery Item 11.14

Perform the sit-and-reach test just described and compare your results with the norms. Are you flexible?

Trunk Extension

Measuring trunk flexion with the sit-and-reach test is prevalent in health-related fitness testing; however, trunk extension assessments are not included in health-related fitness testing, and little research has been conducted on developing a valid field test of this ability. Jensen and Hirst (1980) described a field test in which the subject lies prone with the hands clasped near the small of the back. The subject then raises his or her upper body off the floor as far as possible while an aide holds the legs down. The score is the distance from the suprasternal notch to the floor multiplied by 100 and divided by the trunk length. Safrit (1986) suggested determining the trunk length while the subject is seated; trunk length is then the distance from the suprasternal notch to the seat. There is no reliability and validity information on this test. Furthermore, body weight and back extensor strength would be related to the scores on this test, which would limit it as a valid measure of flexibility. Developing a reliable and valid measure of low back extension requires additional research.

HEALTH-RELATED FITNESS BATTERIES

Several organizations have grouped health-related fitness tests together in a test battery. These test batteries and their documentation provide you with the ability to administer reliable and valid fitness tests, interpret the results, and convey fitness information to your program participants.

YMCA Physical Fitness Test Battery

Throughout this chapter we have highlighted the fitness test battery used by the YMCA (Golding, Myers, and Sinning 1989). The test battery is used in the physical fitness assessments that the YMCA performs on its members. The test is available and easily adaptable to many adult physical fitness testing situations. The entire test battery includes these measurements:

- ▶ Height
- ▶ Weight
- ▶ Resting heart rate
- ▶ Resting blood pressure
- ▶ Body composition
- ▶ Cardiovascular evaluation
- ▶ Flexibility
- ▶ Muscular strength and endurance

We have described and provided norms for the 3-min step test (cardiovascular endurance, p. 233), the skinfold estimation of body composition (p. 240), the sit-and-reach test (flexibility, p. 253), the bench press test (muscular strength, p. 248), and the sit-up test (muscular endurance, p. 251).

Canadian Standardized Test of Fitness

The Canadian Standardized Test of Fitness, a physical fitness test battery, was the result of the Canadian Fitness Survey conducted in 1981 on thousands of subjects to develop an understanding of the fitness level of the Canadian population. The test battery includes the following components:

- Resting heart rate
- Resting blood pressure
- Body composition (skinfolds)
- Cardiorespiratory endurance (A variety of test results can be used, including treadmill time, cycle ergometer results, and distance-run times.)
- Flexibility (sit-and-reach test)
- Abdominal endurance (with the 1 min sit-up test)
- Upper-body strength and endurance (push-up test, p. 249)

PHYSICAL FITNESS ASSESSMENT IN OLDER ADULTS

An older adult is defined as someone aged 65 years and over. In the year 2000 there were approximately 35 million older adults in the United States, about 13% of the entire population, and this number is projected to grow to 69.4 million older adults by the year 2030. The authors of this book may not be alive in 2030, but if we are, we will be members of this dramatically growing segment of the U.S. population. The increase in the number of older adults in the United States is similar to that in other industrialized countries.

Older adults have the highest rates of chronic diseases such as cardiovascular disease (CVD), cancer, diabetes, osteoporosis, and arthritis. Health care costs for older adults are contributing to health financing problems. Medicare, the U.S. national health care program for older adults, is facing an uncertain financial future that may have grave implications for future generations of Americans. As professionals in human performance, we serve an important role in helping improve the physical activity and fitness levels of young, middle-aged, and older adults. Improved health of our population derived from increased physical activity and fitness would help alleviate the health care financing shortages facing the United States.

Aging is related to

- decreased sensations of taste, smell, vision, and hearing;
- decreased mental abilities (e.g., memory, judgment, speech);
- decreased organ function of the digestive system, urinary tract, liver, and kidneys;
- decreased bone mineral content and muscle mass resulting in less lean body weight; and
- decreased physical fitness (e.g., cardiorespiratory endurance, strength, flexibility, muscular endurance, reaction and movement times, balance).

This last factor is of primary concern to anyone entering a health- and fitness-related profession. Studies show that older adults respond to appropriate endurance and strength training programs in a manner similar to younger adults. Hagberg and colleagues (1989) demonstrated a 22% increase in $\dot{V}O_2$max as the result of endurance training in 70- to 79-year-old males and females. Fiatarone and colleagues (1994) conducted a high-intensity strength-training program in older males and females with an average age of 87 years. The subjects were described as frail nursing home residents. Their study demonstrated strength gains of 113% and lean body mass gains of 3%. More important, the subjects dramatically improved their walking speed and stair climbing power, which are clinical measures of functional capacity. **Functional capacity,** the ability to perform the normal activities of daily living, is of prime importance for older adults if they are to maintain independent living status and a high quality of life (USDHHS 1996). The relationships among physical activity, physical fitness, and functional capacity produce the need for reliable and valid measurement of physical activity and physical fitness in older adults.

We have defined health-related physical fitness in adults as including cardiorespiratory endurance, body composition, and musculoskeletal fitness. Are the parameters of health-related physical fitness the same in older adults? Although the same factors are still important, an older individual's health and fitness should also include motor fitness factors such as balance, reaction time, and movement time in order to focus on maintaining functional capacity, activities of daily living, and overall quality of life.

For example, falls are a major health problem in older adults. Many older adults suffer fractures from falls, and the risk for mortality increases after falls occur. Indeed, many older adults become fearful of falling again, and they restrict their movements and lifestyle because of that fear. Strength and balance training in older adults improve fitness parameters that increase lean body and bone mass and lower the risk of falling and of suffering a fracture if one does fall (USDHHS 1996).

In concert with this broader definition of health-related physical fitness in older adults, Rikli and Jones (1999a and 1999b) at the Ruby Gerontology Center at the University of California, Fullerton, have developed a fitness battery for older adults. The test battery incorporates measures of strength, flexibility, cardiorespiratory endurance, motor fitness, and body composition (table 11.24). Local and national advisory panels of health and fitness experts have assisted in selecting and validating the test items. The theme of item selection was the important role that physical fitness has in delaying frailty and maintaining mobility in older adults, both of which are key factors in a healthy older population with a high quality of life. Test items were selected based on whether they met the following criteria. Does the item

- represent major functional fitness components (i.e., key physiologic parameters associated with the functions required for independent living);
- have acceptable test–retest reliability (>.80);
- have acceptable validity, with support for at least two to three types of validation: content-related, criterion-related, or construct-related;
- reflect usual age-related changes in physical performance; and
- require minimum equipment and space?

In addition, is the item

- able to detect physical changes attributable to training or exercise;
- able to be assessed on a continuous scale across wide ranges of functional ability (frail to highly fit);
- easy to administer and score;
- capable of being self- or partner-administered in the home setting;
- safe to perform without medical release for the majority of community-residing older adults;
- socially acceptable and meaningful; and
- reasonably quick to administer, with individual testing time of no more than 30 to 45 min?

Table 11.24 Fitness Parameters and Items of the Older Adult Fitness Test

Physical fitness parameter	Test item
Lower-body strength	30s chair stand
Upper-body strength	Arm curl
Lower-body flexibility	Chair sit-and-reach
Upper-body flexibility	Back scratch
Cardiorespiratory endurance	6-min walk or 2-min step-in-place test
Motor fitness Composite measure of power, speed, agility, and balance	8 ft up-and-go
Body composition	Body mass index (BMI)

Reliability and validity for the test battery are very acceptable. Test–retest reliability estimates exceeded .80 for all tests for older males and females. Criterion-related validity coefficients exceeded .70 for five of the seven performance tests for both males and females, and all seven performance tests demonstrated construct validity. Thus, the battery has been established with content validity and feasibility as guidelines. Further research has established sufficient reliability and validity for the battery.

OLDER ADULT FITNESS BATTERY

By definition, older people would not be determined apparently healthy. But Rikli and Jones (1999a) found that ambulatory, community-residing older adults with "no medical conditions that would be contraindicated for submaximal testing" (ACSM, 1995) who had not been advised to refrain from exercise could be safely tested. Rikli and Jones (1999b) stated that persons who should not take the tests without physician approval are those who

- ▸ have been advised by their doctors not to exercise because of a medical condition;
- ▸ have experienced chest pain, dizziness, or exertional angina (chest tightness, pressure, pain, heaviness) during exercise;
- ▸ have experienced congestive heart failure; or
- ▸ have uncontrolled high blood pressure (greater than 160/100).

Thus, with careful screening and caution, this test battery should be a valid and feasible tool for fitness assessment in older persons.

30-Second Chair Stand

Purpose
To assess lower-body strength.

Objective
To complete as many stands from a sitting position as possible in 30 s.

Equipment
Stopwatch
Straight-back or folding chair (without arms) with seat height approximately 17 in. (43.2 cm)

For safety purposes, place the chair against a wall, or stabilize it in some other way to prevent it from moving during the test. Begin the test with the participant seated in the middle of the chair, back straight, and feet flat on the floor. The participant crosses his or her arms at the wrists and holds them against the chest. On the go signal, the participant rises to a full stand and then returns to a seated position. The participant is encouraged to complete as many full stands as possible within 30 s. Following a demonstration by the tester, the participant should do a practice trial of one or two repetitions as a check for proper form, followed by one 30 s test trial.

Scoring

The score is the total number of stands executed correctly within 30 s. If the participant is more than halfway up at the end of 30 s, it counts as a full stand.

Arm Curl

Purpose

To assess upper-body strength.

Objective

To perform as many correctly executed arm curls as possible in 30 s.

Equipment

Wristwatch with a second hand
Straight-back or folding chair
Hand weight (dumbbell): 5 lb (2.3 kg) for women; 8 lb (3.6 kg) for men

Instructions

The participant sits on the chair, back straight, feet flat on the floor, and with the dominant side of the body close to the edge of the chair. He or she holds the weight at his or her side in the dominant hand using a handshake grip. The test begins with the arm in the down position beside the chair, perpendicular to the floor. At the go signal, the participant turns his or her palm up while curling the arm through a full range of motion and then returns the arm to a fully extended position. At the down position, the weight should return to the handshake grip position.

The examiner kneels (or sits in chair) next to the participant on the dominant arm side, placing his or her fingers on the person's midbiceps to stabilize the upper arm and to ensure that a full curl is made (participant's forearm should squeeze examiner's fingers;

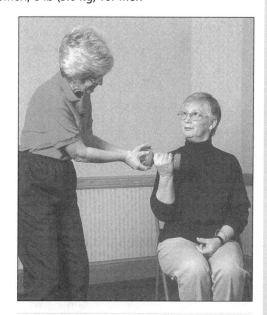

Figure 11.14 Arm curl test.

figure 11.14). The participant's upper arm must remain still throughout the test.

The examiner may also need to position his or her other hand behind the participant's elbow to help gauge when full extension has been reached and to prevent a backward swinging motion of the arm.

(continued)

The participant is encouraged to execute as many curls as possible within the 30 s time limit. Following a demonstration, give a practice trial of one or two repetitions to check for proper form, followed by one 30 s trial.

Scoring

The score is the total number of curls made correctly within 30 s. If the arm is more than halfway curled at the end of the 30 s, it counts as a curl.

6-Minute Walk Test

Purpose

To assess aerobic endurance.

Objective

To assess the maximum distance a participant can walk in 6 min along a 50 yd (45.7 m) course.

Equipment

Stopwatch
Long measuring tape (over 20 yd [18.3 m])
4 cones
20 to 25 Popsicle sticks for each participant
Adhesive name tags (for each participant)
Pen
Chalk or masking tape to mark the course

Set-up

Set up a 50 yd course marked in 5 yd (4.6 m) segments with chalk or tape (figure 11.15). Be sure the walking area is well lit and has a nonslippery, level surface. For safety purposes, position chairs at several points along the outside of the walkway.

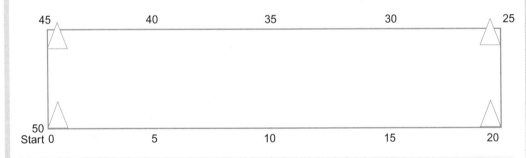

Figure 11.15 The 6-minute walk test uses a 50 yd course measured into 5 yd segments.

Instructions

On the go signal, participants walk as quickly as possible (without running) around the course as many times as they can within the time limit. Give a Popsicle stick (or other similar object) to the participant each time he or she completes one lap, or have someone tally laps. If testing two or more participants at once, stagger starting times 10 s apart so participants do not walk in clusters or pairs. Assign each participant a number to indicate the order of starting and stopping (you can use self-adhesive name tags to number each participant).

During the test, participants may stop and rest (sitting on the provided chairs) if necessary and resume walking. The tester should move to the center of the course after all participants have started and should call out elapsed time when participants are approximately half done, when 2 min are left, and when 1 min is left. At the end of their respective 6 min, instruct participants to stop and move to the right where an assistant will record the score. Conduct a practice test before the test day to assist with proper pacing and to improve scoring accuracy.

Discontinue the test if at any time a participant shows signs of dizziness, pain, nausea, or undue fatigue. At the end of the test, the participant should slowly walk around for about a minute to cool down.

Scoring

The score is the total number of yards walked (to the nearest 5 yd) in 6 min.

2-Minute Step-in-Place

Purpose

An alternative test to assess aerobic endurance.

Objective

To assess the maximum number of steps in place a participant can complete in 2 min.

Equipment

Stopwatch
Tape measure or length of 30 in. (76.2 cm) cord
Masking tape
Mechanical counter (if possible) to ensure accurate counting of steps

Instructions

The proper (minimum) knee-stepping height for each participant is at a level even with the midway point between the

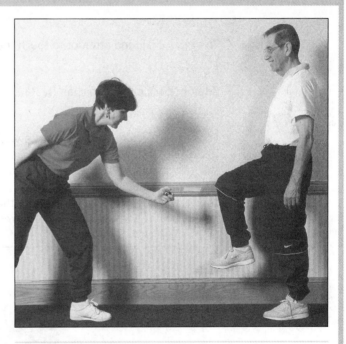

Figure 11.16 2-minute step-in-place.

patella (middle of the knee cap) and the iliac crest (top of the hip bone). You can determine this point using a tape measure or by simply stretching a piece of cord from the patella to the iliac crest and then doubling the cord over to determine the midway point. To monitor correct knee height when stepping, stack books on an adjacent table, or attach a ruler to a chair or wall with masking tape to mark proper knee height.

On the go signal, the participant begins stepping (not running) in place, completing as many steps as possible within 2 min (figure 11.16). The tester counts the number of steps completed, serves as a spotter in case of loss of balance, and ensures that the subject maintains proper knee height. As soon as proper knee height cannot be maintained, the participant is asked to stop or to stop and rest until proper form can be regained. Stepping may be resumed if the 2 min time period has not elapsed. If necessary, the participant can place one hand on a table

(continued)

or chair to assist in maintaining balance. To assist with pacing, participants should be told when 1 min has passed and when there are 30 s to go. At the end of the test, the participant should slowly walk around for about a minute to cool down.

A practice test before the test day can assist with proper pacing and improve scoring accuracy. On test day, the examiner should demonstrate the procedure and allow the participants to practice briefly.

Scoring

The score is the total number of steps taken within 2 min. Count only full steps (i.e., each time the right knee reaches the minimum height).

Chair Sit-and-Reach Test

Purpose

To assess lower-body (primarily hamstring) flexibility.

Objective

To sit in a chair and attempt to touch the toes with the fingers.

Equipment

Straight-back or folding chair (seat height approximately 17 in. [43.2 cm])

Instructions

For safety, place the chair against a wall and check to see that it remains stable when the person sits on the front edge. The participant sits in the chair and moves forward until he or she is sitting on the front edge. The crease between the top of the leg and the buttocks should be even with the edge of the chair seat. Keeping one leg bent and the foot flat on the floor, the participant extends the other leg (the preferred leg) straight in front of the hip, with the heel flat on the floor and the foot flexed at approximately 90° (figure 11.17).

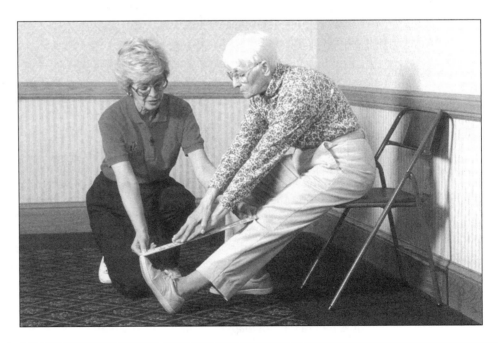

Figure 11.17 Chair sit-and-reach test.

With the leg as straight as possible (but not hyperextended), the participant slowly bends forward at the hip joint, keeping the spine as straight as possible and the head in line with the spine, not tucked. The participant attempts to touch his or her toes by sliding the hands, one on top of the other with the tips of the middle fingers even, down the extended leg. The reach must be held for 2 s. If the extended knee starts to bend, ask the participant to slowly sit back until the knee is straight, before scoring. Remind participants to exhale as they bend forward, to avoid bouncing or rapid, forceful movements, and to never stretch to the point of pain.

Following a demonstration, have the participants determine their preferred leg—the leg that yields the better score. Then give the participant two practice (stretching) trials on that leg, followed by two test trials.

Scoring

Using an 18 in. (45.7 cm) ruler, the scorer records the number of inches a person is short of reaching the toe (minus score) or reaches beyond the toe (plus score). The middle of the toe at the end of the shoe represents a zero score. Record both test scores to the nearest 0.5 in. (1.3 cm) and circle the best score. Be sure to indicate minus or plus on the scorecard.

Back Scratch

Purpose

To assess upper-body (shoulder) flexibility.

Objective

To reach behind the back with the hands to touch or overlap the fingers of both hands as much as possible.

Equipment

18 in. (45.7 cm) ruler (half of a yardstick)

Instructions

In a standing position, the participant places the preferred hand over the same shoulder, palm down and fingers extended, reaching down the middle of the back as far as possible (elbow pointed up). The hand of the other arm is placed behind the back, palm up, reaching up as far as possible in an attempt to touch or overlap the extended middle fingers of both hands.

Without moving the participant's hands, the tester helps to see that the middle fingers of each hand are directed toward each other (figure 11.18). The participants are not allowed to grab their fingers together and pull.

Following a demonstration, the participant determines the preferred hand and is given two stretching trials followed by two test trials.

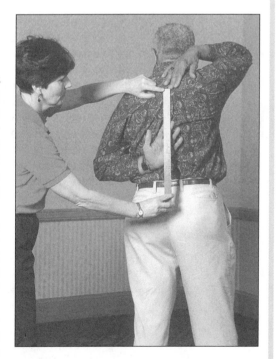

Figure 11.18 Back scratch test.

(continued)

Scoring

Measure the distance of overlap, or distance between the tips of the middle fingers to the nearest 0.5 in. (1.3 cm). Minus scores are given to represent the distance short of touching middle fingers; plus scores represent the degree of overlap of middle fingers. Record both test scores and circle the best score. The best score is used to evaluate performance. Be sure to indicate minus or plus on the scorecard. (It is important to work on flexibility on both sides of the body, but only the better side has been used in developing norms.)

Eight Feet Up-and-Go

Purpose

To assess physical mobility (involves power, speed, agility, and dynamic balance).

Objective

To stand, walk 16 ft (4.9 m), and sit back down in the fastest possible time.

Equipment

Stopwatch
Tape measure
Cone (or similar marker)
Straight-back or folding chair (seat height approximately 17 in. [43.2 cm])

Instructions

Position the chair against a wall or some other way to secure it during testing. Make sure the chair is in a clear unobstructed area, facing a cone marker exactly 8 ft (2.4 m) away (measured from a point on the floor even with the front edge of the chair to the back of the marker). There should be at least 4 ft (1.2 m) of clearance beyond the cone to allow ample turning room for the participant.

Remind the participant that this is a timed test and that the objective is to walk as quickly as possible (without running) around the cone and back to the chair. The participant starts in a seated position in the chair with erect posture, hands on thighs, and feet flat on the floor with one foot slightly in front of the other. On the go signal, the participant gets up from the chair (may push off thighs or chair), walks as quickly as possible around the cone, and returns to the chair and sits down. The tester should serve as a spotter, standing midway between the chair and the cone, ready to assist the participant in case of loss of balance. For reliable scoring, the tester must start the stopwatch on "go," whether or not the person has started to move, and stop the stopwatch at the exact instant the person sits in the chair.

Following a demonstration, the participant should walk through the test one time to practice and then perform 2 test trials. Remind participants that the stopwatch will not be stopped until the participant is fully seated in the chair.

Scoring

The score is the time elapsed from the go signal until the subject returns to a seated position on the chair. Record both test scores to the nearest tenth of a second. Circle the best score (lowest time), which will be the score used to evaluate performance.

Tables 11.25 and 11.26 provide percentile norms for females and males, respectively, on the eight items on the Older Adult Fitness Test Battery (Rikli and Jones 1999b).

Table 11.25   Age-Group Percentile—Women

			60-64 (n = 595)	65-69 (n = 1027)	70-74 (n = 1240)	75-79 (n = 937)	80-84 (n = 502)	85-89 (n = 305)	90-94 (n = 141)
	Chair stand								
percentile	10th	no.	9	9	8	7	6	5	2
	25th		12	11	10	10	9	8	4
	50th		15	14	13	12	11	10	8
	75th		17	16	15	15	14	13	11
	90th		20	18	18	17	16	15	14
	Arm curl		60-64 (n = 598)	65-69 (n = 1034)	70-74 (n = 1258)	75-79 (n = 953)	80-84 (n = 519)	85-89 (n = 329)	90-94 (n = 146)
percentile	10th	no.	10	10	9	8	8	7	6
	25th		13	12	12	11	10	10	8
	50th		16	15	15	14	13	12	11
	75th		19	18	17	17	16	15	13
	90th		22	21	20	20	18	17	16
	6-min walk		60-64 (n = 356)	65-69 (n = 617)	70-74 (n = 728)	75-79 (n = 513)	80-84 (n = 276)	85-89 (n = 152)	90-94 (n = 79)
percentile	10th	yd.	495	440	420	365	310	260	195
	25th		545	500	480	430	385	340	275
	50th		605	570	550	510	460	425	350
	75th		660	635	615	585	540	510	440
	90th		710	695	675	655	610	595	520
	2-min step		60-64 (n = 264)	65-69 (n = 491)	70-74 (n = 597)	75-79 (n = 489)	80-84 (n = 279)	85-89 (n = 167)	90-94 (n = 61)
percentile	10th	no.	60	57	53	52	46	42	31
	25th		75	73	68	68	60	55	44
	50th		91	90	84	84	75	70	58
	75th		107	107	101	100	91	85	72
	90th		122	123	116	115	104	98	85
	Chair sit-and-reach		60-64 (n = 591)	65-69 (n = 1037)	70-74 (n = 1250)	75-79 (n = 954)	80-84 (n = 514)	85-89 (n = 332)	90-94 (n = 151)
percentile	10th	in.	–3.0	–3.0	–3.5	–4.0	–4.5	–4.5	–7.0
	25th		–0.5	–0.5	–1.0	–1.5	–2.0	–2.5	–4.5
	50th		2.0	2.0	1.5	1.0	0.5	–0.5	–2.0
	75th		5.0	4.5	4.0	3.5	3.0	2.5	1.0
	90th		7.0	6.5	6.0	5.5	5.0	4.5	3.5
	Back scratch		60-64 (n = 592)	65-69 (n = 1030)	70-74 (n = 1246)	75-79 (n = 946)	80-84 (n = 517)	85-89 (n = 323)	90-94 (n = 148)
percentile	10th	in.	–5.5	–6.0	–6.5	–7.5	–8.0	–10.0	–11.5
	25th		–3.0	–3.5	–4.0	–5.0	–5.5	–7.0	–8.0
	50th		–0.5	–1.0	–1.5	–2.0	–2.5	–4.0	–4.5
	75th		1.5	1.5	1.0	0.5	0.0	–1.0	–1.0
	90th		4.0	3.5	3.0	3.0	2.5	2.0	2.0

(continued)

Table 12.25   (continued)

	8 ft up-and-go		60-64 (n = 594)	65-69 (n = 1033)	70-74 (n = 1244)	75-79 (n = 938)	80-84 (n = 497)	85-89 (n = 306)	90-94 (n = 142)
percentile	10th	s	6.7	7.1	8.0	8.3	10.0	11.1	13.5
	25th		6.0	6.4	7.1	7.4	8.7	9.6	11.5
	50th		5.2	5.6	6.0	6.3	7.2	7.9	9.4
	75th		4.4	4.8	4.9	5.2	5.7	6.2	7.3
	90th		3.7	4.1	4.0	4.3	4.4	5.1	5.3
	Body mass index		60-64 (n = 572)	65-69 (n = 1016)	70-74 (n = 1213)	75-79 (n = 916)	80-84 (n = 504)	85-89 (n = 337)	90-94 (n = 149)
percentile	10th	kg/m^2	19.6	19.8	20.3	19.8	19.6	19.5	18.3
	25th		22.8	23.0	23.1	22.5	22.0	21.8	21.1
	50th		26.3	26.5	26.1	25.4	24.7	24.3	24.1
	75th		29.8	30.0	29.1	28.3	27.4	26.8	27.1
	90th		33.0	33.2	31.9	31.0	30.0	29.0	29.5

Reprinted from Rikli and Jones (1999b).

Table 11.26   Age-Group Percentile—Men

	Chair stand		60-64 (n = 230)	65-69 (n = 460)	70-74 (n = 498)	75-79 (n = 434)	80-84 (n = 226)	85-89 (n = 108	90-94 (n = 71)
percentile	10th	no.	11	9	9	8	7	6	5
	25th		14	12	12	11	10	8	7
	50th		16	15	15	14	12	11	10
	75th		19	18	17	17	15	14	12
	90th		22	21	20	19	18	17	15
	Arm curl		60-64 (n = 229)	65-69 (n = 458)	70-74 (n = 498)	75-79 (n = 440)	80-84 (n = 232)	85-89 (n = 113)	90-94 (n = 71)
percentile	10th	no.	13	12	11	10	10	8	7
	25th		16	15	14	13	13	11	10
	50th		19	18	17	16	16	14	12
	75th		22	21	21	19	19	17	14
	90th		25	25	24	22	21	19	17
	6-min walk		60-64 (n = 144)	65-69 (n = 281)	70-74 (n = 294)	75-79 (n = 230)	80-84 (n = 130)	85-89 (n = 60)	90-94 (n = 48)
percentile	10th	yd	555	500	480	395	370	295	215
	25th		610	560	545	470	445	380	305
	50th		675	630	610	555	525	475	405
	75th		735	700	680	640	605	570	500
	90th		790	765	745	715	680	660	590

			60-64 (n = 92)	65-69 (n = 211)	70-74 (n = 225)	75-79 (n = 226)	80-84 (n = 119)	85-89 (n = 50)	90-94 (n = 38)
	2-min step								
percentile	10th	no.	74	72	66	56	56	44	36
	25th		87	88	80	73	71	59	52
	50th		101	101	95	91	87	75	69
	75th		115	116	110	109	103	91	86
	90th		128	130	125	125	118	106	102
	Chair sit-and-reach		60-64 (n = 228)	65-69 (n = 461)	70-74 (n = 494)	75-79 (n = 434)	80-84 (n = 231)	85-89 (n = 113)	90-94 (n = 74)
percentile	10th	in.	−6.0	−6.0	−6.5	−7.0	−6.0	−8.0	−9.0
	25th		−2.5	−3.0	−3.5	−4.0	−5.5	−5.5	−6.5
	50th		0.5	0.0	−0.5	−1.0	−2.0	−2.5	−3.5
	75th		4.0	3.0	2.5	2.0	1.5	0.5	0.5
	90th		6.5	6.0	5.5	5.0	4.5	3.0	2.0
	Back scratch		60-64 (n = 228)	65-69 (n = 457)	70-74 (n = 489)	75-79 (n = 430)	80-84 (n = 226)	85-89 (n = 113)	90-94 (n = 73)
percentile	10th	in.	−10.0	−10.5	−11.0	−12.0	−12.5	−12.5	−13.5
	25th		−6.5	−7.5	−8.0	−9.0	−9.5	−10.0	−10.5
	50th		−3.5	−4.0	−4.5	−5.5	−5.5	−6.0	−7.0
	75th		0.0	−1.0	−1.0	−2.0	−2.0	−3.0	−4.0
	90th		2.5	2.0	2.0	1.0	1.0	0.0	−1.0
	8 ft up-and-go		60-64 (n = 229)	65-69 (n = 461)	70-74 (n = 492)	75-79 (n = 436)	80-84 (n = 227)	85-89 (n = 106)	90-94 (n = 72)
percentile	10th	s	6.4	6.5	6.8	8.3	8.7	10.5	11.8
	25th		5.6	5.7	6.0	7.2	7.6	8.9	10.0
	50th		4.7	5.1	5.3	5.9	6.4	7.2	8.1
	75th		3.8	4.3	4.2	4.6	5.2	5.3	6.2
	90th		3.0	3.8	3.6	3.5	4.1	3.9	4.4
	Body mass index		60-64 (n = 228)	65-69 (n = 460)	70-74 (n = 491)	75-79 (n = 429)	80-84 (n = 230)	85-89 (n = 114)	90-94 (n = 69)
percentile	10th	kg/m^2	22.0	22.1	21.6	21.4	21.7	21.8	20.2
	25th		24.6	24.7	24.0	23.8	23.8	23.3	22.4
	50th		27.4	27.5	26.6	26.4	26.1	24.9	24.9
	75th		30.2	30.3	29.2	29.0	28.4	26.5	27.4
	90th		32.8	32.9	31.6	31.4	30.5	28.0	29.6

Adapted from Rikli and Jones (1999b).

SPECIAL POPULATIONS

Special populations include people with physical or mental disabilities or both. Lack of knowledge and understanding has clouded the history of society's care and attention to people with disabilities. Recently, new laws and attitudes have provided a more enlightened approach to society's interactions with disabled people. However, reliable and valid fitness assessment of disabled adults is not a well-researched or understood topic compared with work with other adults. Shephard (1990), in his book *Fitness in Special Populations,* provided a comprehensive and detailed presentation of exercise and fitness issues for disabled persons. He stated that the assessment of fitness in disabled persons should include the following:

▸ Anaerobic capacity and power

▸ Aerobic capacity

▸ Electrocardiographic response to exercise

▸ Muscular fitness, including strength, endurance, and flexibility

▸ Body composition

Appropriate fitness assessment requires accurate classification of the disability of the person and proper test and protocol selection. For example, if you wish to test the aerobic capacity of a person with paraplegia, you can select a wheelchair ergometer test or an arm ergometer test. Test protocols for deaf persons would not be applicable to blind persons. Specific training and education are required for fitness measurement in special populations. Shephard (1990) provided fitness information for people who are wheelchair disabled (e.g., those with paraplegia and amputations), blind, deaf, mentally retarded, or autistic or who have cerebral palsy, muscular dystrophy, or multiple sclerosis.

Meeting the fitness needs of persons with disabilities is a challenge for the future and an area of professional opportunity. From a measurement and evaluation perspective, research on reliable and valid fitness tests in different disabled populations is needed. Research is needed to provide a better understanding of fitness levels and the necessary level of fitness for improved health in different disabled groups.

Mastery Item 11.15

Go to the library and find a book or journal article about fitness testing in special populations. Evaluate the tests' characteristics in terms of reliability, validity, and norm development.

MEASURING PHYSICAL ACTIVITY

The measurement and evaluation of physical fitness have long been an area of research and practice in exercise science. Given the epidemiological evidence supporting the role of **physical activity**—the act of bodily movement that requires the contraction of muscles and the expenditure of energy—in health promotion and chronic disease prevention, the reliable and valid measurement of physical activity is essential in determining

▸ the amount of physical activity individuals get,

▸ the role of physical activity in health status,

▸ factors that relate to physical activity behavior, and

▸ the effect of interventions to promote physical activity.

Of course, physical fitness and physical activity are related in that genetic factors and physical activity behaviors are what determine one's physical fitness. Education and health promotion interventions can be developed to directly improve the quality and quantity of physical activity and as a result improve physical fitness. The Surgeon General's report on physical activity and health concluded that many Americans do not get enough physical activity to promote health and lower the risk of a variety of chronic diseases (USDHHS 1996). Is it the attribute of physical fitness or the behavior of physical activity that is health promoting? This question may never be answered. However, we must be able to measure both physical fitness and physical activity in reliable and valid manners.

Just as the well-known food pyramid presents guidelines for proper nutrition, the physical activity pyramid shown in figure 11.19 presents guidelines for physical activity. The basic concept is for individuals to develop physically active lifestyles that will produce sufficient levels of health-related fitness and will promote physical and mental health. The ACSM suggests that an individual must expend 150 kilocalories (kcal) per day (or 1000 kcal per week) to achieve a health benefit from physical activity. There are a variety of strategies for achieving a lifestyle that includes healthful amounts of physical activity. From a health standpoint, the most important message of the pyramid may be to limit the time a person is inactive because, as the figure indicates, physical inactivity is negatively related to all components of health-related physical fitness.

Physical activity, like other health behaviors, is a difficult behavior to assess with high reliability and validity. It is typically measured through either self-reports (e.g., diaries, logs, recall surveys, retrospective quantitative histories, and global self-reports) or direct monitoring (e.g., behavioral observation by a trained observer; electronic monitoring of heart rate or body motion; physiological monitoring using direct calorimetry in a metabolic chamber;

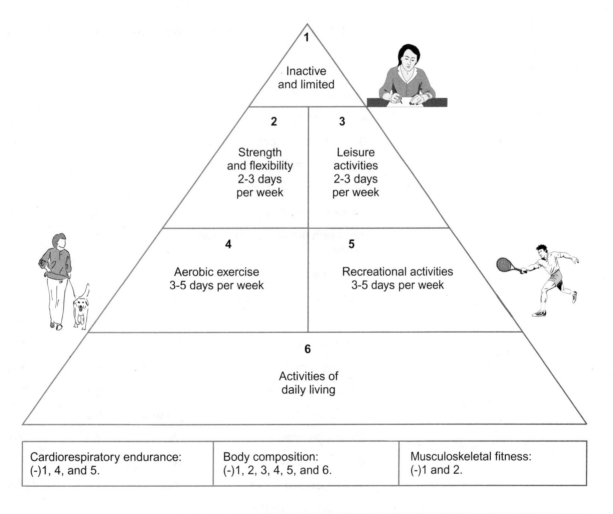

| Cardiorespiratory endurance: (-)1, 4, and 5. | Body composition: (-)1, 2, 3, 4, 5, and 6. | Musculoskeletal fitness: (-)1 and 2. |

Figure 11.19   The relationship of health-related fitness to the physical activity pyramid.

indirect calorimetry using a portable gas analysis system; and doubly labeled water; USDHHS 1996). There have been only limited success and research in establishing the reliability and validity of physical activity measurements, whereas, as the Surgeon General's report states, measures of physical fitness have good to excellent accuracy and reliability (USDHHS 1996). See Montoye et al. (1996) for a review of physical activity assessment tools.

"One of the principal difficulties in establishing the validity of a physical activity measure is the lack of a suitable 'gold-standard' criterion measure for comparison" (USDHHS 1996, 35). Because of this problem, many validation studies have used the physical fitness attribute of cardiorespiratory endurance as the criterion for concurrent validity estimation. The median correlation or concurrent validity coefficient between measures of cardiorespiratory fitness and self-report survey measures of physical activity was .41 in 12 research studies that used two popular physical activity survey instruments (USDHHS 1996, 35). Although the magnitude of this validity coefficient may seem low, remember that genetics accounts for at least 30% of cardiorespiratory endurance and that physical activity is still the most important predictor of cardiorespiratory endurance (Blair, Kannel, Kohl, and Goodyear 1989; Perusse, Tremblay, Leblanc, and Bouchard 1989). Not all measures of physical activity are appropriate for all populations. Table 11.27 provides a list of self-report and direct monitoring assessment procedures and the age groups for which they are suited.

Reliability of measures of physical activity is obviously an important issue. In a comprehensive study, Jacobs, Ainsworth, Hartman, and Leon (1993) examined the reliability and validity of 10 commonly used physical activity questionnaires and estimated the reliability

Table 11.27   **Physical Activity Assessment in Age Groups of the Population**

Type	Instrument	Children	Adults	Older persons
Self-report survey	Task specific diary	No	Yes	Yes
	Recall questionnaire	No	Yes	Yes
	Quantitative history	No	Yes	Yes
	Global self-report	No	Yes	Yes
Direct monitoring	Behavioral observation	No	Yes	Yes
	Job classification	No	Yes	No
	Heart rate monitor and motion sensor	Yes	Yes	Yes
	Heart rate monitor	Yes	Yes	Yes
	Electronic motion sensor	No	Yes	Yes
	Pedometer	No	Yes	Yes
	Gait assessment	Yes	Yes	Yes
	Accelerometers	Yes	Yes	Yes
	Horizontal time monitor	Yes	Yes	Yes
	Stabilometers	Infants	No	No
	Direct calorimetry	Yes	Yes	Yes
	Indirect calorimetry	No	Yes	Yes
	Doubly labeled water	Yes	Yes	Yes

Reprinted from Laporte, Montoye, and Caspersen (1985).

Accurately assessing physical activity requires measuring the frequency, duration, and intensity.

of measures of total physical activity and subsets of physical activity (i.e., light, moderate, heavy, leisure, and work). Test–retest reliability estimates across a 12-month interval for all questionnaires and 1-month interval for two of the questionnaires were reported. As expected, the reliability estimates for the 12-month interval were extremely variable, ranging from a low of .12 to a high of .93. The estimates for the month interval were much higher, ranging from .63 to .95.

An accurate assessment of physical activity includes measuring not only frequency and duration of the activity but also the intensity of the activity. Accurate determination of energy expenditure requires measures of all three components. Ainsworth and colleagues (1993, 2000) developed the Compendium of Physical Activities to aid in the assessment of energy expenditure attributable to physical activity. The compendium is a coding scheme for classifying a wide variety of physical activities into energy expenditure levels. The coding scheme uses five digits to classify physical activity by purpose, specific type, and intensity. The compendium standardizes measurements of physical activity and allows a better comparison and evaluation of different research findings.

Recently a very simple self-report method of assessing physical activity has demonstrated excellent reliability and criterion-related validity comparable to more complicated assessment techniques. The method uses a single-response scale to assess physical activity behaviors. Figure 11.20 depicts the five-level version of this single-response approach. The

- ▷ I don't exercise/walk regularly now and I do not plan to start in the near future.
- ▷ I don't exercise/walk regularly but I have been thinking of starting.
- ▷ I am doing moderate physical activity fewer than 5 times a week, or vigorous ones fewer than 3 times a week.
- ▷ I have been doing moderate physical activities 5 or more times a week, or vigorous ones at least 3 times a week, for the last 1 to 6 months.
- ▷ I have been doing moderate physical activities 5 more more times a week, or vigorous exercise at least 3 times a week, for 7 months or longer.

Figure 11.20   The single-response physical activity scale (five levels).

scale is based on the stages of change in physical activity behavior. Test–retest reliability estimates have exceeded .90 for the scale, and the correlation ($r = .48$) between the scale responses and treadmill assessments of aerobic capacity was consistent with validity coefficients for other self-report techniques.

Reliable and valid measurement of physical activity is a difficult but not impossible task. Selecting appropriate instruments, using standardized procedures, and performing pilot studies are necessary steps to achieve reliable and valid measurements. A simple construct validity concept holds true for assessments of physical activity: People classified as physically active with accepted measures in the epidemiological literature have lower rates of morbidity and mortality attributable to chronic disease than those who are inactive (USDHHS 1996). This consistent and persistent finding in the research literature supports the reachable goal of reliable and valid assessment of physical activity. Welk (2002) compiled detailed presentations of the variety of measurement techniques in the assessment of physical activity. This is an excellent source for developing a comprehensive picture of this challenging measurement area.

CERTIFICATION PROGRAMS

The ACSM is an American organization that leads in the research and promotion of all areas of exercise science. Any professional who is serious about a career in adult fitness programs that include reliable and valid fitness assessment should be an active member in this organization.

The ACSM offers a certification program that includes the following:

- ▶ ACSM Health/Fitness Instructor
- ▶ ACSM Exercise Specialist
- ▶ ACSM Registered Clinical Exercise Physiologist

The National Strength and Conditioning Association (NSCA) offers two certification programs: Certified Strength and Conditioning Specialist (CSCS) and Certified Personal Trainer (NSCA-CPT). By directing your education and preparation toward achieving these certifications, you will have verified, from a measurement and evaluation perspective, your ability to conduct reliable and valid fitness assessments. Howley and Franks (2003) and the ACSM (2005) provide an excellent and comprehensive resource for the professional in adult fitness and evaluation programs.

PRACTICAL HEALTH-RELATED FITNESS BATTERY

A large number of physical fitness tests have been presented in this chapter. People often want a quick, feasible, yet valid means of self-testing their fitness. An easily measured fitness test battery that can satisfy this desire can be developed from the tests presented in this chapter. Such a battery would include the following:

- ▶ Cardiovascular endurance—the Rockport 1-Mile Walk Test
- ▶ Body composition—the BMI and the waist–hip girth ratio
- ▶ Abdominal power—the YMCA 1-Minute Sit-Up Test
- ▶ Upper-body strength and endurance—the Canadian Standardized Test of Fitness Push-Up Test
- ▶ Hamstring flexibility—the YMCA Sit-and-Reach Test

Measurement and Evaluation Challenge

Jim had two major tasks before meeting with the YMCA executive director: to outline a suggested adult fitness assessment program and to investigate the certification he would seek. One of them is easily solved, given the resources the YMCA already has available: He can suggest implementing the YMCA's fitness testing program. The other, deciding the certification he should seek, is also fairly easily accomplished. Because he will be working with apparently healthy people who are not athletes, he probably should seek the ACSM's Health/Fitness Instructor certification. The fitness staff he may employ should have or seek certifications such as the ACSM's exercise leader or the NSCA's certified personal trainer. Additionally, in his research Jim has learned that the specific test that he might administer depends on the purposes of the test and the age and gender of the participants. He is on his way to having an effective fitness promotion and evaluation program with qualified personnel.

SUMMARY

The material presented in this chapter should provide you with a sound basis for understanding the various factors involved in assessing adult fitness and physical activity reliably and validly. Mastering the material does not make you qualified to administer fitness tests in all adult situations but is an important step to achieving those qualifications. If adult fitness programs are to be an important part of your professional career, seek appropriate education and training to obtain ACSM certification. Adult fitness testing, as with any other testing situation, requires test selection, preparation, practice trials, and attention to detail for sound measurement and evaluation of performance.

12

Physical Fitness and Activity Assessment in Youth

Measurement and Evaluation Challenge

Jo is an elementary school physical education teacher. For years, her school district has been using a locally developed fitness test to recognize children with a physical fitness award. The district's test included 600 yd walk–run, shuttle run, 50 yd dash, standing long jump, and sit-ups. The new physical education coordinator for the district has instituted a policy that requires the use of the FITNESSGRAM across the district because it is a "health-related" fitness test. Jo is unfamiliar with the FITNESSGRAM— its philosophy, components, and administration. Additionally, she has heard that the FITNESSGRAM assesses physical activity levels as well as physical fitness. The award system for the FITNESSGRAM is reportedly different from the one with which she is familiar. Previously, student awards were based on percentile performance, but the FITNESSGRAM uses health-related criterion-referenced standards. Jo sees that she has much to do to prepare for the new school year.

Objectives

After studying this chapter, you will be able to

▶ discuss the status of youth fitness in the United States;

▶ differentiate athletic, or motor fitness, and health-related fitness in youth;

▶ describe similarities and differences in youth fitness batteries;

▶ administer fitness tests for youth in a reliable and valid manner;

▶ address issues in physical fitness assessment in special populations; and

▶ assess physical activity in youth.

The levels of physical activity and fitness of youth in the United States are a controversial issue in the field of human performance. Various political leaders, physical educators, and fitness experts proclaim that the nation's youth are dangerously inactive and physically unfit. Historically, awareness of and concern for poor physical fitness levels in youth can be tracked to World Wars I and II, when many young men were deemed not physically fit enough to serve in the armed services during times of national emergency. In the early 1950s, physical fitness testing indicated that European children had higher levels of physical fitness than American children. This led President Eisenhower to establish what has become the President's Council on Physical Fitness and Sports. The council, along with the American Alliance of Health, Physical Education and Recreation (AAHPER), established a national youth fitness testing program, which included the Presidential Physical Fitness Award. This program also developed the AAHPER Youth Fitness Test. This test is an example of a **youth fitness battery,** which combines several fitness tests to provide an overall assessment of physical fitness. Several youth fitness test batteries are now available that form the basis for youth fitness assessment in the United States. In terms of measurement and evaluation, there has been a strong shift toward an increased emphasis on the promotion and assessment of physical activity (the behavior) instead of on physical fitness assessment (the outcome).

Mastery Item 12.1

Were you given physical fitness tests during your youth? What tests did you take? What items did you complete? Did the test results provide a reliable and valid estimate of your physical fitness?

The National School Population Fitness Survey, conducted and published by the President's Council on Physical Fitness and Sports (Reiff et al. 1985), concluded in part that

> the physical performance of children and youth in 1985 was not much different from that of youth in 1975. Extrapolated to the entire population, the study data show there is still a low level of performance in important components of physical fitness by millions of our youth. (p. 45)

The same survey indicated that "large numbers of boys and girls" had a low level of performance on cardiorespiratory endurance tests. The Office of Disease Prevention and Health Promotion, an agency of the U.S. Public Health Service, conducted the National Children and Youth Fitness Study (NCYFS; Pate, Ross, Dotson, and Gilbert 1985) to evaluate the fitness levels of children. The skinfold measures in 1985 were significantly greater than skinfold measurements of children from the 1960s (Pate, Ross, Dotson, and Gilbert 1985). Thus, the more recently born children appeared to have higher percent body fat. This information indicated that many youth in the United States are unfit and that fitness levels may be decreasing.

However, conflicting information indicates that youth fitness levels are not necessarily poor and may not be decreasing. Looney and Plowman (1990) conducted a study in which they applied the criterion-referenced standards of the FITNESSGRAM youth fitness battery to the test results of the NCYFS I (1985) and NCYFS II (1987). They found that 80% to 90% of the children could pass the body composition standards, depending on the age of the child. Boys and girls demonstrated high passing rates for the 1 mi (1.6 km) run test in some age groups (figure 12.1).

Rowland (1990), a respected researcher in exercise science with children, summarized this issue, stating that there is conflicting information. The conclusion that youth fitness levels are low and declining is not clearly supported in the literature. In 1992, a forum on youth fitness was published in *Research Quarterly for Exercise and Sport* that included seven papers by recognized experts. These papers did not present a consensus on the levels of youth fitness and the trend of those fitness levels. Blair (1992) estimated that 20% of American youth, or between 8 and 9 million children, have potentially unhealthy fitness levels. The

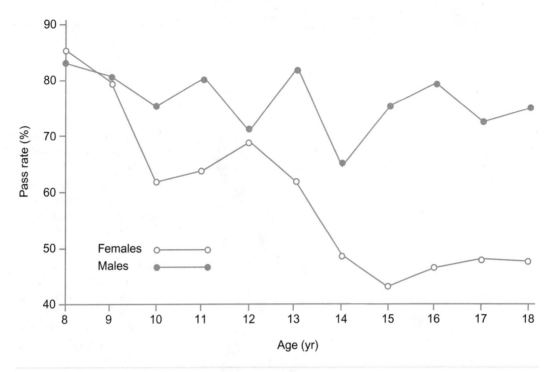

Figure 12.1 Passing rates for the FITNESSGRAM 1 mi run.

National Center for Health Statistics (NCHS) has presented an alarming trend in overweight children and adolescents in the United States (NCHS 2002). This trend indicates a growth in the percentage of overweight children and adolescents from 4% to 5% in the 1960s to more than 15% in 1999-2000 (NCHS 2002). In relation to physical activity, some of the most recent data have been provided by the Centers for Disease Control and Prevention (CDC) through the Youth Risk Behavior Surveillance System. In 2003, the CDC conducted a national school-based Youth Risk Factor Behavior Survey for grades 9 through 12 that resulted in a sample size of 15,240 students from 152 schools across the United States (Grunbaum et al. 2004). Figure 12.2 shows that many of the students did not participate in vigorous, moderate, and strength-related physical activities. It is also apparent that the majority of students are not

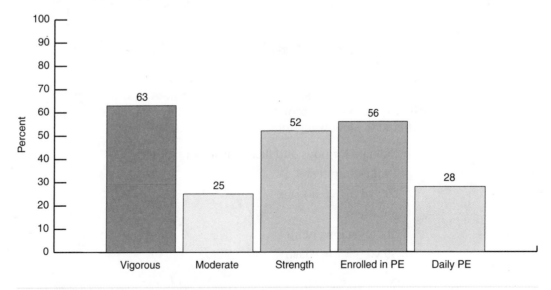

Figure 12.2 Physical activity in grades 9 through 12 in the United States.

If a child passes a standard on one test but fails it on another, is he or she physically fit?

enrolled in and attending daily physical education classes. Only a minority are sufficiently physically active during physical education classes. We also know that for three of the variables, the percentages of children from the Youth Risk Factor Behavior Survey are far below the objectives set by the Healthy People 2010 national health initiative. In the spring of 2003, President Bush announced the Healthier US Initiative. The initiative had four major components: physical fitness, nutrition, prevention (preventive medical screenings), and make healthy choices (avoid risky behaviors). As a professional in human performance, you should know that there are many children with low levels of fitness and poor physical activity levels. National and international leaders and public health professionals are calling for improved physical activity and fitness for those children who are inactive and unfit. However, most children are physically active and can pass fitness test standards. As you will learn, different tests have different standards. A child may pass the standard on one test and yet fail it on another. Whether he or she is considered physically fit depends on the standards and the tests used in the evaluation.

Mastery Item 12.2

How would you describe the status of physical fitness of American youth?

HEALTH-RELATED FITNESS AND MOTOR FITNESS

The original AAHPER Youth Fitness Test developed in the 1950s and revised in the 1970s included not only tests of health-related fitness described in chapter 11 but also tests of **athletic fitness,** or **motor fitness**—physical fitness related to sport performance. The test included the following items:

- ▸ Pull-ups for boys and flexed arm hang for girls
- ▸ 1 min sit-up test
- ▸ 600 yd (548.6 m) run
- ▸ Shuttle run
- ▸ Standing long jump
- ▸ 50 yd (45.7 m) dash

As you can see, only the first three items might be considered health-related based on the definition of the ACSM (1991) given in chapter 11 (p. 225). However, even the 600 yd (548.6 m) run is not a good measure of aerobic capacity. The other items relate to the ability to participate in sports and athletics.

In the 1970s, another youth fitness test was developed for the American Alliance of Health, Physical Education, Recreation and Dance (AAHPERD). This test was called the AAHPERD Health-Related Physical Fitness Test (AAHPERD 1980) and included measures of the following components:

- Cardiovascular endurance (distance runs)
- Body composition (skinfolds)
- Musculoskeletal function
    - Abdominal function (1 min sit-up test)
    - Low back and hamstring function (sit-and-reach test)

The test items in this battery are components consistent with the present definition of health-related fitness. In the future, youth fitness evaluation will focus on administering health-related fitness batteries. The principal programs in the United States are the FITNESSGRAM (Cooper Institute for Aerobics Research 1987, 1992, 1999, and 2004) and the President's Challenge (PCPFS 1999). We will discuss each of these testing programs later in the chapter.

Mastery Item 12.3

Write your own definitions of health-related and athletic, or motor, fitness. What items would be on test batteries for each of these?

NORMS VERSUS CRITERION-REFERENCED STANDARDS

Another important change in youth fitness testing has been the change from **norm-referenced fitness standards**—that is, levels of performance relative to a specifically defined group—to **criterion-referenced fitness standards**—that is, specific predetermined levels of performance—for determining fitness achievement. The AAHPERD Health-Related Physical Fitness Test was published with percentile norms (AAHPERD 1980). Children who scored below the 50th percentile on a test or tests were encouraged to train to be able to achieve that level. The FITNESSGRAM was the first nationally recognized fitness test battery that used health-related criterion-referenced fitness standards (Cooper Institute for Aerobics Research 1987, 1992, 1999). The President's Council on Physical Fitness and Sports (PCPFS 1999) uses criterion-referenced standards for evaluation of youth fitness scores in the President's Challenge program. As you will see later in the chapter, although these programs all use criterion-referenced standards, the passing standards are different for the same tests in the different programs. Safrit and Looney (1992) and others have taken the position that health-related criterion-referenced standards are useful and appropriate but that normative data on youth fitness scores are also useful for (a) evaluating a program, (b) identifying excellence in achievement, and (c) identifying the current status of individuals either locally or nationally.

Mastery Item 12.4

Why do you think these different programs use different standards for evaluating the youth fitness test scores? How would you explain the standards to your students, their parents, and your supervisors?

NORMATIVE DATA

Results of the NCYFS I and II, mentioned earlier, were published in the *Journal of Physical Education, Recreation and Dance* in two segments (NCYFS I: Ross, Dotson, Gilbert, and Katz 1985; NCYFS II: Ross et al. 1987). Supported by the U.S. Public Health Service, the NCYFS strove to develop the best normative data on youth fitness. The children in the study represented a national probability sample in order to provide a sample representative of children in the United States. Previous databases had been considered questionable because they represented large "convenience samples" that may have biased estimates or not fully represented the population of children in the United States. NCYFS I resulted in norms for youth aged 10 to 18; NCYFS II produced norms for children aged 6 to 9.

Although youth fitness evaluation has moved away from the use of norm-referenced standards to criterion-referenced standards, we agree with Safrit and Looney (1992) that the availability of normative data has value in youth fitness assessment. Tables 12.1 through 12.5 provide you with percentile norms for the various tests and measurements gathered in NCYFS I and II. This gives you a good summary of the best normative data available on youth in the United States.

Table 12.1 Percentile Norms for Males and Females on Distance Runs (min:sec)

Male percentile	Age (years)												
	6	7	8	9	10	11	12	13	14	15	16	17	18
90	4:27	4:11	8:46	8:10	8:13	7:25	7:13	6:48	6:27	6:23	6:13	6:08	6:10
75	4:52	4:33	9:29	9:00	8:48	8:02	7:53	7:14	7:08	6:52	6:39	6:40	6:42
50	5:23	5:00	10:39	10:10	9:52	9:03	8:48	8:04	7:51	7:30	7:27	7:31	7:35
25	5:58	5:35	12:14	11:44	11:00	10:32	10:13	9:06	9:10	8:30	8:18	8:37	8:34
10	6:40	6:20	14:05	13:37	12:27	12:07	11:48	10:38	10:34	10:13	9:36	10:43	10:50
Female percentile													
90	4:46	4:32	9:39	9:08	9:09	8:45	8:34	8:27	8:11	8:23	8:28	8:20	8:22
75	5:13	4:54	10:23	9:50	10:09	9:56	9:52	9:30	9:16	9:28	9:25	9:26	9:31
50	5:44	5:25	11:32	11:13	11:14	11:15	10:58	10:52	10:32	10:46	10:34	10:34	10:51
25	6:14	6:01	12:59	12:45	12:52	12:54	12:33	12:17	11:49	12:18	12:10	12:03	12:14
10	6:51	6:38	14:48	14:31	14:20	14:35	14:07	13:45	13:13	14:07	13:42	13:46	15:18

Note: Subjects aged 6 to 7 run 1/2 mi; others run 1 mi.

Table 12.2 Percentile Norms for Males and Females for Sum of Skinfolds (mm)

Male percentile	Age (years)												
	6	7	8	9	10	11	12	13	14	15	16	17	18
90	12	12	12	12	12	12	12	11	12	12	12	13	13
75	14	14	14	15	14	14	14	13	13	14	14	14	15
50	16	17	18	21	17	18	17	17	17	17	17	17	18
25	20	22	24	29	24	25	24	23	22	22	22	22	24
10	27	32	37	40	35	36	38	34	33	32	30	30	30
Female percentile													
90	15	15	15	16	13	14	15	15	17	19	19	20	19
75	18	18	19	20	16	17	18	19	20	23	22	23	22
50	21	22	24	26	20	21	22	24	26	28	26	28	27
25	27	28	33	35	27	30	29	31	33	34	33	36	34
10	33	37	43	45	36	40	40	43	40	43	42	42	42

Note: For subjects aged 6 to 9, values represent the sum of triceps and calf skinfolds; for other subjects, values are the sum of triceps and subscapular skinfolds.

Table 12.3 Percentile Norms for Males and Females on the 1-Minute Bent-Knee Sit-Up Test (number completed)

Male percentile	Age (years)												
	6	7	8	9	10	11	12	13	14	15	16	17	18
90	28	32	35	39	47	48	50	52	52	53	55	56	54
75	24	28	30	33	40	41	44	46	47	48	49	50	50
50	19	23	26	28	34	36	38	40	41	42	43	43	43
25	14	18	20	23	28	30	32	32	35	36	38	37	36
10	9	12	15	16	22	22	25	28	30	31	32	31	31
Female percentile													
90	28	33	34	36	43	42	46	46	47	45	49	47	47
75	23	27	29	31	37	37	40	40	41	40	40	40	40
50	18	21	25	26	31	32	33	33	35	35	35	36	35
25	14	16	19	21	25	26	28	27	29	30	30	30	30
10	6	11	13	15	20	20	21	21	23	24	23	24	24

Table 12.4 Percentile Norms for Males and Females on the Pull-Up Test (number completed)

Male percentile	Age (years)												
	6	7	8	9	10	11	12	13	14	15	16	17	18
90	15	19	20	20	8	8	8	10	12	14	14	15	16
75	10	13	14	15	4	5	5	7	8	10	12	12	13
50	6	8	10	10	1	2	3	4	5	7	9	9	10
25	3	4	6	6	0	0	0	1	2	4	6	5	6
10	1	1	3	3	0	0	0	0	0	1	2	2	3
Female percentile													
90	2	13	16	17	17	3	3	2	2	2	2	2	2
75	1	9	11	11	12	1	1	1	1	1	1	1	1
50	0	6	7	8	9	0	0	0	0	0	0	0	0
25	0	3	4	4	4	0	0	0	0	0	0	0	0
10	0	0	1	1	1	0	0	0	0	0	0	0	0

Table 12.5 Percentile Norms for Males and Females on the Sit-and-Reach Test (inches)
Note: Subjects aged 6 tp 9 performed modified pull-ups.

Male percentile	Age (years)												
	6	7	8	9	10	11	12	13	14	15	16	17	18
90	16.0	16.0	16.0	15.5	16.0	16.5	16.0	16.5	17.5	18.0	19.0	19.5	19.5
75	15.0	15.0	14.5	14.5	14.5	15.0	15.0	15.0	15.5	16.5	17.0	17.5	17.5
50	13.5	13.5	13.5	13.0	13.5	13.0	13.0	13.0	13.5	14.0	15.0	15.5	15.0
25	12.0	11.5	11.5	11.0	11.5	11.5	11.0	11.0	11.0	12.0	13.0	13.0	13.0
10	10.5	10.0	9.5	9.5	10.0	9.5	8.5	9.0	9.0	9.5	10.0	10.5	10.0
Female percentile													
90	16.5	17.0	17.0	17.0	17.5	18.0	19.0	20.0	19.5	20.0	20.5	20.5	20.5
75	15.5	16.0	16.0	16.0	16.5	16.5	17.0	18.0	18.5	19.0	19.0	19.0	19.0
50	14.0	14.5	14.0	14.0	14.5	15.0	15.5	16.0	17.0	17.0	17.5	18.0	17.5
25	12.5	13.0	12.5	12.5	13.0	13.0	14.0	14.0	15.0	15.5	16.0	15.5	15.5
10	11.5	11.5	11.0	11.0	10.5	11.5	12.0	12.0	12.5	13.5	14.0	13.5	13.0

YOUTH FITNESS TEST BATTERIES

As we have discussed, assessing physical fitness in youth has changed from a motor fitness emphasis to a health-related emphasis in the United States and abroad. Youth fitness test batteries are used to measure and evaluate physical fitness. Currently, several test batteries are available for youth fitness assessment. Table 12.6 lists four different youth fitness batteries, the items contained on each, and the organization where you can obtain the battery. Three elements are present in all the test batteries:

1. Health-related fitness items
2. Criterion-referenced standards for each test
3. Motivational awards

All batteries use criterion-referenced standards for passing or failing a test and achieving the fitness awards. The FITNESSGRAM and the President's Challenge use two sets of standards for each test: a lower passing standard for achievement of a minimally accepted level of fitness and a higher standard to provide motivation and a greater challenge to the student. Safrit and Pemberton (1995) provided a complete guide to youth fitness testing.

Youth Fitness Test Contact Information

FITNESSGRAM

Human Kinetics
P.O. Box 5076
Champaign, IL 61825

The President's Challenge

Poplars Research Center
400 E. 7th Street
Bloomington, IN 47405-3085

Eurofit

Vrije Universiteit en Universiteit van Amsterdam
Meibergdreef 15-1105
AZ Amsterdam

Mastery Item 12.5

What differences exist between the President's Challenge Health Fitness Award and the Presidential Award?

The President's Challenge program offers a flexible approach to evaluating the physical fitness test performance of the child. The following delineates the evaluation and award scheme:

▶ The Presidential Physical Fitness Award—for children who reach a high standard of performance (85th percentile on 1985 School Population Fitness Survey) on five required test items listed in table 12.6.

▶ The National Physical Fitness Award—for children who reach a moderate standard of performance (50th percentile on 1985 School Population Fitness Survey) on five required test items listed in table 12.6.

▶ The Participant Physical Fitness Award—for students whose scores fall below the 50th percentile on one or more of the test items.

▶ The President's Challenge Health Fitness Award—for children who achieve a specific health-related criterion on five required test items listed in table 12.6. Qualifying standards for the President's Challenge are presented in tables 12.7 to 12.9.

Table 12.6   **Youth Fitness Test Batteries**

	FITNESSGRAM	President's Challenge (Presidential, National, or Participant Physical Fitness Award)	President's Challenge (Health Fitness Award)	Eurofit
Aerobic capacity	PACER (r) Mile run Walk test	1 mi run 1/4 mi run (6-7 years) 1/2 mi run (8-9 years)	1 mi run 1/4 mi run (6-7 years) 1/2 mi run (8-9 years)	Endurance shuttle run Bike ergometer test
Body composition	Skinfolds (r) Body mass index		Body mass index	Height Weight Skinfolds
Abdominal strength and endurance	Curl-ups (r)	Curl-ups Partial curl-ups	Partial curl-ups	Sit-ups
Upper-body strength and endurance	Push-ups (r) Modified pull-ups Pull-ups Flexed arm hang	Pull-ups Push-ups	Push-ups Pull-ups	Hand grip Bent arm hang
Trunk extensor strength and endurance	Trunk lift (r)			
Flexibility	Back saver sit-and-reach Shoulder stretch	V-sit reach Sit-and-reach	V-sit reach Sit-and-reach	Sit-and-reach
Running speed and agility		Shuttle run		Shuttle run
Speed of limb movement				Plate tapping
Power				Standing long jump
Balance				Flamingo balance

Note: r = recommended; PACER = progressive aerobic cardiovascular endurance run. Test battery contact information is in the highlight box on page 283.

Table 12.7 Presidential Physical Fitness Award (85th Percentile)

Age	Curl-ups (*n* in 1 min)	Shuttle run (s)	V-sit reach (in.)	Sit-and-reach (cm)	One-mi run (min:sec)	Distance option 1/4 mi (min:sec)	Distance option 1/2 mi (min:sec)	Pull-ups (*n*)
Boys								
6	33	12.1	+3.5	31	10:15	1:55		2
7	36	11.5	+3.5	30	9:22	1:48		4
8	40	11.1	+3.0	31	8:48		3:30	5
9	41	10.9	+3.0	31	8:31		3:30	5
10	45	10.3	+4.0	30	7:57			6
11	57	10.0	+4.0	31	7:32			6
12	50	9.8	+4.0	31	7:11			7
13	53	9.5	+3.5	33	6:50			7
14	56	9.1	+4.5	36	6:26			10
15	57	9.0	+5.0	37	6:20			11
16	56	8.7	+6.0	38	6:08			11
17	55	8.7	+7.0	41	6:06			13
Girls								
6	32	12.4	+5.5	32	11:20	2:00		2
7	34	12.1	+5.0	32	10:36	1:55		2
8	38	11.8	+4.5	33	10:02		3:58	2
9	39	11.1	+5.5	33	9:30		3:53	2
10	40	10.8	+6.0	33	9:19			3
11	42	10.5	+6.5	34	9:02			3
12	45	10.4	+7.0	36	8:23			2
13	46	10.2	+7.0	38	8:13			2
14	47	10.1	+8.0	40	7:59			2
15	48	10.0	+8.0	43	8:08			2
16	45	10.1	+9.0	42	8:23			1
17	44	10.0	+8.0	42	8:15			1

From the Presidential Physical Fitness Award.

1/4 and 1/2 norms are reprinted, by permission, from Amaateur Athletic Union, 1998, *Amateur Athletic Union Physical Fitness Program*.

Additional norms information can be seen on http://www.presidentschallenge.org/misc/downloads.aspx.

Table 12.8 National Physical Fitness Award (50th Percentile)

Age	Curl-ups (n in 1 min)	Shuttle run (s)	V-sit reach (in.)	Sit-and-reach (cm)	One-mi run (min:sec)	Distance option 1/4 mi (min:sec)	Distance option 1/2 mi (min:sec)	Pull-ups (n)	Flexed arm hang (s)
Boys									
6	22	13.3	+1.0	26	12:36	2:21		1	6
7	28	12.8	+1.0	25	11:40	2:10		1	8
8	31	12.2	+0.5	25	11:05		4:22	1	10
9	32	11.9	+1.0	25	10:30		4:14	2	10
10	35	11.5	+1.0	25	9:48			2	12
11	37	11.1	+1.0	25	9:20			2	11
12	40	10.6	+1.0	26	8:40			2	12
13	42	10.2	+.05	26	8:06			3	14
14	45	9.9	+1.0	28	7:44			5	20
15	45	9.7	+2.0	30	7:30			6	30
16	45	9.4	+3.0	30	7:10			7	28
17	44	9.4	+3.0	34	7:04			8	30
Girls									
6	23	13.8	+2.5	27	13:12	2:26		1	5
7	25	13.2	+2.0	27	12:56	2:21	4:56	1	6
8	29	12.9	+2.0	28	12:30		4:50	1	8
9	30	12.5	+2.0	28	11:52			1	8
10	30	12.1	+3.0	28	11:22			1	8
11	32	11.5	+3.0	29	11:17			1	7
12	35	11.3	+3.5	30	11:05			1	7
13	37	11.1	+3.5	31	10.23			1	8
14	37	11.2	+4.5	33	10:06			1	9
15	36	11.0	+5.0	36	9:58			1	7
16	35	10.9	+5.5	34	10:31			1	7
17	34	11.0	+4.5	35	10:22			1	7

From the Presidential Physical Fitness Award.

1/4 and 1/2 norms are reprinted, by permission, from Amaateur Athletic Union, 1998, *Amateur Athletic Union Physical Fitness Program*.

Additional norms information can be seen on http://www.presidentschallenge.org/misc/downloads.aspx.

Table 12.9 Health Fitness Award

Age	Partial curl-ups (n)	One-mi run (min:sec)	Distance option 1/4 mi (min:sec)	Distance option 1/2 mi (min:sec)	V-sit reach (in.)	Sit-and-reach (cm)	Right-angle push-ups (n)	Pull-ups (n)	BMI (range)
Boys									
6	12	13:00	2:30		1	21	3	1	13.3-19.5
7	12	12:00	2:20		1	21	4	1	13.3-19.5
8	15	11:00		4:45	1	21	5	1	13.4-20.5
9	15	10:00		4:35	1	21	6	1	13.7-21.4
10	20	9:30			1	21	7	1	14.0-22.5
11	20	9:00			1	21	8	2	14.0-23.7
12	20	9:00			1	21	9	2	14.8-24.1
13	25	8:00			1	21	10	2	15.4-24.7
14	25	8:00			1	21	12	3	16.1-25.4
15	30	7:30			1	21	14	4	16.6-26.4
16	30	7:30			1	21	16	5	17.2-26.8
17	30	7:30			1	21	18	6	17.5-27.5
Girls									
6	12	13:00	2:50		2	23	3	1	13.1-19.6
7	12	12:00	2:40		2	23	4	1	13.1-19.6
8	15	11:00		5:35	2	23	5	1	13.2-20.7
9	15	10:00		5:25	2	23	6	1	13.5-21.4
10	20	10:00			2	23	7	1	13.8-22.5
11	20	10:00			2	23	7	1	14.1-23.2
12	20	10:00			2	23	8	1	14.7-24.2
13	25	10:00			3	23	7	1	15.5-25.3
14	25	10:00			3	23	7	1	16.2-25.3
15	30	10:00			3	23	7	1	16.6-26.5
16	30	10:00			3	23	7	1	16.8-26.5
17	30	10:00			3	23	7	1	17.1-26.9

From the Presidential Physical Fitness Award.

Note: BMI = body mass index. Criterion standards listed are adapted from Amateur Athletic Union Physical Fitness Program; AAHPERD Physical Best; Cooper Institute for Aerobic Research, FITNESSGRAM; Corbin, C., & Lindsey, R., *Fitness for Life,* 4th edition; and YMCA Youth Fitness Test.

The Eurofit physical fitness test battery, which is a combination of health-related and motor fitness components, is also presented in table 12.6. The Eurofit is widely used on the European continent. Eurofit combines health-related and performance-related items with additional motor ability items (e.g., plate tapping, power, and balance).

The standards themselves are quite different among the test batteries. We examine this issue in more detail later in this chapter. Another important point about the standards is that some of the passing standards are associated with a health-related criterion, whereas others are based on norms with the assumption that a higher fitness level is healthier than a lower one and that performance evaluation should be based on comparison with peers. The validity of these specific criterion-referenced standards in indicating a reduced risk of a specific disease in childhood or in adulthood has not been demonstrated in a scientific manner (Plowman 1992a).

Jo, our teacher from the measurement challenge, will need to talk with her colleagues and discuss the implications for testing and evaluation that will result from using the FIT-NESSGRAM.

FITNESSGRAM

We highlight the FITNESSGRAM physical fitness battery because it has been validated and is used by millions of children. The FITNESSGRAM is a physical fitness battery that includes health-related criterion-referenced standards. Each item has two standards of achievement: one reflecting a minimally acceptable level for health, and a higher one to motivate students and provide them with a fitness challenge. This results in a "Healthy Fitness Zone." The program includes optional tests, effective computer software and support, and an awards program. Following are five items from the battery. The criterion-referenced passing standards for each test are provided in table 12.10.

The one-mile run and the one-mile walk test are alternatives to FITNESSGRAM's PACER.

Table 12.10   FITNESSGRAM Health-Related Criterion-Referenced Standards

Age	PACER (no. of laps)		Skinfolds (percent fat)		Curl-ups (number completed)		Trunk lift (in.)		Push-ups (number completed)	
Boys										
5			25	10	2	10	6	12	3	8
6			25	10	2	10	6	12	3	8
7			25	10	4	14	6	12	4	10
8			25	10	6	20	6	12	5	13
9			25	10	9	24	6	12	6	15
10	23	61	25	10	12	24	9	12	7	20
11	23	72	25	10	15	28	9	12	8	20
12	32	72	25	10	18	36	9	12	10	20
13	41	72	25	10	21	40	9	12	12	25
14	41	83	25	10	24	45	9	12	14	30
15	51	94	25	10	24	47	9	13	16	35
16	61	94	25	10	24	47	9	12	18	35
17	61	94	25	10	24	47	9	12	18	35
17+	61	94	25	10	24	47	9	12	18	35
Girls										
5			32	17	2	10	6	12	3	8
6			32	17	2	10	6	12	3	8
7			32	17	4	14	6	12	4	10
8			32	17	6	20	6	12	5	13
9			32	17	9	22	6	12	6	15
10	15	41	32	17	12	26	9	12	7	15
11	15	41	32	17	15	29	9	12	7	15
12	23	41	32	17	18	32	9	12	7	15
13	23	51	32	17	18	32	9	12	7	15
14	23	51	32	17	18	32	9	12	7	15
15	23	51	32	17	18	35	9	12	7	15
16	32	61	32	17	18	35	9	12	7	15
17	41	61	32	17	18	35	9	12	7	15
17+	41	61	32	17	18	35	9	12	7	15

Note: Upper and lower limits of healthy fitness zones are included.

Adapted from The Cooper Institute for Aerobics Research (1999).

PACER

Purpose

To measure aerobic capacity. (PACER stands for progressive aerobic cardiovascular endurance run.)

Objective

To run as long as possible back and forth across a 20 m (21.9 yd) space at a specified pace that gets faster each minute.

Equipment and facilities

A flat, nonslippery surface at least 20 m long that allows a width of 40 to 60 in. (1.02 to 1.52 m) for each student.
CD or cassette player with adequate volume
PACER CD or audiocassette
Measuring tape
8 or more marker cones
Pencil
Copies of scoresheet A or B

Note

Students should wear shoes with nonslip soles.

Instructions

Mark the 20 m (21.4 yd) course with marker cones to divide lanes and a tape or chalk line at each end. If using the audiotape, calibrate it by timing the 1 min test interval at the beginning of the tape. If the tape has stretched and the timing is off by more than 0.5 s, obtain another copy of the tape. Make copies of scoresheet A or B for each group of students to be tested.

Before the test day, allow students to listen to several minutes of the tape so that they know what to expect. Students should then be allowed at least two practice sessions.

Allow students to select a partner. Have students who are being tested line up behind the start line.

The PACER CD has a music version and one with only the beeps. The PACER tape has two music versions and one beep-only version. Each version of the test will give a 5 s countdown and tell the students when to start.

Students should run across the 20 m distance and touch the line with their foot by the time the beep sounds. At the sound of the beep, they turn around and run back to the other end. If some students get to the line before the beep, they must wait for the beep before running the other direction. Students continue in this manner until they fail to reach the line before the beep for a second time.

A single beep will sound at the end of the time for each lap. A triple beep sounds at the end of each minute. The triple beep serves the same function as the single beep and also alerts the runners that the pace will get faster.

When to stop

The first time a student does not reach the line by the beep, she or he reverses direction immediately. Allow a student to attempt to catch up with the pace. The test is completed for a student when she or he fails to reach the line by the beep for the second time. Students just completing the test should continue to walk and stretch in the cool-down area. Figure 12.3 provides diagrams of testing procedures.

Scoring

In the PACER test, a lap is one 20 m distance (from one end to the other). Have one student record the lap number (cross off each lap number on a PACER score sheet). The recorded score is the total number of laps completed by the student. For ease in administration, count the first lap that the student does not reach the line by the beep. Be consistent with all of the students and classes.

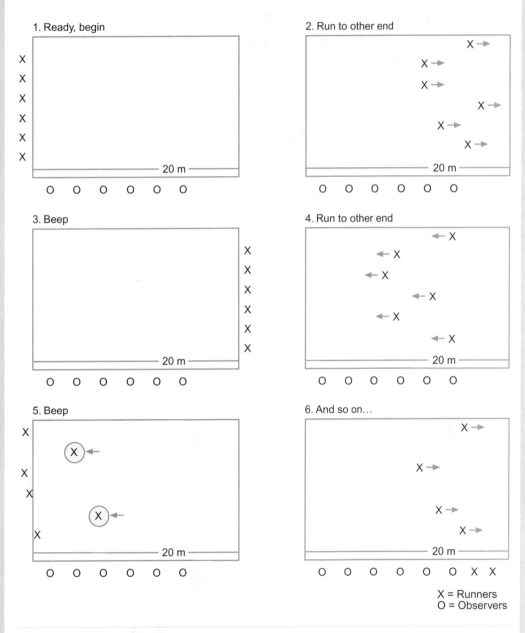

Figure 12.3 Diagram of PACER.

Reprinted from The Cooper Institute for Aerobics Research 1999.

Skinfold Measurements

Purpose

To measure percent body fat.

Objective

The thickness of the skinfolds at the triceps and medial calf sites is measured and used to estimate percent body fat.

Equipment

Skinfold calipers (Both expensive and inexpensive calipers have been shown to provide reliable and valid measures.)

Instructions

The triceps skinfold is taken on the right arm over the triceps muscle. The skinfold is vertical and midway between the acromion process of the scapula and the elbow (see figure 11.8c, page 239). The medial calf skinfold is a vertical skinfold taken at the level of maximal girth of the calf on the right leg (figure 12.4). The foot rests on a stool or other device so that the knee is at a 90° angle.

Scoring

Measure each skinfold three times and record the median value. Record the skinfold thickness to the nearest 0.5 mm.

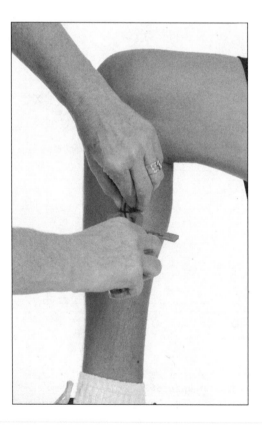

Figure 12.4 FITNESSGRAM calf skinfold measurement.

Mastery Item 12.6

Review the material provided in chapter 11 (pp. 238-239) on obtaining reliable and valid skinfold measurements.

Curl-Up

Purpose

To measure abdominal strength and endurance.

Objective

The child performs as many curl-ups as possible to a maximum of 75.

Equipment

Gym mats
Cardboard strips (a narrow strip [30 in. (76.2 cm) × 3 in. (7.62 cm)] for grades K to 4 and a wider strip [30 in. × 4.5 in. (11.4 cm)] for older children)
Tape player and audiotape (for controlling cadence)

Instructions

Students perform this test in groups of three. One student performs the curl-ups, the second supports the head of the performer, and the third secures the strip so that it does not move. The student being tested lies in a supine position on the mat with knees bent at an approximate 140° angle, legs apart, arms straight and parallel to the trunk with the palms resting on the mat (figure 12.5). The fingers are extended and the head is in the partner's hand resting on the mat. The cardboard strip is placed under the knees with the fingers touching the nearer edge. The third student stands on the strip so that it does not move during the test. The student curls up so that the fingers slide to the other side of the strip. The student performs as many curl-ups as possible while maintaining a cadence of one curl-up every 3 s but stops at a maximum number of 75.

Scoring

Record the number of curl-ups to a maximum of 75.

Figure 12.5 FITNESSGRAM curl-up test.

Trunk Lift

Purpose

To measure trunk extensor strength and flexibility.

Objective

The student lifts the upper body from the ground using the muscles of the back and holds the position to allow an accurate measurement from chin to the ground.

Equipment

Gym mats and yardstick or ruler

Instructions

The student being tested lies face down on a mat (figure 12.6). The toes are pointed and the hands placed under the thighs. The student lifts the upper body, in a slow and controlled pace, to a maximum of 12 in. (30.5 cm). The student holds the position until a ruler is placed in front of him or her and the measurement taken from the floor to the chin.

Scoring

Perform two trials and record the highest score to the nearest inch, up to a maximum of 12 in.

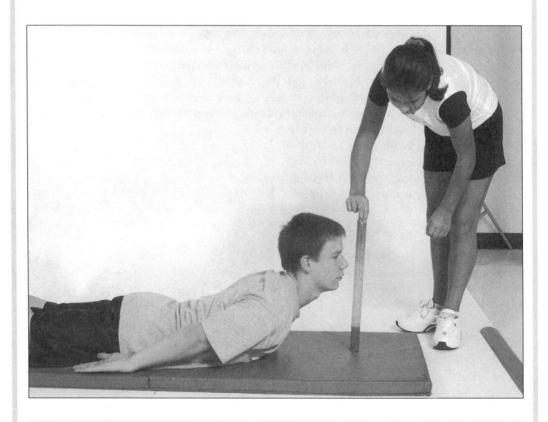

Figure 12.6 FITNESSGRAM trunk lift test.

Push-Ups

Purpose

To measure upper-body strength and endurance.

Objective

The students complete as many push-ups as possible.

Equipment

Tape player
Audiotape to control the cadence of 1 push-up every 3 s.

Instructions

The students work in pairs, with one counting while the other is tested. The student being tested lies face down on the ground with the hands under the shoulders; fingers stretched; legs straight, parallel, and slightly apart; and the toes tucked under the feet (figure 12.7). At a start command, the student pushes up with the arms until they are straight. The legs and back should be kept straight throughout the test. The student lowers the body using the arms until the arms bend to a 90° angle and the upper arms are parallel to the floor. This action is repeated as many times as possible following the cadence of one repetition every 3 s. The test is continued until the student cannot maintain the pace or demonstrates poor form.

Scoring

Record the number of push-ups performed.

Figure 12.7 FITNESSGRAM push-up test.

The muscular fitness protocols of FITNESSGRAM are comparatively new. Work with a fellow student and administer the curl-up, trunk lift, and push-up tests to each other. Perform several trials on different days to examine the consistency of your test performances.

VARIABLE STANDARDS IN YOUTH FITNESS TESTS

As previously mentioned, different test batteries use different standards for passing or failing the same fitness test. Consequently, interpretation of test data can vary greatly, depending on the standard used. Table 12.11 displays the 1 mi run passing standards (ages 6-11) for the FITNESSGRAM and for the President's Challenge National and Presidential Awards. As you can see, the passing standards are very different. For example, the passing times for an 8-year-old boy range from 13:00 to 8:48. This is quite a large range! Logically, the percentage of students passing this test of cardiovascular endurance would vary greatly depending on which set of standards is used. Ezzell, Smith, and Jackson (1991) examined the 1 mi run test results of 992 boys and 1028 girls between the ages of 6 and 10. The researchers applied the three sets of standards noted in table 12.11 to determine the different passing rates (figure 12.8). As expected, there was a wide range in passing rates, which matched the wide range in passing standards. If you were a teacher discussing the passing rates for cardiovascular endurance in your school with your principal, the message could be different depending on your objective. For example, if you want to get more time for physical education and activity, you might indicate to the principal that only 21% of the boys and 20% of the girls

Table 12.11 **1 Mile Run Standards of Youth Fitness Batteries (min:sec)**

Group/age	FITNESSGRAM	President's Presidential	President's National
Boys			
6	15:00	10:15	12:36
7	14:00	9:22	11:40
8	13:00	8:48	11:05
9	12:00	8:31	10:30
10	11:00	7:57	9:48
11	11:00	7:32	9:20
Girls			
6	16:00	11:20	13:12
7	15:00	10:36	12:56
8	14:00	10:02	12:30
9	13:00	9:30	11:52
10	12:00	9:19	11:22
11	12:00	9:02	11:17

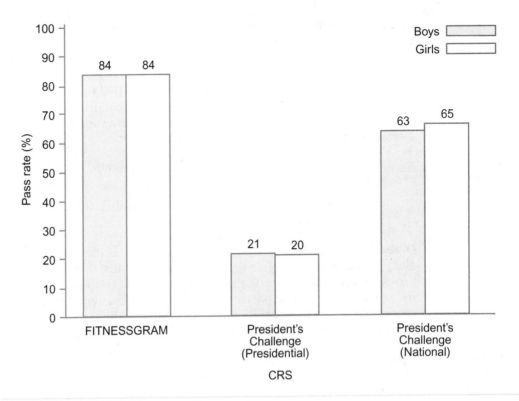

Figure 12.8 Comparison of the 1 mi walk/run passing rates.

had sufficient cardiovascular endurance, based on the President's Challenge Presidential Award, and that more time for endurance training is needed. Or if you want to get a pay raise, you might use the FITNESSGRAM passing rates and tell the principal that your students have high levels of cardiovascular endurance because of your effective physical education program. The results of Ezzell, Smith, and Jackson's (1991) study provide clear insight into why there is disagreement on the fitness levels of American youth. Clearly the choice of testing protocol and standards used will influence your interpretation of results. You need to consider these differences carefully when choosing a testing protocol. If Jo, the elementary school physical education teacher mentioned at the beginning of the chapter, plans to use FITNESSGRAM, she should understand these issues.

Mastery Item 12.8

| Why is the President's Challenge Presidential Award so difficult to receive?

ENHANCING RELIABLE AND VALID FITNESS TEST RESULTS WITH CHILDREN

The tests presented in the fitness batteries in table 12.6 have different levels of evidence to support their validity. Distance runs and skinfolds have demonstrated criterion-related validity in research (Lohman 1989; Safrit et al. 1988). The tests of muscular fitness, pull-ups, push-ups, modified pull-ups, flexed arm hang, sit-ups, and curl-ups, and the sit-and-reach test have limited criterion-related validity support but are generally accepted on the basis of content or logical validity. Research has questioned the assumption of content validity for some tests: sit-and-reach, pull-ups, and sit-ups (Engelman and Morrow 1991; Hall, Hetzler, Perrin, and Weltman 1992; Jackson and Baker 1986). However, you can assume sufficient validity of your tests provided that you are using one of the fitness batteries presented in this chapter.

The practical issue is for you to take the appropriate steps during fitness test administration that will produce reliable test results. This is an important issue in youth fitness testing because research has shown that field tests of fitness in children can demonstrate poor reliability, especially in younger children (Rikli, Petray, and Baumgartner 1992). To enhance reliability ($r_{xx'} > .80$), you can take a variety of practical steps to minimize measurement error, including the following:

1. Attain adequate knowledge of test descriptions.
2. Give proper demonstrations and instructions.
3. Develop good student and teacher preparation through adequate practice trials.
4. Conduct reliability studies.

Of these steps, providing adequate practice trials is the most important. Children, especially younger ones, need several trials of a fitness test to learn how to take the test and to provide consistent results. For example, children need to practice the PACER several times to learn the appropriate pace and to provide a consistent time that is a reliable and valid representation of their cardiovascular endurance. If you are going to take skinfolds, you need to practice the measurement techniques discussed in chapter 11 (p. 240).

Mastery Item 12.9

What steps would you take to demonstrate and ensure the reliability of the sit-and-reach test in an elementary school setting?

SPECIAL CHILDREN

One of the biggest measurement challenges that you may confront as a professional in human performance is the assessment of physical fitness in children with physical or mental disabilities. Keep in mind that the fitness test batteries discussed in this chapter would exclude or be biased against many **physically disabled** children (i.e., those having physical or organic limitations, such as cerebral palsy) and **mentally disabled** children (i.e., those with mental or psychological limitations, such as autism) because of their specific disabilities. *Before you can administer or evaluate fitness test results, you must consider the subjects' physical and organic limitations, neural and emotional capacity, interfering reflexes, and acquisition of prerequisite functions, responses, and abilities (Seaman and DePauw 1989).* You can develop the basic knowledge and competence that you need for assessing the fitness and activity of disabled children from your instruction and learning experiences in **adapted physical education,** that is, physical education adjusted to accommodate children with physical or mental limitations. The physical fitness tests selected should be appropriate for the individual student based on his or her disability as well as on the fitness capacity to be measured. Seaman and DePauw (1989) and Winnick and Short (1999) are excellent sources for detailed information on fitness assessment of special children.

The Brockport Physical Fitness Test (Winnick and Short 1999), a health-related physical fitness test for youths aged 10 through 17 with various disabilities, was developed through a research study, Project Target, funded by the U.S. Department of Education. The test battery includes criterion-referenced standards for 25 tests. The test manual helps the professional consider the child's disability and select the most appropriate test and test protocol. The test comes with Fitness Challenge software to help the professional administer the test and develop a database. Table 12.12 provides potential items for fitness assessment, the appropriate disabled population for the test, and reliability and validity comments for each test. The complete test kit includes a manual, software, a demonstration video, and a fitness training guide.

Table 12.12 Brockport Physical Fitness Test Items

Test item	Mental retardation	Blind with assistance	Cerebral palsy	Spinal cord injury	Congenital anomalies/ amputation	Reliability	Validity
PACER	R	R			O	Acceptable	Content Concurrent
1 mi run/walk		O			R	Acceptable	Concurrent
Target aerobic movement test	R		R	R	R	Acceptable	Content
Skinfolds	R	R	R	R	R	Acceptable	Concurrent
BMI	O	O	O			Acceptable	Concurrent
Reverse curl				R		Not reported	Content
Seated push-up			R	R	R	Not reported	Content
40 m push/walk			R			Not reported	Content
Wheelchair ramp test			R/O			Not reported	Content
Bench press	O			O	R	Acceptable	Content
Dumbbell press			R/O	O	R/O	Acceptable	Content Concurrent
Extended arm hang	R					Acceptable	Content
Dominant grip strength	O		O	R	O	Acceptable	Construct
Isometric push-up	O					Acceptable	Content
Push-up		O				Acceptable	Content
Pull-up		O				Acceptable	
Modified pull-up		O				Acceptable	Content
Curl-up		R			R	Acceptable	Content
Modified curl-up		R				Acceptable	Content
Trunk lift	R	R			R	Acceptable	Content
Modified Apley test			R	R	R	Not reported	Content
Shoulder stretch	O	O				Acceptable	Content
Modified Thomas stretch			R	R		Not reported	Content
Back saver sit-and-reach test	R	R			R	Acceptable	Content
Target stretch test			R/O	R	R	Acceptable	Content Concurrent

Note: O = Option; R = Recommended.

Mastery Item 12.10

Instructing special children and evaluating their physical fitness can be challenging. List specific preparations you would make to address this challenge.

MEASURING PHYSICAL ACTIVITY IN YOUTH

The concerns about youth fitness discussed earlier in this chapter are matched by concerns about youth physical activity levels. Public health officials and physical activity researchers need reliable and valid measures of physical activity in children and youth for effective research studies that will increase youth activity and fitness levels (Sallis et al. 1993). Issues regarding the physical activity levels of youth include the following:

▶ Physical activity improves the overall health of children.

▶ Physically inactive children tend to become inactive adults with higher risks for chronic diseases.

▶ Conversely, physically active children tend to become active adults with lower risks for chronic diseases.

For example, Dennison, Straus, Mellits, and Charney (1988) found that children with the poorest performances on distance-run tests had the highest risk of being physically inactive as young adults. Telama, Yang, Laasko, and Viikari (1997) also reported that children and adolescents who are active tend to be more physically active as young adults. These kinds of longitudinal studies, examining the relationship between youth physical activity and adult physical activity, are impossible without accurate assessments of physical activity while the subjects are children.

In Canada, the national approach has shifted from assessing physical fitness in youth to assessing and promoting physical activity. The Canadian Active Living Challenge (CALC) has replaced the Canada Fitness Awards as a national school- and community-based educational intervention aimed at positively influencing knowledge, beliefs, and attitudes about physical activity and healthy lifestyles. The intervention implements a behavioral approach to enhancing these variables, one that involves self-monitoring and reinforcement.

As discussed in chapter 11 regarding adults, assessing physical activity levels of children is most reliably and validly done via direct monitoring (e.g., accelerometers, pedometers, and direct observation). But lack of feasibility limits applying such procedures in large-scale studies. In a 1993 article, three different self-report instruments—a 7-day recall interview, a self-administered survey, and a simple activity rating—were examined for test-retest reliability and validity (Sallis et al. 1993). Reliability ranged from .77 to .89 on the three self-reports over all subjects. However, as would be expected, reliability was higher for older children than for younger children. The 7-day recall had moderate concurrent validity ($r = .44$-$.53$) when related to a criterion of monitored heart rate. Again, validity improved as the age of the children increased. Sallis and colleagues concluded that self-report techniques could be used with youth of high school age, but caution was necessary with younger children.

The President's Council on Physical Fitness and Sports has recognized the importance of physical activity. In 2002, the Council added the Presidential Active Lifestyle Award.

The award is based on the following requirements:

▶ Physical activities recorded

  ▸ 60 min per day or 11,000 pedometer steps for girls or 13,000 steps for boys

  ▸ 5 days per week

  ▸ 6 weeks

▶ Weekly signatures of participant. Final signature of supervising adult attesting to qualification.

The President's Council has also established a Web site where people of all ages can track their physical activity behaviors. You can use the site to track all of your physical activities, including the use of pedometers. Individuals can receive the Presidential Active Lifestyle Award (PALA) based on maintaining a physically active lifestyle. The site is http:\\www.presidentschallenge.org\.

Mastery Item 12.11

In preparation for entering data into the President's Challenge Web site, a group of elementary school children wore pedometers for a week to measure their physical activity levels. The mean daily number of steps was calculated for each student. Using the data provided on the World Wide Web for MI 12.11, determine if there is a significant difference in the mean number of steps taken between boys and girls (use an independent *t* test). Also confirm that the correlation between the number of steps taken and their weight is .306.

The Cooper Institute has developed an assessment of physical activity to accompany the FITNESSGRAM physical fitness testing program. The ACTIVITYGRAM assessment has two different approaches. The first, the FITNESSGRAM Physical Activity Questionnaire, is based on three questions from the Centers for Disease Control and Prevention's Youth Risk Factor Behavior Survey, a surveillance instrument that tracks national trends in physical activity.

1. Aerobic activity question: On how many of the past 7 days did you participate in physical activity for a total of 30 to 60 min, or more, over the course of a day? This includes moderate activities (walking, slow bicycling, or outdoor play) as well as vigorous activities (jogging, active games, or active sports such as basketball, tennis, or soccer). (0, 1, 2, 3, 4, 5, 6, 7 days)

2. Strength activity question: On how many of the past 7 days did you do exercises to strengthen or tone your muscles? This includes exercises such as push-ups, sit-ups, or weightlifting. (0, 1, 2, 3, 4, 5, 6, 7 days)

3. Flexibility activity question: On how many of the past 7 days did you do stretching exercises to loosen up or relax your muscles? This includes exercises such as toe touches, knee bending, or leg stretching. (0, 1, 2, 3, 4, 5, 6, 7 days)

Two questions regarding participation in physical activity during physical education class are also asked.

1. In an average week when you are in school, on how many days do you go to physical education classes? (0, 1, 2, 3, 4, 5 days)

2. During an average physical education class, how many minutes do you spend actually exercising or playing sports? (I do not take physical education, <10 min, 10-20 min, 21-30 min)

Using the child's responses to the questions, FITNESSGRAM/ACTIVITYGRAM computer software provides individualized feedback to the student on the standard FITNESSGRAM/ACTIVITYGRAM assessment shown in figure 12.9.

The second ACTIVITYGRAM approach to physical activity assessment is based on the Previous Day Physical Activity Recall (Weston, Petosa, and Pate 1997). This "segmented day" approach requires the child to report activities in 30 min blocks from 7:00 A.M. to 11:00 P.M. The child reports the frequency, intensity, time, and type of physical activity. This provides content validity for the activity recall in that the relevant factors of exercise prescription and participation are assessed. The logging chart that the student completes is illustrated in figure 12.10. The data from the chart are entered into the FITNESSGRAM computer software, and the ACTIVITYGRAM report is generated and given to the student. An example of the report is presented in figure 12.11.

FITNESSGRAM®

Charlie Brown
Grade: 5   Age: 11
Madison County Elementary School
Instructor: Kathy Read

	Test Date	Height	Weight
Current	07/15/99	5'01"	105
Past	07/13/99	5'3"	122

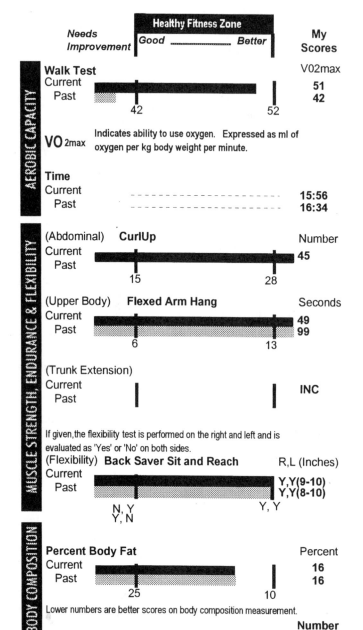

AEROBIC CAPACITY

Healthy Fitness Zone
Needs Improvement | Good _____ Better

My Scores

Walk Test   VO2max
Current   **51**
Past   **42**
42   52

VO2max Indicates ability to use oxygen. Expressed as ml of oxygen per kg body weight per minute.

Time
Current   **15:56**
Past   **16:34**

MUSCLE STRENGTH, ENDURANCE & FLEXIBILITY

(Abdominal) CurlUp   Number
Current   **45**
Past
15   28

(Upper Body) Flexed Arm Hang   Seconds
Current   **49**
Past   **99**
6   13

(Trunk Extension)
Current   **INC**
Past

If given, the flexibility test is performed on the right and left and is evaluated as 'Yes' or 'No' on both sides.

(Flexibility) Back Saver Sit and Reach   R,L (Inches)
Current   **Y,Y(9-10)**
Past   **Y,Y(8-10)**
N, Y   Y, Y
Y, N

BODY COMPOSITION

Percent Body Fat   Percent
Current   **16**
Past   **16**
25   10

Lower numbers are better scores on body composition measurement.

ACTIVITY

Number of Days

On how many of the past 7 days did you participate in physical activity for a total of 30-60 minutes, or more, over the course of a day?   **4**

On how many of the past 7 days did you do exercises to strengthen or tone your muscles?   **3**

On how many of the past 7 days did you do stretching exercises to loosen up or relax your muscles?   **2**

MESSAGES

Charlie, your scores on all test items were in or above the Healthy Fitness Zone. You are also doing strength and flexibility exercises. However, you need to play active games, sports or other activities at least 5 days each week.

Although your aerobic capacity score is in the Healthy Fitness Zone now, you are not doing enough physical activity. You should try to play very actively at least 60 minutes at least five days each week to look and feel good.

Your abdominal strength was very good. To maintain your fitness level be sure that your strength activities include curl-ups 3 to 5 days each week. Remember to keep your knees bent. Avoid having someone hold your feet.

Your upper body strength was very good, Charlie. To maintain your fitness level be sure that your strength activities include arm exercises such as push-ups, modified push-ups or climbing activities 2 to 3 days each week.

Charlie, your flexibility is in the Healthy Fitness Zone. To maintain your fitness, stretch slowly 3 or 4 days each week, holding the stretch 20 - 30 seconds. Don't forget that you need to stretch all areas of the body.

Charlie, your body composition is in the Healthy Fitness Zone. If you will be active most days each week, it may help to maintain your level of body composition.

To be healthy and fit it is important to do some physical activity almost every day. Aerobic exercise is good for your heart and body composition. Strength and flexibilty exercises are good for your muscles and joints.

Good job, you are doing enough physical activity for your health. Additional vigorous activity would help to promote higher levels of fitness.

©The Cooper Institute for Aerobics Research

Figure 12.9 FITNESSGRAM report.

FITNESSGRAM

ACTIVITYGRAM Logging Chart

Name_____ Age _____ Teacher _____ Grade _____

Record the primary activity you did during each 30-minute interval during the day using the list at the bottom of the page. Then select and intensity level that best describes how it felt (Light: "Easy"; Moderate: "Not too tiring"; Vigorous: "Very tiring"). Note: all time periods of rest should have Rest checked for intensity level.

Time	Activity	Rest	Light	Mod.	Vig.	Time	Activity	Rest	Light	Mod.	Vig.
7:00						3:00					
7:30						3:30					
8:00						4:00					
8:30						4:30					
9:00						5:00					
9:30						5:30					
10:00						6:00					
10:30						6:30					
11:00						7:00					
11:30						7:30					
12:00						8:00					
12:30						8:30					
1:00						9:00					
1:30						9:30					
2:00						10:00					
2:30						10:30					

Categories of Physical Activities

Lifestyle activity	Active aerobics	Active sports	Muscle fitness activities	Flexibility exercises	Rest and inactivity
"Activities that I do as a part of my normal day"	"Activities that I do for aerobic fitness"	"Activities that I do for sports and recreation"	"Activities that I do for muscular fitness"	"Activities that I do for flexibility and fun"	"Things I do when I am not active"
1. Walking, bicycling, or skateboarding	11. Aerobic dance activity	21. Field sports (baseball, softball, football, soccer, etc.)	31. Gymnastics or cheer, dance, or drill teams	41. Martial arts (Tai Chi)	51. Schoolwork, homework, or reading
2. Housework or yardwork	12. Aerobic gym equipment (stairclimber, treadmill, etc.)	22. Court sports (basketball, volleyball, soccer, hockey, etc.)	32. Track-and-field sports (jumping, throwing, etc.)	42. Stretching	52. Computer games or TV/videos
3. Playing active games or dancing	13. Aerobic activity (bicycling, running, skating, etc.)	23. Racquet sports (tennis, racquetball, etc.)	33. Weightlifting or calisthenics (push-ups, sit-ups, etc.)	43. Yoga	53. Eating or resting
4. Work-active job	14. Aerobic activity in physical education	24. Sports during physical education	34. Wrestling or martial arts (Karate, Aikido)	44. Ballet dancing	54. Sleeping
5. Other _____	15. Other _____	25. Other _____	35. Other _____	45. Other _____	55. Other _____

Figure 12.10 ACTIVITYGRAM logging chart.
Reprinted from The Cooper Institute for Aerobics Research 1999.

ACTIVITYGRAM

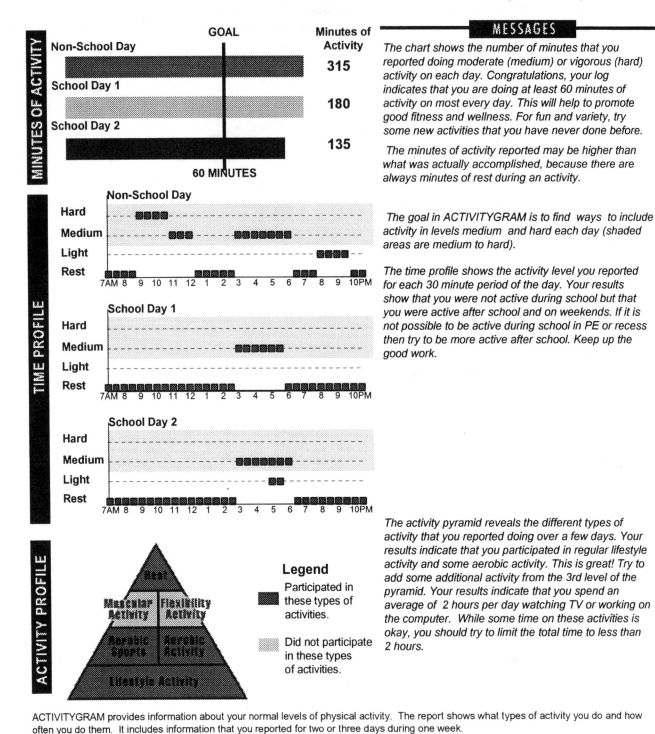

MESSAGES

The chart shows the number of minutes that you reported doing moderate (medium) or vigorous (hard) activity on each day. Congratulations, your log indicates that you are doing at least 60 minutes of activity on most every day. This will help to promote good fitness and wellness. For fun and variety, try some new activities that you have never done before.

The minutes of activity reported may be higher than what was actually accomplished, because there are always minutes of rest during an activity.

The goal in ACTIVITYGRAM is to find ways to include activity in levels medium and hard each day (shaded areas are medium to hard).

The time profile shows the activity level you reported for each 30 minute period of the day. Your results show that you were not active during school but that you were active after school and on weekends. If it is not possible to be active during school in PE or recess then try to be more active after school. Keep up the good work.

The activity pyramid reveals the different types of activity that you reported doing over a few days. Your results indicate that you participated in regular lifestyle activity and some aerobic activity. This is great! Try to add some additional activity from the 3rd level of the pyramid. Your results indicate that you spend an average of 2 hours per day watching TV or working on the computer. While some time on these activities is okay, you should try to limit the total time to less than 2 hours.

ACTIVITYGRAM provides information about your normal levels of physical activity. The report shows what types of activity you do and how often you do them. It includes information that you reported for two or three days during one week.

ACTIVITYGRAM is a module within FITNESSGRAM 6.0 software. FITNESSGRAM materials are distributed by the American Fitness Alliance, a division of Human Kinetics. www.americanfitness.net

© The Cooper Institute for Aerobics Research

Figure 12.11 ACTIVITYGRAM report.

Direct observation of physical activity behaviors can be expensive in terms of time and labor. However, new technologies including portable computers and measurement systems like BEACHES, SOFIT, and SOPLAY have allowed the reliable, valid, and feasible measurements using direct observation of physical activity. BEACHES (McKenzie, Sallis, Patterson, et al. 1991) was designed to record physical activity and eating behaviors and related environmental factors. SOFIT (McKenzie, Sallis, Nader 1991) was developed to record physical activity, lesson context, and teacher behavior during physical education classes. SOPLAY (McKenzie, Marshall, Sallis, and Conway 2000) was designed to measure the physical activity of groups of people instead of individuals, as was the case for BEACHES and SOFIT. SOPLAY allows recording information about the environment that may play a role in the observed physical activity behavior of the group being studied.

Measurement and Evaluation Challenge

Jo has learned a great deal about youth fitness and physical activity assessment. She is now familiar with the difference between a number of factors. She recognizes the difference between health-related fitness and motor fitness. She understands the benefits of criterion-referenced standards compared with the norm-referenced standards the district previously used. She is also now aware of the importance of measuring physical activity in addition to measuring physical fitness. She sees that the FITNESSGRAM/ACTIVITYGRAM provides many opportunities to assess fitness and activity that were unavailable to her with the district's previous fitness assessment.

SUMMARY

The reliable and valid assessment of youth physical activity and fitness is a major objective of the assessment of human performance. This is especially true of the public school or private school physical educator. Although there is not universal agreement about the levels of poor fitness in youth, it is clear that there are many children whose fitness and physical activity levels are not consistent with good health (Blair 1992). Reliable and valid fitness testing of youth involves selecting tests related to program objectives, appropriate training, and administering sufficient practice trials to ensure stable test results. A good resource with detailed information about fitness and activity assessment for the professional is *Measurement in Pediatric Exercise Science* (Docherty 1996). Testing special children requires special attention to test selection and administration. The professional physical educator needs appropriate education for training and testing special children.

Although the validity of a test and its passing standard cannot be associated with a reduced risk of a specific disease, the tests that are administered communicate an important message. When we test for cardiovascular endurance, we are telling the child that this is an important attribute. This is true for the other health-related fitness tests as well. In general, it has been scientifically documented that higher levels of physical activity and fitness are associated with reduced risks of certain diseases. Effectively measuring and evaluating the physical fitness and activity levels of American youth will aid in developing a healthy and physically educated population.

chapter **13**

Assessment of Sport Skills and Motor Abilities

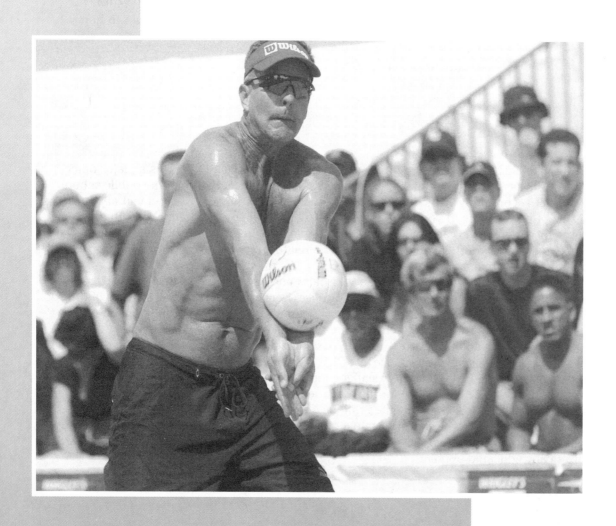

Outline

Measurement and Evaluation Challenge

Scott is a graduate assistant for the men's volleyball team. He is currently enrolled in a graduate measurement class. He has been asked to complete a project for the class, and he's interested in tying in the project with his work with the volleyball team. The team has some very good athletes that are not especially tall, as well as some members that are very tall but not especially quick. Team members have been tested on a series of motor performance tests related to volleyball playing potential. Scott would like to use this test battery to design individualized physical training programs for the players. He is not sure how to go about this. He has decided to talk with his measurement and evaluation instructor.

Objectives

After studying this chapter, you will be able to

- differentiate between skills and abilities;
- apply sound testing procedures in ability and skill assessment;
- develop psychomotor tests with sufficient reliability and validity;
- differentiate among and be able to use the different types of sport skills tests;
- define and delineate the basic motor abilities; and
- develop tests and test batteries that will be able to select, classify, and diagnose athletes based on psychomotor tests.

he measurement of sport skills and motor abilities is one of the fundamental aspects of the measurement of human performance. Fleishman (1964) provided the modern foundation for work in this area, including a delineation between skills and abilities. According to Fleishman, a **skill** is a learned trait based on the abilities that a person has; **abilities** are more innate than skills; skills are more sport specific, whereas abilities are more general. Battinelli (1984) summarized this relationship in an article tracing the history of the debate over generality versus specificity of motor ability, stating the following:

> The evident trend that emerges from the literature . . . over the years seems to demonstrate that the acquirement of motor abilities and motor skills through motor learning processes is dependent upon general as well as specific factors. The general components of motor ability (muscle strength, muscle endurance, power, speed, balance, flexibility, agility and cardiovascular endurance) have become the practical physical supports to motor learning, while skill specificities have been shown to be representative of the neural physiological processes exemplified in such learning. (p. 111)

The definition of the trait to be measured is important in determining the manner in which it will be measured. In chapters 11 and 12, we noted that not only the definition of physical fitness has changed but also the manner in which it is measured. In sport skills and motor performance testing, the distinction emphasized by Fleishman and the advent of high-speed computers and advanced statistical techniques have also brought about changes. These changes, as well as the accepted practices associated with the measurement of sport skills and motor abilities, are presented in this chapter.

GUIDELINES FOR SPORT SKILLS AND MOTOR PERFORMANCE TESTS

When selecting or devising tests to measure sport skills or motor ability, you should follow certain accepted testing procedures. Whether you use standardized tests or your own tests depends on your expertise and the specific use of the test. If you will be making comparisons with other groups, use some type of standardized test. If the information is simply for you, then you can modify a standard test or develop a new one to suit your purposes.

The American Alliance for Health, Physical Education, Recreation and Dance (AAHPERD) has provided guidelines for skills test development (Hensley 1989); these are the basis for the development of the AAHPERD skills test series and also apply to the measurement of motor abilities.

The guidelines state that skills tests should

- ▶ have at least minimally acceptable reliability and validity,
- ▶ be simple to administer and to take,
- ▶ have instructions that are easy to understand,
- ▶ require neither expensive nor extensive equipment,
- ▶ be reasonable in terms of preparation and administration time,
- ▶ encourage correct form and be gamelike but involve only one performer,
- ▶ be of suitable difficulty (neither so difficult that they cause discouragement nor so simple that they are not challenging),
- ▶ be interesting and meaningful to the performer,
- ▶ exclude extraneous variables as much as possible,
- ▶ provide for accurate scoring by using the most precise and meaningful measure,
- ▶ follow specific guidelines if a target is the basis for scoring,

▶ require a sufficient number of trials to obtain a reasonable measure of performance (tests that have accuracy as a principal component require more trials than tests measuring other characteristics), and

▶ yield scores that provide for diagnostic interpretation whenever possible (p. 2).

The guidelines also indicate that if a target is the basis for scoring, it should have the capacity to encompass 90% of the attempts; near misses may need to be awarded points. Determining target placement should be based on two principal factors: (a) the developmental level of the student (e.g., the height of a given target may be appropriate for a 17-year-old but not for a 10-year-old) and (b) the allocation of points for various strategic aspects of the performance (e.g., the placement of a badminton serve to an opponent's backhand should score higher than an equally accurate placement to the opponent's forehand).

Because most tests of sport skills or motor abilities can be objectively measured, you can expect reliability and validity coefficients higher than those associated with written instruments. AAHPERD suggests that reliability and validity coefficients exceed .70. However, although many reliability coefficients exceed this lower bound value, it is difficult to develop tests whose validity coefficients reach it. A validity coefficient (r) of .70 would mean that nearly 50% of the variance of the test would be associated with the criterion ($r^2 = .49$). When examining the validity of a test, consider not only its statistical relationship with the criterion but also its practical relevance. Also consider test feasibility as you select instruments to use.

EFFECTIVE TESTING PROCEDURES

The procedures for psychomotor testing are the same as with written testing. They may be classified as pretest, testing, and posttest duties. It is imperative that you think through all aspects of the testing procedures in detail so that you can gather consistent, accurate results.

Pretest Duties

Pretest planning is the first element of preparing to administer tests. The tester must be completely familiar with the test, the items to be administered within the test, the facilities available, and the necessary equipment and markings. For a physical performance test, those being tested need to be exposed to the exact test items and allowed time to practice so that they can learn how to pace themselves and how to perform the test in the most efficient manner. This way, when the test is finally administered, the measures are accurate estimates of the actual amount of learning that has taken place rather than of the students' abilities to perform in an unfamiliar situation.

With batteries of performance tests, there are other elements to consider. The first is the order of administration. If there are several physically taxing items in a test battery, spread them out across several days so that the test subjects are not unduly fatigued when they participate or perform the test. Also, pair time-consuming items with items that do not take as long.

Consider whether you will need assistants to help record the data. This is important when you determine the type of score sheets to use. Two types of recording sheets are used when multiple testers and multiple stations are available. With one form, all the data for a test subject are recorded on the subject's sheet (figure 13.1), and these sheets are carried from station to station by the subjects. One problem with this method is that the sheets can be misplaced or damaged as they go from station to station. The second method is to provide each tester with a master list of all those taking the test; as the subjects rotate through the stations, the testers simply record the data on the master list (figure 13.2). The problem with the second approach is the lengthy transcription process in the posttest recording stage. If only one tester is available, we recommend using the master list method.

Soccer skills test score sheet

Name: _____

Class: _____

Speed dribble (# of touch violations in parentheses)

T1 _____   (   )

T2 _____   (   )

Control dribble

T1 _____

T2 _____

Passing		Left foot	Right foot	Shooting		
10 yd	1	_____	_____	Left target	1	_____
	2	_____	_____		2	_____
15 yd	1	_____	_____		3	_____
	2	_____	_____	Right target	1	_____
					2	_____
					3	_____

Figure 13.1   Sample individual score sheet.

Fielding, throwing, and running

Grade _____

Name	Gender	Fielding trials					Throwing trials			Running trials		
		1	2	3	4	Total	1	2	Total	1	2	Total
1. _____												
2. _____												
3. _____												
4. _____												
5. _____												
6. _____												
7. _____												
etc. _____												

Figure 13.2   Example of a tester's master sheet.

You will need to develop and write standardized instructions so that each tester knows exactly what to do. Testers themselves should be familiar with the test. The instructions should be read to the test subjects so that each is given exactly the same information for the test. Consider giving each student a written copy of the instructions prior to testing. The students should be familiar with the test procedures and should have even practiced the exact test that they will take prior to the actual test.

Testing Duties

The second phase is the actual testing. Prepare the test location as early as possible. Make sure that the surfaces are clean so that the students will be able to do their best. Also, address any safety concerns at this time, such as objects close to the testing area or unsafe surfaces or equipment. Give students the opportunity to warm up for physical performance tests, to allow them to reach their potential. Present the instructions to the students, being sure that any hints or "motivational" comments are consistent across all groups.

Posttest Duties

The final phase of the testing procedure involves transcribing the test results and analyzing the scores. The method of transcription depends upon how the data were actually gathered, and the data need to be verified any time they are transferred from one medium to another. Each time the data are transcribed, they need to be proofed by another person. This is best done by involving another tester, or test subject helpers can call out scores.

The analysis you use depends on the purpose of the testing. If individual performance within the group is important, use norm-referenced analytical procedures. On the other hand, if the purpose of the test is to compare an individual's performance with a standard, use criterion-referenced analysis.

Keep all test subjects' scores as confidential as possible. If test subjects help record the data, position them directly next to the tester. This can alleviate the problem of having to call out scores.

DEVELOPING PSYCHOMOTOR TESTS

Often, teachers, coaches, and researchers are interested in administering test batteries for a certain sport or motor performance area. In the analysis of athletic performance, tests of sport skills and motor performance are often used in conjunction with each other to give the teacher or coach additional information about prospective athletes. Strand and Wilson (1993) presented a 10-step flowchart for the construction of test batteries of this nature; we present a modified version of their flowchart in figure 13.3. Although the flowchart was developed primarily for sport skills, it is also applicable to motor ability testing. The steps are as follows:

> **Step 1.** Review the criteria for a good test. Basically, these criteria concern the statistical aspects of reliability, validity, and objectivity. The tester needs to be familiar with the equipment, personnel, space, and time available for testing. A test must be appropriate for the age and gender of the students and must be closely associated with the skill in question. Consider safety aspects also at this time.

> **Step 2.** Analyze the sport to determine the skills or abilities to be measured. If you are trying to evaluate current performance, then skills tests are most appropriate, but if you are trying to determine student potential for playing, then motor ability tests might be more useful. Combination batteries can also be useful, depending on the specific purpose of the administration of the battery.

▶ **Step 3.** Review the literature. Once you have analyzed the sport for its salient areas, review previous test batteries and literature associated with the specific skills or motor performance areas. Consult experts (such as colleagues, textbooks, professionals, teachers) at this point.

▶ **Step 4.** Select or construct test items. Ensure that the test items are (a) representative of the performances to be analyzed, (b) administered with relative ease, (c) as closely associated with the actual performance as possible, and (d) practically important. Each test or item should measure an independent area. It is not an efficient use of time to have several tests that measure the same basic skill or ability.

▶ **Step 5.** Determine the exact testing procedures. This includes selecting the number of trials necessary for the testing, the trials that will be used to establish the criterion score, and the order of the tests.

▶ **Step 6.** Peer review. Have experts such as other teachers or coaches who are familiar with the activity examine the test battery.

▶ **Step 7.** Pilot testing. If the selected tests are based on experts' opinions and a review of the literature, the pilot study analysis will help you determine the appropriateness of the test. Pilot testing is an important step before the test is finalized; it helps determine the total time of administration and the clarity of instructions and points out possible flaws in the test.

▶ **Step 8.** Determine the statistical qualities of the test—its reliability, validity, and objectivity. The reliability coefficients are estimates and are accurate only for the groups tested. They can be group specific, especially with adolescents, so it is important that both genders and all ages and skill levels included in the normative sample are tested. The methods for validating skills or motor ability tests are presented in chapters 6 and 7. Content validity is important before the statistical evaluation of the tests. You can determine concurrent validity or construct validity at this time by performing the appropriate procedures. Remember that validity coefficients are estimates and are only appropriate for groups comparable to those tested. The stability of the motor performance tests is normally established with reliability coefficients that involve repeated measures. This is the one phase of motor performance testing that separates it from the preparation phases of many psychological or educational tests. Eliminate or modify potential items at this point if they have poor reliability or validity.

▶ **Step 9.** Establish norms for norm-referenced tests or determine standards for criterion-referenced tests.

▶ **Step 10.** Construct the test manual to fully describe the test, the scoring procedures, and its statistical qualities (reliability and validity). Follow the guidelines suggested by the American Psychological Association (APA 1999) in developing a test manual.

▶ **Step 11.** Reevaluate the instrument from time to time. As time passes, your students may have different preparation levels; thus, norms and standards that were appropriate at one time may not be appropriate at another.

ISSUES IN SKILLS TESTING

The usual measurement issues of reliability and validity are important in skills testing. However, there are two other issues that we highlight in this chapter. One is feasibility. Skills tests typically take time to administer. You have to ask yourself, *Is it more important for me to be teaching the skills and having the students perform the skills or for me to spend the time evaluating the skills?* The second issue is determining the best way to evaluate the skill. Will

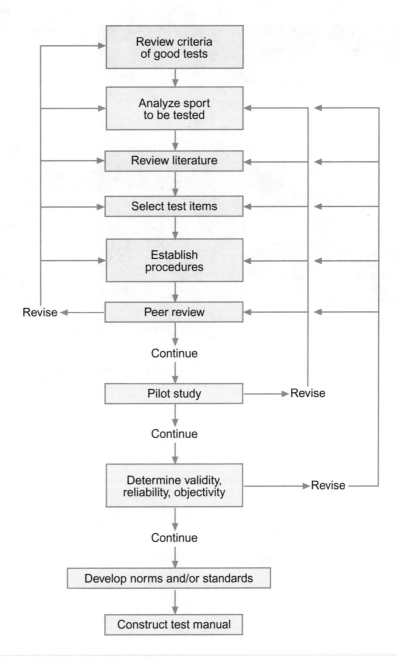

Figure 13.3 Flowchart for constructing motor performance tests.

you need to have highly objective tests that may not be as gamelike, or is it possible to use subjective tests that are more gamelike but may have lower reliabilities? If you prefer the latter approach, see chapter 8 for more on alternative assessment.

With skills testing, there often has to be a compromise between selecting a test that is extremely objective and reliable but not gamelike and selecting one that is highly valid but less objective and more time consuming. For example, consider Scott's challenge at the beginning of this chapter. He might choose to measure several skills primarily associated with volleyball performance, along with conducting some motor performance tests. One such skill is the forearm pass. He initially selected three tests to measure forearm passing: the self-pass, the wall pass, and the court pass. The self-pass and the wall pass are both simple tests that can be administered to large groups and require a minimal amount of space. They both produce consistently high degrees of reliability. The problem with these tests

The AAHPERD tennis battery is designed for the beginning-level player and includes a backhand return test.

is that they are not gamelike. For the self-pass, performers pass the ball repeatedly to themselves, the minimum criterion being that the ball must be passed at least 10 ft (3.0 m) off the ground on every repetition. For the wall pass, the performer passes the ball above a target line, again 10 ft above the ground, while staying behind a restraining line 6 ft (1.8 m) from the wall. In both cases, the performers can practice the test by themselves and can administer a self-test. The tests are designed for group administration and partner scoring; they are feasible, involving minimal court markings and class time. However, most volleyball experts would agree that the tests may not readily transfer to game skills—the ability to pass a serve or a hard-driven spike. The court pass test, on the other hand, places performers in a position in the backcourt, where they are most likely to receive the serve. They are asked to pass to a target area a ball that has been "served" or tossed by a coach or a tester. A point system is used to determine the accuracy and skill involved in the passing of 10 balls. This test has good reliability and is valid in terms of content validity. However, it requires lengthy administration time; it takes much longer to administer this gamelike court pass test than it does to administer the self-pass or wall pass tests. Also, the testers must produce reasonably consistent serves or tosses. Scott chose to go with the court pass to measure his team because of the small number of participants and their skill level. Had Scott desired to test beginners, either the self-pass or wall pass would test as effectively as the court pass.

Mastery Item 13.1

| When might you choose a simple, highly reliable, but not gamelike test?

Mastery Item 13.2

| In what situations might a gamelike test be used?

Over the past 20 years, the AAHPERD basketball test (Hopkins, Shick, and Plack 1984) and softball test (Rikli 1991) have been revised, and a tennis test (Hensley 1989) has been developed (table 13.1). AAHPERD's comprehensive testing program provides physical educa-

Table 13.1 **AAHPERD Skills Test Batteries**

Basketball	Tennis	Softball
Speed spot shooting	Ground stroke	Batting
Passing	Forehand/backhand	Fielding ground balls
Control dribble	Serve	Overhand throwing
Defensive movement	Volley	Base running

tion instructors and coaches with a battery of reliable and valid tests that can be administered in a minimal amount of time with a minimal amount of court markings. The AAHPERD guidelines indicate that skills tests should include the primary skills involved in performing a sport, usually not to exceed four. The tests of a battery should have acceptable reliability and validity and not have high intercorrelations. The battery in general should be able to validly discriminate among performance levels, thus demonstrating construct validity. The AAHPERD tests are designed primarily for the beginning-level performer.

Mastery Item 13.3

Select one of the AAHPERD skills test manuals: basketball, tennis, or softball. Before examining the manual, think of ways in which you would measure the areas delineated by AAHPERD. Examine the manual and see how your choices compare to AAHPERD's.

SKILLS TEST CLASSIFICATION

When you construct skills tests, consider whether to use objective or subjective procedures. Be sure that you are measuring only one trait at a time. The AAHPERD skills test batteries include only objective tests; however, subjective ratings are included as alternate forms for the tennis and softball batteries. You must consider a number of factors, including time, facilities, number of testers, and number and type of tests to be used, when deciding on a specific approach to skills testing.

Objective Tests

There are four primary classifications for objective skills tests:

- Accuracy-based skills tests
- Repetitive-performance tests
- Total body movement tests
- Distance or power performance tests

Some tests may be combinations of two of these classifications. Each classification involves specific measurement issues.

Accuracy-Based Tests

Accuracy-based skills tests usually involve the skill of serving an object, such as a volleyball, tennis ball, or badminton shuttlecock. They may also involve some other test of accuracy: throwing in football or baseball, free throws or other shots in basketball, or kicking goals in soccer. *The primary measurement issue associated with accuracy tests is the development of a scoring system that will provide reliable yet valid results.* Consider a soccer goal-scoring test. The test is set up so that the performer must kick the ball into the goal from 12 yd (11.0 m) away. For the performer to get the maximum amount of points, the ball must go between the goal posts and a restraining rope 3 ft (0.9 m) inside them. A performer is allowed six kicks: three kicks to the left side of the goal and three kicks to the right side. Two points are awarded if the ball goes into the goal in the target area, 1 point if it simply goes into the goal. The problem with this test is that good shooters often try to put the ball in the target area but end up being just slightly off, for a score of 0, whereas less competent performers may shoot every shot for the middle of the goal to ensure that they receive at least 1 point. Good shooters may thus obtain lower scores than shooters not quite as skilled. This reduces the reliability and the validity of the test.

Mastery Item 13.4

Can you detect another weakness with this soccer goal-scoring test that might affect its validity?

In volleyball tests, the court is often marked to delineate where the most difficult serves should be aimed. The problem with the soccer goal-scoring test just mentioned is inherent in this test as well, because serves that fall outside the court are given a score of 0. To rectify this problem, point values slightly lower than that of the target area can be assigned to serves that fall within some area slightly outside the target. This can, however, create a problem of feasibility—that is, of marking the court for this area.

Another consideration with accuracy-based tests is the number of repetitions necessary to produce reliable scores. When the AAHPERD Tennis Skills Test was first developed, the serving component did not attain the reliability values necessary for inclusion in the battery. The Spearman–Brown prophecy formula, when applied to the tennis-serving data, indicated that to improve the reliability of the test significantly, the number of trials would need to be more than doubled. This would affect the feasibility of the test. Therefore, AAHPERD developed a modified scoring system that enabled the reliability to increase to an acceptable level. When using the multiple-trial approach in serving, testers must maximize reliability while minimizing the number of trials.

An example of a currently used accuracy-based skills test is the North Carolina State University (NCSU) Volleyball Service Test (Bartlett, Smith, Davis, and Peel 1991). Content validity was claimed because serving is a basic volleyball skill. An intraclass reliability coefficient of .65 for college students was reported. Procedures for the administration of this test are presented here:

NCSU Volleyball Service Test

Purpose

To evaluate serving in volleyball.

Equipment

Volleyballs
Marking tape
Rope
Scorecards or recording sheets
Pencils

Instructions

Use regulation-sized volleyball courts for the serve test; prepare them as shown in figure 13.4. Mark point values on the floor. The test subject stands in the serving area and serves 10 times either underhand or overhand.

Scoring

Give scores of 0 to balls that contact the net or antennas or land out of bounds. Give the higher point value to balls that land on a line.

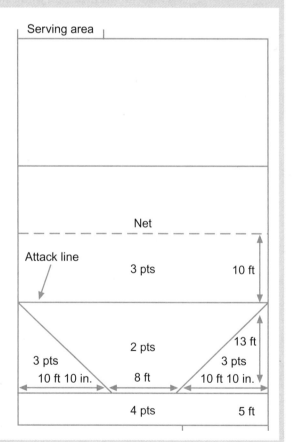

Figure 13.4   Floor plan for the NCSU Volleyball Service Test.
Reprinted from Bartlett et al. 1991.

Repetitive-Performance Tests

Repetitive-performance tests are tests that involve continuous performance of an activity (e.g., volleying) for a specified period of time. They are commonly called wall volleys or self-volleys and can be used to measure the strokes of racket sports, such as the forehand or backhand stroke in tennis and volleying and passing in volleyball. Repetitive-performance tests usually have a high degree of reliability, but unless they are constructed carefully they may not approximate the same type of action that is used in the game, in which case validity is reduced. Furthermore, because they are not gamelike, they may not transfer into actual game performance as well as some other court tests. Therefore, it is extremely important when using repetitive-performance tests that the testers make sure that the test subjects use the correct form.

An example of a repetitive-performance test is the Short Wall Volley Test for racquetball (Hensley, East, and Stillwell 1979). The test was originally administered to college students but is considered appropriate for junior and senior high school students as well. Test–retest reliability coefficients are .86 for women and .76 for men. A validity coefficient of .86 has been obtained by using the instructor's rating of students as the criterion measure. Procedures for the administration of this test are presented here:

Short Wall Volley Test for Racquetball

Purpose

To evaluate short wall volley skill.

Equipment

Rackets
Eye protection
Four balls at each testing station
Measuring tape
Marking tape
Stopwatches
Scorecards or recording sheets
Pencils

Instructions

The test subject stands behind the short service line, holding two balls. An assistant, located within the court but near the back wall, holds two additional balls. To begin, the subject drops a ball and volleys it against the front wall as many times as possible in 30 s. All legal hits must be made from behind the short line. The ball may be hit in the air after rebounding from the front wall or after bouncing on the floor. The ball may bounce as many times as the hitter wishes before it is volleyed back to the front wall. The subject may step into the front court to retrieve balls that fail to return past the short line but must return behind the short line for the next stroke. If a ball is missed, a second ball may be put into play in the same manner as the first. (The missed ball may be put back into play, or a new ball can be obtained from the assistant.) Each time a volley is interrupted, a new ball must be put into play by being bounced behind the short line. Any stroke may be used to keep the ball in play. The scorer may be located either inside the court or in an adjacent viewing area.

Scoring

The 30 s count should commence the instant the student drops the first ball. Trial 2 should begin immediately after trial 1. The score recorded is the sum of the two trials for balls legally hitting the front wall.

Mastery Item 13.5

What else could be done to increase the reliability of the Short Wall Volley Test?

Total Body Movement Tests

Total body movement tests are often called speed tests, because they assess the speed at which a performer completes a task that involves movement of the whole body in a restricted area. A dribbling test in basketball or soccer measures this skill; base running tests in baseball and softball are also tests of this type. *These tests usually have a high degree of reliability because a large amount of interindividual variability is associated with timed performances.* These tests can be administered quickly, but they have two inherent problems. First, a test must approximate the game performance, and in many cases flat-out speed of movement is not always required in the game. For example, in basketball, even on the fast break there must be some degree of controlled speed to allow for control of the ball. Therefore, to measure basketball dribbling skill, AAHPERD has selected a control-dribble test involving dribbling around a course of cones (Hopkins, Shick, and Plack 1984; figure 13.5). When this type of test is used, the course must be set up to validly

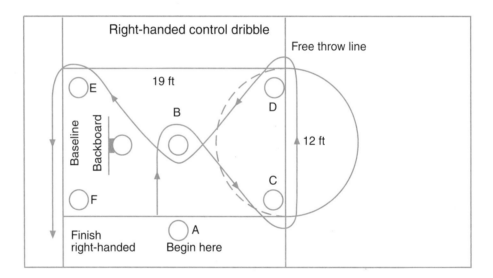

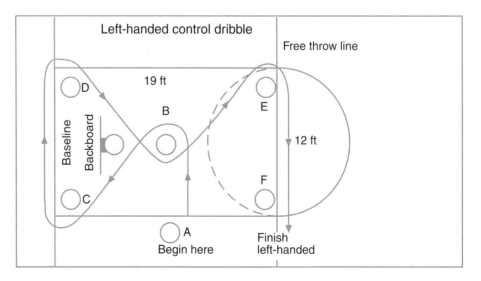

Figure 13.5   AAHPERD control-dribble test.

Reprinted from Hopkins, Shick, and Plack 1984.

approximate the skill used in a game. Second, when such tests are used for evaluation, performance time is the valid criterion, but if you are interested in efficiency of performance, these tests would be highly related to the performer's speed. Obviously, a faster performer can complete the test faster than a slower one, even if the slower performer has more skill in actual ball handling.

A way to eliminate the speed problem inherent in total body movement tests is to create performance ratios. You create a **performance ratio** by dividing a performance time by a movement time for the same subject. For example, you could compare a student's dribbling time to his or her movement time over the same course. That is, dribbling efficiency (a performance ratio) is equal to dribble time divided by movement time. Performance ratios can be extremely effective motivational tools for both highly skilled and less skilled performers because both are performing against themselves, trying to reduce the ratio to as close to 1 as possible. Theoretically, a ratio of 1 would be the minimum value on this type of test. Conversely, because movement time is measured by this type of ratio, faster performers could be unduly penalized. Also, if those being tested are aware of the manner in which these ratios are created, they might not give their maximum effort on the movement time test. This would allow them to achieve a better ratio score than their performance warrants. These ratios are not appropriate for use with a sports team because absolute performance is the primary measure. They are, however, an effective approach for measuring skill proficiency in a classroom setting.

Mastery Item 13.6

Describe how you could create a performance ratio for the control-dribble test in basketball.

Mastery Item 13.7

Can you describe a potential problem with performance ratios?

Another example of a total body movement test is the defensive movement test from the AAHPERD Basketball Skills Test (Hopkins, Shick, and Plack 1984). Intraclass reliability coefficients above .90 were reported, and concurrent validity established for the full test battery ranged from .65 to .95. The procedures for administering this test are presented here:

Defensive Movement Test for Basketball

Purpose

To measure performance of basic defensive movement.

Equipment

Stopwatch
Standard basketball lane
Tape for marking change-of-direction points

Instructions

Mark the test perimeters by the free-throw boundary line behind the basket, and mark the rebound lane lines into sections by a square and two lines (figure 13.6). Only the middle line—rebound lane marker (C in figure 13.6)—is a target point for this test. Mark with tape the additional spots outside the four corners of the area at points A, B, D, and E. There are three trials to the test. The first is a practice trial, and the last two are scored for the record. The performer starts at A and faces away from the basket. On the signal, "Ready, Go!" the

(continued)

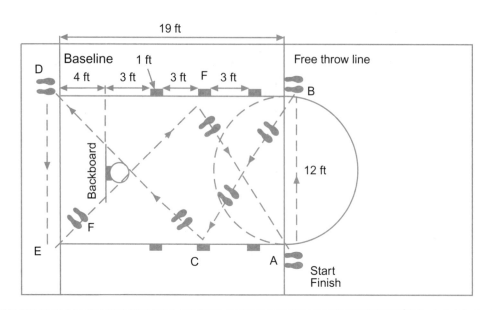

Figure 13.6   Perimeters and setup for the AAHPERD defensive movement test for basketball.
Reprinted from Hopkins, Schick, and Plack 1984.

performer slides to the left without crossing the feet and continues to point B, touches the floor outside the lane with the left hand, executes a drop step, slides to point C, and touches the floor outside the lane with the right hand. The performer continues the course as diagrammed. The performer completes the course when both feet have crossed the finish line. Violations include foot faults (crossing of the feet during slides or turns and runs), failure of the hand to touch the floor outside the lane, and executing the drop step before the hand has touched the floor. If a performer violates the instructions, the trial stops and timing must begin again from the start.

Scoring

The score for each trial is the elapsed time required to legally complete the course. Record scores to the nearest tenth of a second for each trial; the final score is the sum of the two latter trials.

Distance or Power Tests

The final classification for objective skills tests is **distance or power performance tests**, which assess one's ability to project an object for maximum displacement or force. The badminton drive test for distance and the racquetball power serve test are examples of this type (Strand and Wilson 1993), as are distance throws in softball and baseball and punt, pass, and kick competitions. One problem with these tests is ensuring that they are performed in a gamelike manner. Another question with tests of distance is whether to take accuracy into account. For example, in the punt, pass, and kick contest, the distance off the line of projection is subtracted from the distance of projection. Because of this, a performer might hold back from using maximum force for fear of losing accuracy. In contrast, in track events such as the discus and shot put, these corrections are not made as long as the object is projected within a given area. Consequently, before using a test it is important to consider whether the test requires or accounts for any correction for accuracy. In the volleyball arm power test using the basketball throw (Disch 1978), accuracy is not considered an important component because arm power associated with the ability to spike a volleyball is being measured; therefore, absolute distance is the criterion. However, in other sport skills tests, such as throwing a football or baseball, the accuracy aspect may be important, so some correction needs to be used. The simplest correction is subtracting the distance off the line of projection from the total distance.

An example of a distance or power performance test is the overhand throwing test from the AAHPERD Softball Skills Test (Rikli 1991). Intraclass reliability coefficients were found to exceed .90 for all samples, and concurrent validity coefficients were found to range from .64 to .94. The procedures for administering this test are presented here:

Overhand Throwing Test for Softball

Purpose

To measure the skill involved in throwing a softball for distance and accuracy.

Equipment

A smooth grass field that can be marked off in feet
2 measuring tapes
Softballs
2 small cones or marking stakes

Instructions

Mark off in feet or meters a throwing line (or position a measuring tape) down the center of a large, open field area, with a restraining line marked at one end perpendicular to the throwing line. A back boundary line is marked off 10 ft (3.0 m) behind the restraining line. The person being tested stands between the restraining line and the back boundary line, back far enough to take one or more steps before throwing. The subject, after 3 to 4 min of short-throw warm-ups, has two trials to throw the softball as far and as straight as possible down the throwing line, without stepping over the restraining line. Position assistants or other waiting subjects in the field to indicate, using a cone or marking stake, the spot where each ball first touches the ground. If a player steps on or over the restraining line before releasing the ball, he or she must repeat the trial.

Scoring

The net throwing score equals the throwing distance, measured at a point on the throwing line perpendicular to the spot where the ball landed, minus the error distance—the number of feet the ball landed away from the throwing line. The player's score is the better of the two throws. Measure both the distance and the error score to the nearest foot or meter.

Mastery Item 13.8

What are some factors that could reduce the reliability of the Overhand Throwing Test for Softball?

Subjective Ratings

Objective tests are attractive because they usually have a high degree of reliability, can be developed to produce good validity, and measure specific components of skill performance. However, **subjective ratings,** the value a rater places on a skill or performance based on personal observation, offer attractive alternatives for physical education teachers, coaches, and others interested in human performance analysis. Subjective ratings can be developed for individual skills that are often process oriented, making the ratings attractive from a teaching standpoint. A process-oriented skill is one in which the form of the skill is evaluated (e.g., in diving and gymnastics). Performers are evaluated on their preliminary position, force production phase, and follow through. Specific cues can be given to them about where they may be losing efficiency in performance. Ellenbrand (1973), for example, developed a general rating scale for the performance of gymnastic stunts (table 13.2). This scale could easily be adapted to evaluate other process-oriented performances. The alternative evaluation procedures presented in chapter 8 (pp. 133-135) are closely aligned with subjective ratings. The same reliability, validity, fairness, and feasibility issues described there are important here.

Table 13.2 Ellenbrand Gymnastic Rating Scale

The score for each item (event) is the product of a difficulty value and the executing rating. The sum of all test items in each event is the score for that event. The final test score is the sum of the scores for all events or all test items.

3 points	Correct performance. Proper mechanics. Executed in good form. Performer shows balance, control, and amplitude in movements.
2 points	Average performance. Errors evident in either mechanics or form. May show some lack of balance, control, or amplitude in movements.
1 point	Poor performance. Errors in both mechanics and form. Performer shows little balance, control, or amplitude in movement.
0 points	Improper or no performance. Incorrect mechanics or complete lack of form. No display of balance, control, or amplitude in movement.

There is no deduction for falls or repeated skills. However, a stunt that is executed with assistance receives a rating of 0.

Reprinted from Ellenbrand (1973).

Another application of subjective ratings is to observe participants in the activity and give them a **global rating** based on their overall performance in a competitive situation. This allows the tester to evaluate several performers concurrently and possibly evaluate some intangible aspects related to game performance that are not identified by performance of individual skills in a nongame setting. The problem with this approach lies in the number of observations possible. In one game, a student may have the opportunity to contact the ball several times, but in another game that student may make very few contacts. Other problems with subjective ratings are defining criteria and ensuring consistency among raters. In most physical education situations, the teacher is the only available person in class to do the rating, so the evaluation may be biased by preconceived notions about the performance or the performer. This is also true in many coaching situations. If several ratings could be obtained, reliability would be increased. However, this would also decrease the feasibility of the testing by either requiring more raters or increasing the number of viewing sessions necessary.

Mastery Item 13.9

Select a sport in which a serve is involved. Create a rating scale to evaluate the serving process. Create a rating scale to evaluate serving effectiveness in a game situation. Discuss which is better.

Types of Rating Scales

Verducci (1980) delineated two basic types of scales: relative scales and absolute scales. **Relative scales** are scored by comparing performance to that of others in the same group. This normative approach has the virtue of distinguishing ability well within the group but creates problems if performers are to be compared with those in other groups. Relative scales are classified as follows: rank order, equal-appearing intervals, and paired comparisons.

The most widely used of these approaches is the rank-order method. In this approach, all performers within a group are ranked on a given skill. If you are evaluating more than one skill, evaluate all the subjects on one skill before moving on to the next. Rank-ordering forces differentiation among all performers, but it does not account for the degree of difference between them.

The equal-appearing intervals method is often used when ranking groups of 20 or more subjects. Using this technique, several categories are assumed to be equally spaced from one another, and the rater places subjects with similar performances into the same categories. For example, categories such as best, good, average, poor, or worst might be used. A rater may decide to divide the group into the five categories. Typically, a larger percentage of subjects fall into the middle categories than the extreme categories.

In the paired-comparison method, the rater compares each subject to every other subject and determines which of each pair is better than the other on the trait being assessed. When this is done for all possible pairs of subjects, the results can be used to establish a relative ranking of all the subjects in the group. This technique works well with groups of fewer than 10 subjects.

With **absolute ratings,** the performer is evaluated on a fixed scale; performance is compared with a predetermined standard. This approach is not affected by the group in which a person is tested, and several people may end up with the same rating. Absolute scales can be classified into four types: numerical scales, descriptive scales, graphic scales, and checklists—the most widely used being numerical scales and checklists, which we will discuss in some detail. For discussions of less popular types of relative and absolute scales, see Verducci (1980, chapter 13).

Numerical scales are common. The numerical scales in tables 13.2 and 13.3 describe the levels of performance needed to earn a certain number of points. The Ellenbrand scale (table 13.2) ranges from 0 to 3, whereas the Hensley scale (table 13.3) ranges from 1 to 5. In general, numerical scales range from 1 up to 9 points; it is usually difficult to accurately discriminate more than nine performance levels. Numerical scales are most useful when performers can be classified into a limited number of ordered categories and there is consistent agreement about the characteristics of each category.

Checklists are useful if both the process and the outcome are being evaluated. **Checklists** usually represent a dichotomous rating of a trait—it is either absent or present. The following 15 basic skills, necessary for a person to be considered competent at swimming level I,

Table 13.3   Tennis Rating Scale: Forehand and Backhand

5 = excellent	Proper grip, good balance, footwork, and near-perfect form. Demonstrates consistent stroke mechanics. Anticipates opponent's shots. Placement appropriate for opponent's weaknesses or position.
4 = good	Proper grip, good balance, adequate footwork, and acceptable but not perfect form. Demonstrates above-average consistency of stroke mechanics. Anticipates opponent's shots. Consistent placement within court area.
3 = average	Proper grip and acceptable balance, but footwork is poor. Form is somewhat erratic and inefficient, resulting in inconsistency in shot placement. Style of play may be defensive. Little anticipation of opponent's shots.
2 = fair	Uses improper grip at times, poor footwork, and basically incorrect form. Inconsistent stroke mechanics. Defensive style of play, merely trying to get ball over net. Little anticipation of opponent's shots. Unable to sustain a rally.
1 = poor	Incorrect grip, off balance, with poor footwork. Form is very poor and erratic. Inaccurate shot placement. No anticipation of opponent's shots. Experiences difficulty in getting ball over net.

Reprinted from Hensley (1989).

have been taken from the American Red Cross Water Safety Instructor's Manual (American Red Cross 1992):

- Fully submerge the face for 3 s.
- Experience buoyancy: Bounce up and down in chest-deep water maintaining an upright position for 10 bounces, or bob to chin level with support—10 times.
- Demonstrate the support float on your front.
- Demonstrate the support float on your back.
- Demonstrate bubble blowing.
- Enter and exit water independently using a ladder, ramp, steps, or the side of the pool.
- Walk 5 yd in chest-deep water maintaining balance, or move 5 yd along the side of the pool maintaining contact with the wall.
- Demonstrate kicking on your front.
- Demonstrate kicking on your back.
- Walk 5 yd in chest-deep water using an alternating arm stroke or demonstrate alternating arm action for 10 s while holding the overflow gutter.
- Discuss the importance of swimming pool rules.
- Discuss the role of safety personnel and emergency medical services.
- Demonstrate reaching assists without equipment.
- Demonstrate how to relieve a cramp.
- Demonstrate wearing a life jacket on deck and entering shallow water.

Each skill is checked as completed or not completed. This criterion-referenced approach provides a concrete evaluation of given performance levels and gives specific feedback to the performer. It is easy for the instructor to determine which competencies are lacking and to provide extra practice in those areas.

Common Errors in Rating Scales

Several common errors occur with rating scales. The most common of these is termed the **halo effect,** which is the tendency of a rater to elevate a person's score because of bias. This can occur in two ways. First, a rater may have a strong attitude about a performer that biases the evaluation in the direction of the attitude. Second, the rater may believe that the performance being evaluated is not indicative of the performer's normal level and rate the performer based on previous performances. The halo effect may also work in reverse; the rater may reduce a person's score because of negative bias.

Another common error, termed **standard error,** occurs when a rater has a standard different from that of other raters. Consider the judges' ratings in table 13.4. Inspection of

Table 13.4 **Standard Error Exemplified by Three Judges' Ratings**

	Judge		
Performer	A	B	C
1	9	8	4
2	8	9	4
3	7	7	3
4	5	6	1
5	5	5	1

the rating indicates that all three judges ordered the performance similarly; however, the numerical ratings from judge C were substantially below those from judges A and B, indicating that judge C was functioning with a standard vastly different from that used by judges A and B. This could be a major problem if all subjects were not rated by all judges.

A third error, called **central-tendency error,** reflects the hesitancy of raters to assign extreme ratings, either high or low. Assume that you are rating people on a scale of 1 to 5. There is a common tendency not to use the extreme categories, which reduces the effective scale to three categories (i.e., 2, 3, and 4). This not only causes scores to bunch around the mean but also reduces the variability in the data, which can reduce reliability.

Suggestions for Improving Rating Scales

You can take several steps to alleviate many of the problems associated with rating scales.

1. Develop well-constructed scales. Here are several suggestions for doing this:
   - State objectives in terms of observable behavior.
   - Select traits that determine success.
   - Define selected traits with observable behaviors.
   - Determine the value of each trait in relation to success.
   - Select and develop the appropriate scale for the rating instrument.
   - Select the degrees of success or attainment for each trait and define them with observable behaviors.
   - Try out and revise the rating scale.
   - Use the rating scale in an actual test situation.

Can we measure athletic potential and predict athletic success? The answer may lie in developing test batteries that discriminate among performance levels.

2. Thoroughly train the raters. They should have a clear understanding of the traits measured and be able to differentiate thoroughly the differences in performance levels.

3. Explain common rating errors to the raters. If they are aware of these pitfalls, they may avoid them.

4. Allow raters ample time to observe behaviors. This will increase the sampling unit of the performances.

5. Use multiple raters whenever possible. If this is not possible, test several raters on a common group to check their objectivity. Finally, raters should rate one trait at a time, then move on to the next trait. This improves consistency.

Other Tests

In addition to objective tests and subjective ratings, there are other classifications of tests for the measurement of human performance. Performance-based testing and trials-to-criterion testing are described briefly in the following paragraphs. Performance-based testing involves actual performance of the activity that is being assessed. Trials-to-criterion testing offers instructors a way to reduce the amount of time spent testing and to increase the amount of time available for teaching.

Performance-Based Testing

A third classification for evaluation of skill performance is the performance producing the score. In this situation, a concrete criterion exists. This occurs in sports such as archery, bowling, golf, and swimming. In archery, the score for a round of arrows indicates how well the performer did for that performance. Scores can be evaluated to examine the stability of performance at a given distance, or performance at different distances can be evaluated to determine the concurrent validity of shooting across distance.

Bowling and golf provide a unique situation: Although you can use the total score for a game of bowling or a round of golf for assessment purposes, this does not evaluate specific elements of the game. Picking up a specific pin for a spare in bowling, or your drive, your short game, and your putting in golf, are elements that can be evaluated separately. These performances often focus on the process rather than the product, and rating scales often have a good degree of validity in these areas. The virtue of developing rating scales in these areas is that you have a concrete criterion against which to validate them. On the other hand, correlating a rating scale against an objective skills test may or may not represent validity. You may be simply getting a congruency of the two performances, neither of which may be valid.

Trials-to-Criterion Testing

An alternative approach to the measurement of skills involves the use of **trials-to-criterion** testing (Shifflett and Shuman 1982), in which the student performs a skill until he or she reaches a certain criterion performance. For example, consider a conventional free throw test involving 20 attempts. A class of 30 students would have 600 free throws to complete. Using a trials-to-criterion approach, the students would be given some number of successes to complete. For example, instead of shooting 20 free throws and counting the number made in 20 attempts, the students might be instructed to shoot until they make 8 free throws. As soon as they make 8, they are to report the number of attempts it took to make the 8; the best score would be 8 attempts to make 8 shots. (At some point the test would have to be discontinued for those students who simply cannot make the 8 free throws, but for a given class, most of the shooters can probably make 8 out of 20 attempts.) If the average number of trials needed to make 8 free throws is 12 in a class of 30 students, only 360 free throws would need to be attempted (not 600), thus saving a great amount of time. If the correlation between the trials-to-criterion test scores and the scores on the conventional free throw test (involving 20 attempts) is high, you would have an attractive way of reducing the time of testing. Furthermore, in a teaching situation, this would allow the teacher to spend additional time with the weaker shooters, while the stronger shooters move on to other skills.

Mastery Item 13.10

Scott wants to add a spiking test to his volleyball performance battery. Develop a trials-to-criterion test to help him measure spiking.

TESTING MOTOR ABILITIES

Prediction in human performance and sport has long been a popular topic of debate. Is there such a thing as a "natural athlete"? What physical attributes are most important for high levels of athletic performance? Is it possible to measure athletic potential and predict future athletic success?

With the advent of high-speed computers and the application of multivariate statistical techniques to the analysis of human performance, researchers are now able to explore these questions in a manner that was not possible 40 years ago. Although the statistics involved are relatively complex, the theory is basic: Attempts are made to develop test batteries that discriminate among performance levels.

Historical Overview

Early researchers operated on the theory that just as there were tests for assessing the innate ability of intelligence in the cognitive domain, there must also be a way to measure innate motor ability in the psychomotor domain. These early researchers—Rogers, Brace, Cozens, McCloy, Scott, and others—concentrated from the early 1920s through the early 1940s on determining the physical components that are basic to and necessary for successful human performance.

One of the initial attempts was the development of classification indexes for categorizing students according to ability. This was to allow physical education classes to be formed homogeneously so that they could be taught with increased efficiency. The earliest classification indexes focused on predicting ability by age, height, and weight information. C.H. McCloy (1932) developed three classification indexes in his early work that could adequately classify students:

Elementary: $(10 \times age) + weight$

Junior/senior high school: $(20 \times age) + (6 \times height) + weight$

College: $(6 \times height) + weight$

where age is in years, height is in inches, and weight is in pounds. By inspecting these formulas, McCloy found that at the college level age was no longer an important factor in classification and that at the elementary level height contributed little. This was one of the first attempts to predict performance, but at this point no motor performance tests were actually used.

Next, Neilson and Cozens (1934) developed a classification index based on the same principle; however, they used powers (exponents) of height in inches, age in years, and weight in pounds. The McCloy and the Neilson and Cozens classification systems were so similar that the correlation between them was .98.

At about the same time, researchers began classifying by motor ability testing. The term **general motor ability (GMA)**—referring to overall proficiency in performing a wide range of sport-related tasks—was coined. Rogers and McCloy developed strength indexes that included some power tests, the standing long jump, and the vertical jump, which had been found to correlate moderately with ability in a variety of activities (Clarke and Bonesteel 1935). To increase the accuracy of the prediction, test batteries were designed on the premise that certain motor abilities, such as agility, balance, coordination, endurance, power, speed, and strength, were the basis of physical performance.

An example of an early test of general motor ability is the Barrow Motor Ability Test (Barrow 1954). Although the test was initially designed for college men, norms were developed later for both junior and senior high school boys. Originally, the test included 29 test items measuring eight theoretical factors: agility, arm–shoulder coordination, balance, flexibility, hand–eye coordination, power, speed, and strength. Barrow constructed an eight-item test battery to examine the validity of the factors and obtained a multiple correlation coefficient of .92 between the composite score for the eight-item battery and a composite score based on all 29 items. Test–retest reliability coefficients ranged from .79 to .89. From a measurement standpoint, the problem associated with this validation procedure is the composite criterion: Because both the predictor tests (a subbattery) and the criterion test (the full battery) included the same tests, Barrow was simply predicting the whole from a part, giving a spuriously high correlation of .92. It is statistically invalid to use this approach. The importance of this early motor ability test battery was that Barrow started to examine the theoretical structure of the various components of the athletic ability necessary to perform a variety of sports.

Mastery Item 13.11
| What could Barrow have selected to validate his battery?

Larson (1941) provided a different measurement approach to the examination of motor ability. He analyzed the factors underlying 27 test items and six batteries and gathered further statistical evidence on other test batteries related to measuring motor ability. This represented an attempt to examine the construct validity of motor ability tests through the use of a statistical technique called factor analysis. However, he also developed these batteries using the composite criterion approach, which reduced the validity of his findings. He did find reliability coefficients in excess of .86.

The next step in the use of motor ability tests was the development of tests to measure **motor educability**—the ability to learn a variety of motor skills. These tests included track-and-field items as well as a number of novel stuntlike tests. One of the stunts involved having students grab one foot with the opposite hand and then jumping to try to bring the other leg through the opening formed by the leg and the arm. Another item had the students jump up, spin around, and try to land facing exactly the same direction as they faced at the start. David Brace (1927) developed the Iowa Brace Test, involving 20 different stunts, each scored on a pass–fail basis. McCloy later used the Iowa Brace Test as a starting point in developing a test for motor educability. McCloy's revision investigated 40 stunts and selected 21 of them. Six different combinations of 10 stunts were ultimately selected to measure motor educability in the following categories: upper elementary school boys, upper elementary school girls, junior high boys, junior high girls, senior high boys, and senior high girls. A problem with tests of motor educability was the reliability of the measures. Because they were all pass–fail items, it was an "either/or" evaluation, which tended to reduce reliability. Furthermore, the tests tended to correlate poorly with most sports performance measures, thus casting doubt on their validity in any realistic situation.

Mastery Item 13.12
| How could the reliability of these motor educability items be improved?

The Sargent Jump was one of the first motor performance tests used to examine a person's potential for athletic performance (Sargent 1921). Named after Dudley Sargent, the test was a vertical jump purported to be a measure of leg power. This test was widely used and did measure a trait important to many power-based sports. Whereas this is a reliable test that is validly related to certain aspects of performance in a large number of sports, this single measure obviously gives an incomplete picture of overall athletic ability.

In the years that followed, the concept of **specificity**—motor abilities unique to individual psychomotor tasks—arose through the works of Franklin Henry (1956, 1958). Using correlational analysis, Henry stated that traits that have more than 50% shared variance ($r^2 > .50$), or, in other words, that have correlations above .70, were said to be general in nature. Any two tests that had correlations of .70 or less were said to be specific. Because of the magnitude of the correlations that Henry chose, most motor ability tests were found to be specific in nature. The cause-and-effect implication from this was that traits should be trained specifically. Also, anyone proficient athletically would have a high degree of many of these specific traits.

From the 1940s to the 1970s, other researchers, such as Seashore, Fleishman, Cumbee, Meyer, Peterson, and Guilford, developed the notion that motor ability is specific rather than general in nature. Factors most often cited by these investigators included muscular strength and endurance, speed, power, cardiovascular endurance, flexibility, agility, and balance. In general, their theories were based on the correlations between physical factors. High correlation suggests that items have much in common, but low correlation suggests that items measure different traits. Thus, specificity of tasks can be viewed as a concurrent validity approach.

During this period, Fleishman (1964) was developing what he termed the **theory of basic abilities.** This theory serves as the basis for most of the scientific research that has been conducted subsequently in this area. Fleishman distinguished between skill and ability in the following manner: *Skills are learned traits based on the abilities that the person has, whereas abilities are more general and innate in nature than skills.* For example, the tennis serve, badminton serve, and volleyball spike are all specific skills that involve similar overarm patterns. The overarm pattern is considered the ability. Fleishman was also one of the first to examine his theory using factor analysis. His work is classic in the field.

Measurement Aspects of the Domain of Human Performance

A multitude of motor performance factors affect an individual's ability to perform specific sport skills. These factors are the underlying bases for maximized human performance. You have just read about the historical development of the research in this area. Fleishman and Quaintance (1984) further examined these factors (domains) from a construct-validity perspective. They improved on earlier works by expanding the taxonomies of human performance.

Using Fleishman's work as their basis, a number of other researchers began examining the construct validity of various areas of human performance. The broad area of human performance is referred to as the *domain of human performance*. The structure of the areas within the broad domain—the subdomains—has been the topic of research for a number of scholars in our field. Examining the subdomains of human performance allows you to understand the qualities necessary to perform various tasks. It is important to examine these subdomains from a variety of different standpoints, because factors such as age and ability level could alter the structure of these domains.

The primary subdomains of human performance are as follows:

- Muscular strength
- Speed
- Agility
- Anaerobic power
- Flexibility
- Balance
- Kinesthetic perception

Researchers have examined the construct validity of most of these subdomains.

Muscular strength can be classified in relation to either the body segment isolated or the method of measurement. Jackson and Frankiewicz (1975) examined the factor structure of muscular strength and concluded that there were two general dimensions: upper body and leg. To measure strength comprehensively, at least one test from each dimension should be selected. The manner in which strength is measured also affects the measurement situation. Clarke spent years examining the various aspects of isometric (static) strength (Clarke and Munroe 1970). Isotonic (now often referred to as dynamic) strength was the subject of investigation by a number of researchers, most notably Berger (1966). With the implementation of machine-based strength, the examination of isokinetic strength has become an area of research interest. Brown (2000) edited a book detailing the relationships between isokinetic strength and a variety of athletic performances and human movements. All these types of strength present different criteria for how strength is best measured. It is important when measuring strength that you select the method of measurement and the body region tested based directly on the task being evaluated.

The subdomain of speed is important to a number of athletic competitions and is most commonly measured by a short sprint. Disch (1979) found the subdomain of speed to consist of four dimensions: sprinting speed, controlled speed (commonly called agility), arm speed, and leg speed. Because these are distinct factors, it is necessary to measure them all if a comprehensive measure of speed is required.

Mastery Item 13.13

What are some sports in which tests of arm speed or leg speed might be important predictors of playing ability?

From a mechanical standpoint, power is physical work divided by time, where work is defined as weight times the distance the weight is moved. Using this definition, Barlow (1970) and Considine (1970) examined the subdomain of anaerobic power, finding it to be unique to certain body regions, specifically the arms and legs. Selected tests to measure this domain are as follows:

Arm power

- One-hand shot put
- Two-hand shot put over head
- Medicine-ball pitch
- Basketball throw

Leg power

- Margaria–Kalamen Leg Power Test
- Incline run

Flexibility has been found to be body-area specific. Harris (1969) examined the factor structure of flexibility and found that two types of flexibility exist: movements of a limb that require only one joint action, and movements that require more than one joint action. She found 13 distinct factors and concluded that there are a number of types of flexibility within the body.

Fleishman (1964) identified a factor that he called *dynamic flexibility,* which involved a person's ability to change direction with the body parts rapidly and efficiently. This can be thought of as agility that does not involve running. This basic ability is also specific to the sport with which it is associated.

Balance is a multidimensional subdomain. Bass (1939) worked extensively with balance. Balance can be classified as either static or dynamic. Static balance is the ability to maintain total body equilibrium while standing in one spot, whereas dynamic balance involves the ability to maintain equilibrium while moving from one point to another. The types of static bal-

Remember, a test is reliable and valid only under particular circumstances (e.g., for a given gender, age, or test environment) and is not generally reliable or valid.

ance can be influenced by the restrictions of the balancing task as well as by whether the balance is performed with the eyes open or closed. Dynamic balance can be divided into simple or complex tasks based on the planes of balance involved. An example of a simple balance task involves balance on a platform stabilometer. The person has to balance in only one plane of movement. If a ball-and-socket stabilometer is used, complex balance is required in more than one plane.

Kinesthetic perception is the ability to perceive the body's position in space (Singer 1968). Although tests of it are the least objective of the subdomains of human performance, it is very well accepted as an area that must be considered. It is by far the most difficult to measure in terms of reliability and validity.

Perhaps the most important issue to keep in mind when you are testing in the human performance subdomains is specificity of task. This specificity relates directly to the reliability and validity of particular measurements. Remember, a test is reliable and valid only under particular circumstances (e.g., for a given gender, age, or test environment) and is not generally reliable or valid.

PURPOSES OF HUMAN PERFORMANCE ANALYSIS

Human performance analysis can be applied to several research questions. *The primary purposes of human performance analysis are selection, classification, and diagnosis.* Analysis questions apply not only to athletic assessment but also to job performance. Selection refers to the ability of a test to discriminate among levels of ability and thus allow someone to make choices. From an athletic standpoint, this could involve making cuts from or assigning players to teams. In job performance, selection is used in hiring. Classification involves clustering subjects into groups for which they are best suited. In athletic situations, players are assigned to positions or events; job classification involves assignment to a task. Diagnosis concerns determining a person's shortcomings based on tests that are validly related to performance in a given area. Diagnosis is used in sport to design individualized training programs to help improve performance. In the working world, diagnostic tests could be used to examine job performance.

MacDougall and Wenger (1991) discussed benefits that the athlete can derive from motor performance testing. In their work for the Canadian Association of Sport Science, they noted the following beneficial functions of performance testing:

▶ Indicates athletes' strengths and weaknesses within the sport and provides baseline data for individualized training

▶ Provides feedback for the athlete and coach about effectiveness of the training

▶ Provides information about the athlete's current performance status

▶ Is an educational process to help the athlete and coach better monitor performance

They further stated that for testing to be effective, evaluators should follow these procedures:

▶ Include variables relevant to the sport.

▶ Select reliable and valid tests.

- ▶ Develop sport-specific test protocols.
- ▶ Control test administration rigidly.
- ▶ Maintain the athletes' right to respect.
- ▶ Repeat the testing periodically.
- ▶ Interpret the results to both the coach and the athlete directly.

There are two major statistical approaches to the analysis of the subdomains of human performance. The first is the correlational approach to examine the various relationships across groups. You can use multiple regression for this approach if only two groups are involved. The second approach is the divergent group approach using discriminant analysis. The divergent group approach examines groups (can be more than two) known to differ by using variables that are logically related to performance of a skill. Tests found to discriminate between performance levels in the divergent groups are then said to be predictors from a construct validity standpoint. The inferential statistical procedures presented in chapter 5 can be used for this validation process also.

The following section presents examples of analysis of human performance for the purposes of selection, classification, and diagnosis.

Selection Example

An interesting study of the use of predictive validity for selection was conducted by Grove (2001). He studied 74 male baseball players who were competing at the college level: junior college (JUCO) or Division I (D1). A third group of players was formed by 16 who were drafted by the professional league within 24 months of the testing (Pro). The players were measured on run times (30 and 60 yd [27.4 and 54.9 m]), throwing speed, vertical jump, and medicine ball throw. The data were analyzed using analysis of covariance with age used as the covariate. Significant differences ($p < .001$) were found for vertical jump, medicine ball, and throwing speed. Post hoc analysis (using Bonferroni t tests) indicated that the JUCO group was lower than the D1 and Pro groups on the vertical jump and medicine ball. The Pro group performed better on the throwing speed test than the other two groups. The means and standard deviations for all of the tests are presented in table 13.5. The battery was simple to administer, and with the exception of the vertical jump and medicine ball, the tests are widely used by baseball coaches and scouts at all levels. Grove concluded that the test battery used has merit as a cost-effective screening test for talent identification in baseball. He further stated that more pronounced differences may exist when players are grouped according to position as well as playing level.

A second example of the use of selection can be seen in the study by Thissen-Milder and Mayhew (1991). The purpose of their study was to determine the accuracy of general and specific tests for identifying freshman, junior varsity, and varsity high school volleyball players. The specific tests (skills tests) were an overhead volley, a forearm pass, a wall spike, and a bump-set test. The general tests (motor performance and anthropometric tests) were height, weight, percent body fat, agility run, vertical jump, and two flexibility tests. The authors found that 68% of the players could be properly classified according to team based on the following variables: forearm pass, overhead volley, vertical jump, and weight. The group means for the various tests are presented in table 13.6. Further examination of the data indicated that 78% of the starters and nonstarters could be correctly classified by the bump-set test, height, weight, and shoulder flexibility. This type of information is very helpful to coaches not only for making decisions about who plays on which team but also in providing corroborating evidence for their decisions. Such data can also be used diagnostically to help individuals train in specific areas.

Mastery Item 13.14

Describe how you would set up a selection study in an area of interest to you. Discuss the variables that you would use as predictors. How would your groups be formed?

Table 13.5   **Descriptive Statistics and Subgroup Comparisons for the Performance Tests by Grove (2001)**

Measure	JUCO players		Division 1 players		Pro	
	Mean	SD	Mean	SD	Mean	SD
30 yd [27.4 m] time (s)	4.04	.025	4.02	0.17	3.97	0.16
60 yd [54.9 m] time (s)	7.39	.045	7.34	0.28	7.27	0.24
Radar gun (mph)*	78.14	3.45	80.61	4.63	84.40	3.49
Vertical jump (kgm)**	54.96	7.53	56.04	8.32	57.33	10.70
Medicine ball (kgm)**	14.50	1.61	16.47	1.37	16.18	1.80

Note: JUCO = junior college. Group sizes for the JUCO, Div 1, and Pro players were 32, 26, and 16, respectively.
$*p < .0005; **p < .001.$

Table 13.6   **Volleyball Group Data**

Variable	Freshman ($n = 12$)		Junior varsity ($n = 14$)		Varsity ($n = 24$)		*F
	Mean	SD	Mean	SD	Mean	SD	
Age (yr)	14.12	0.61	15.65	0.63	16.4	0.64	47.64
Height (cm)	167.1	6.7	167.0	7.4	168.7	7.8	0.22
Weight (kg)	58.8	6.4	50.7	8.0	58.6	10.5	0.24
Sum of skinfolds (mm)	69.7	13.0	75.4	18.3	66.5	16.6	1.03
Percent fat	18.1	2.6	19.6	3.4	17.2	3.8	1.65
Shoulder flexibility (in.)	0.3	0.7	0.5	1.0	0.3	0.9	0.55
Sit-and-reach flexibility (in.)	6.5	1.8	6.6	2.5	6.6	2.2	0.01
Agility (s)	33.8	2.0	31.6	2.3	41.9	1.8	13.42
Vertical jump (cm)	37.8	7.1	35.6	5.9	43.6	5.6	5.29
Anaerobic power (W)	782.2	152.7	783.6	112.4	836.9	79.9	0.82
Forearm pass (cts/min)	21.1	8.0	25.5	8.4	40.8	6.9	25.63
Overhead volley (cts/min)	22.9	7.2	22.3	7.2	34.8	6.8	12.57
Wall spike (cts/min)	5.2	4.6	7.1	8.8	13.7	7.3	6.65
Bump-set (cts/min)	27.6	9.2	31.4	7.8	41.9	3.6	13.42

Note: $*F = 5.10$ significant at $p < .01.$

Classification Example

Leone, Lariviere, and Comtois (2002) provided an excellent example of classifying athletes based on anthropometric and biomotor variables. The subjects for their study were elite adolescent female athletes with a mean age of 14.3 (standard deviation of 1.3 years). The athletes competed in tennis ($n = 15$), swimming ($n = 23$), figure skating ($n = 46$), and volleyball ($n = 16$). The descriptive values for the anthropometric and biomotor variables are presented in table 13.7.

Discriminant analysis of the tests revealed three significant functions ($p < .05$). The maximum number of significant functions possible is $K - 1$, where K is the number of groups; therefore, this test battery maximally discriminated across the four groups. Inspection

Table 13.7   Physical Characteristics of the Athletes by Sport (Mean ± SD)

	Tennis (n = 15)	Skating (n = 46)	Swimming (n = 23)	Volleyball (n = 16)
Age (years)	13.9 ± 1.3	14.7 ± 1.5	14.3 ± 1.3	13.8 ± 1.3
Body mass (kg)	50.6 ± 8.3	46.6 ± 8.0	54.3 ± 6.9	57.7 ± 8.3
Height (m)	1.61 ± 0.06	1.54 ± 0.07	1.62 ± 0.06	1.63 ± 0.05
Elbow (cm)	6.12 ± 0.30	5.87 ± 0.35	6.29 ± 0.26	6.40 ± 0.33
Knee (cm)	8.81 ± 0.43	8.63 ± 0.76	8.77 ± 0.34	9.31 ± 0.50
Biceps (cm)	25.5 ± 2.8	24.4 ± 2.3	27.8 ± 1.8	26.6 ± 2.2
Calf (cm)	34.0 ± 2.8	33.0 ± 2.7	34.4 ± 1.6	34.4 ± 2.2
Skinfolds (mm)	57.4 ± 17.8	47.7 ± 12.3	56.0 ± 15.0	63.1 ± 15.5
Push-ups (n)	57.8 ± 14.4	36.7 ± 13.5	62.1 ± 16.0	50.2 ± 13.5
Burpees (n)	46.1 ± 23.8	64.6 ± 33.2	52.5 ± 32.7	56.0 ± 28.4
Flexibility (cm)	37.3 ± 5.0	42.6 ± 5.1	41.0 ± 6.0	39.1 ± 6.9
$\dot{V}O_2max$ (ml · kg^{-1} · min^{-1})	49.5 ± 4.4	48.3 ± 4.0	47.6 ± 3.1	48.9 ± 3.6

of the functions revealed that function 1 discriminated between the figure skaters and all of the other sports grouped together. The variables that accounted for this discrimination were body mass, height, push-ups, and biceps girth. Function 2 reflected differences between volleyball players and swimmers. The variables accounting for this discrimination were body mass, biceps girth, calf girth, and height. The third function differentiated the swimmers and tennis players. The variables responsible for this discrimination were body mass, biceps girth, calf girth, sum of skinfolds, and height. The significant discriminant functions were able to classify 88% of the players into their correct sports. The classification summary table is presented in table 13.8. The results of this study indicated that elite adolescent female athletes could be properly classified into their respective sports based on the selected test battery. The anthropometric variables were primarily responsible for most significant classifications. Obviously, physical and anthropometric tests alone will not perfectly classify participants. Other factors, such as motivation and desire, could affect ultimate sport performance. These factors are presented in chapter 14.

Table 13.8   Classification for All Significant Discriminant Functions After Validation

		Predicted group membership, n (%)			
Groups	n	Tennis	Skating	Swimming	Volleyball
Tennis	15	11 (73.3)	0 (0.0)	3 (20.0)	1 (6.7)
Skating	46	0 (0.0)	46 (100)	0 (0.0)	0 (0.0)
Swimming	23	3 (13.0)	0 (0.0)	18 (78.3)	2 (8.7)
Volleyball	16	1 (6.3)	0 (0.0)	2 (12.6)	13 (81.3)

Mastery Item 13.15

Examine the misclassifications found in table 13.8. How might you interpret these findings?

Diagnostic Example

A study by Doyle and Parfitt (1996) based on the principles of personal construct theory (Kelly 1955) examined the possibility of performance profiling athletes. This study is interesting because it presented a unique quantitative performance profiling technique that involves not only motor performance factors but also psychological parameters. The profiling technique is presented in figure 13.7. Participants in the study were 39 track-and-field athletes, 22 males and 17 females, with a mean age of 20.9 years (standard deviation = 2.26). The unique aspect of this study is that the profiling technique used was to examine how each athlete currently felt about his or her preparation for competition. Instead of actually completing performance tests, the athletes were asked to respond to questions rating themselves on the various parameters displayed in the profile. The athletes were asked to respond on a

Figure 13.7 Sample performance profile.

scale of 1 to 10, with 1 being not important at all and 10 being of crucial importance. Their reported scores were then correlated with their performance in three upcoming competitions. To establish a criterion for success, the individual's performance was recorded as a percentage of his or her personal best time divided by his or her performance time. This allowed all the athletes to be compared across various events. Multiple correlations were calculated between the performance and the event competition scores. The results of the analysis indicated that the profiling technique could validly predict competition scores. Progressively stronger relationships were found between profile scores and performance measures from competition one to competition three. It was concluded that there may be a learning process involved in the ability to rate current state more precisely.

Mastery Item 13.16

A high school basketball coach is interested in the free-throw shooting ability of his players. He wants to know if there are differences between the varsity, junior varsity, and freshman teams. Each player attempted 100 free throws per day for 4 weeks of preseason practice. Their average score for the 4 weeks was used as their criterion score. Download and use the data from Mastery Item 13.16 to calculate a one-way ANOVA to determine if there are significant differences among the teams.

Measurement and Evaluation Challenge

Scott found that the information gathered from these tests could be used as a diagnostic technique to help him develop individualized training programs for his athletes. He calculated percentiles for the data and generated performance profiles for all of the players. An example of this type of table is presented in table 13.9 (Disch and Disch 1979). By examining the profiles, he could see which motor performance areas needed to be developed for each player.

In table 13.9, three players are examined. Player 1 had high percentile scores on all the motor performance tests and the anthropometric measures. This player was selected as an All-American setter at the U.S. Volleyball Association Collegiate Championships. The profile of player 2 includes very high scores in the motor performance characteristics but a lack of high scores on the anthropometric characteristics. This player was an excellent setter but did not quite reach the level of performance of player 1. Player 3 was found to have very favorable anthropometric characteristics, but his motor performance profile was much below that of the first two players. The data indicate that player 3 needs to concentrate on improving his motor performance characteristics. This should improve his performance on the volleyball court.

Table 13.9   **Men's Volleyball Performance Profile**

Percentile	Weight (lb)	Height (in.)	Reach (in.)	Percent body fat	Vertical jump (in.)	Triple hop (in.)	Agility run (s)	20 yd dash (s)
99	(200)	78		5.70		344	(7.7)	2.5
95	189	77	100	5.94	(29)	(341)		(2.7)
90	188		(99)	(6.15)	(27)	333	(7.8)	
85	(185)	(76)	(94.5)	(6.58)	26	(330)		(2.8)
80	183	75.5	97	6.86	25	319	7.9	
75	182		96	(6.99)		313		
70	181	75		7.30	24	303	8.1	
65	(180)		95.5	7.41		302		2.9
60	179	74	95	7.55		297		
55	174		94.5	7.6	(23)	296	8.3	
50	172		(94)	7.74		(295)	(8.5)	
45	169	(73)		8.09		292		
40	162		93.5	8.21		287		
35	161		93	8.47		285		(3.0)
30	158	72	92.5	9.68		279	8.8	
25	157	71		9.88	22	276		
20	156	70.5		10.15				
15	154		90	10.88	21	272	8.9	3.1
10	151	70	89	11.63		266	9.4	3.3
5	136	69	88	11.63	20	254	9.5	3.4

Note: Player 1: solid line; Player 2: dotted line; Player 3: dashed line.

SUMMARY

Reliable and valid measurement of sport skills and basic physical abilities has a prominent place in human performance testing. Assessment of psychomotor ability is an essential task that may confront you as a physical therapist, personal trainer, physiologist, physical education instructor, athletic coach, or other human performance professional. Sport skills testing will be important to you as a physical education instructor, athletic coach, physical therapist, exercise scientist, or other human performance professional. A reliable and valid testing program will help you become a respected professional in human performance.

Skills testing comprises a variety of test methods, including objective procedures, subjective ratings, and direct performance assessment. An extensive presentation of a wide range of current sport skills tests is beyond the scope of this text. Those of you who are interested can find a thorough compilation of sport skills tests in Strand and Wilson (1993). Motor ability testing, as you have seen, has a long history in human performance and will take on increased importance in athletics and employment testing. *The most important consideration is to select valid tests that meet your test objectives and are feasible in terms of time and effort.* Kirby (1991) is an excellent resource for descriptions and critiques of motor performance tests.

14

Psychological Measurements in Sport and Exercise

Robert S. Weinberg, Miami University

Measurement and Evaluation Challenge

Bill Keller has just been hired to coach a team in the National Football League. Bill was a football player in college, but because he didn't have as much natural ability as many of his opponents or teammates, he always believed that his mental skills and competitiveness really helped him achieve the high level he attained. For example, Bill believed that he was a self-motivated athlete, was usually able to control his emotions, was able to keep focused throughout the game, was generally confident in his abilities, and didn't let a series of bad losses get him down. Therefore, as he prepares to take over his first head coaching assignment, he believes that it's important not only to assess his players' physical abilities but also to evaluate their mental skills. However, Bill has little background in assessing players' personality and mental skills and thus he has a lot of questions. For example, what psychological inventories should he use to evaluate his players? Should he be administering and interpreting these tests, or should he hire a qualified sport psychologist to do so? When during the season should these psychological inventories be administered? Should he use interviews to find out about his players' mental skills? These are difficult questions, but if he can get the answers, Bill believes that the information derived from these psychological assessments will be invaluable to both him and his players in understanding and improving their mental skills.

Objectives

After studying this chapter, you will be able to

- define and identify the scope of the field of sport psychology,
- illustrate the differences between the performance enhancement aspect and mental health aspect of sport psychology,
- differentiate between psychological states and traits,
- explain the differences between general and sport-specific psychological tests,
- discuss ethics and cautions in the use of psychological testing with athletes,
- describe the qualifications necessary for using and interpreting psychological tests,
- illustrate the feedback process involved in psychological testing of athletes,
- discuss the use and abuse of psychological tests for team selection,
- identify factors related to the reliability and validity of general and sport-specific psychological inventories typically used in sport and exercise settings, and
- differentiate between the research and application perspectives of psychological inventories used in sport and exercise.

The purpose of this chapter is to provide you with an introduction to the evolving field of sport psychology and to highlight the measurement techniques and instruments typically used to assess psychological attitudes, states, and traits. In addition, we discuss issues related to the measurement and interpretation of psychological tests along with ethical considerations when using psychological tests with athletes and sport participants.

SPORT PSYCHOLOGY: PERFORMANCE ENHANCEMENT AND MENTAL HEALTH

The field of sport psychology has developed so rapidly in the last 30 years that many people do not have a clear understanding of what the field encompasses. Most definitions of sport psychology clearly underscore two main areas: *performance enhancement* and *mental health*. Many people see the field of sport psychology as narrow, when in fact it has a wide scope and applications to many areas of our lives.

The performance enhancement aspect of sport psychology refers to the effects of psychological factors on sport performance. These psychological factors include anxiety, concentration, confidence, motivation, mental preparation, and personality (the totality of a person's distinctive psychological traits). The performance enhancement aspect of sport psychology is not restricted to elite athletes. Rather, it spans a continuum from young athletes participating in youth sports all the way to older adults playing in recreational or competitive leagues. *The key point is that the mind affects the body; therefore, the way we think and feel has a strong impact on how we physically perform.* In competitive sports, participants' physical skills are often comparable, and the difference between winning and losing is in their mental skills.

The other main focus of sport psychology is enhancing mental health and well-being through participation in sport, exercise, and physical activity—that is, enhancing the psychological effects of participation in sport, exercise, and physical activity. We just noted that our minds can have an important effect on our bodies; conversely, our bodies (i.e., the way we feel physically) can have an important effect on our minds. For example, research indicates that vigorous physical activity is related to a reduction in anxiety (distress and tension caused by apprehension) and depression (a mental condition of gloom, sadness, or dejection). Similarly, sport participation has been related to increases in self-esteem and self-confidence. *In essence, sport, exercise, and physical activity have been shown to have the capacity to increase our feelings of psychological well-being and thus exert a positive influence on our mental health.*

Participating in competitive sports can also sometimes be frustrating and upsetting when we lose or don't perform up to expectations; this can lead to increases in anxiety, depression, and aggressiveness. Sport psychologists attempt to accentuate the positive aspects of sport participation so that individuals will receive positive psychological benefits.

Sport psychologists attempt to accentuate the positive aspects of sport participation so that individuals will receive positive psychological benefits.

Mastery Item 14.1

Describe the difference between the performance enhancement and mental health aspects of sport psychology. How might these two aspects be associated with health-related benefits?

Researchers studying the performance enhancement aspect and those studying the mental health aspect of sport psychology have different objectives, so it is not surprising that their measurement objectives and the types of psychological tests they use differ considerably (although there is some overlap). For example, in studying performance enhancement, researchers are typically interested in measuring psychological factors that affect performance. Their tests might measure attentional focus, confidence, precompetitive anxiety, self-motivation, and imagery. For researchers interested in studying mental health, tests that measure anxiety, depression, self-esteem, self-concept, mood, and anger are appropriate.

Let's discuss some of the issues that sport psychologists face when measuring psychological factors that affect performance and the psychological results of participation in sport and exercise.

TRAIT VERSUS STATE MEASURES

When we are assessing personality and psychological variables in sport psychology, it is important to distinguish between trait and state measures. Trait psychology was the first scientific approach that evolved for the study of personality. Trait psychology is based on the assumption that personality **traits**—the fundamental aspects of personality—are relatively stable, consistent attributes that shape the way people behave. *In essence, the trait approach considers the general source of variability to reside within the person; it minimizes the role of situational or environmental factors.* This means that a person high on the trait of aggressiveness would tend to act more aggressively in most situations than someone who is low on the trait of aggressiveness.

Traits, or predispositions, may be acquired through learning or be constitutionally (genetically) inherent. A well-known trait typology is extroversion–introversion. These traits relate to a person's general tendency to respond in an outgoing or a shy manner, without regard for the situation. For example, an extrovert placed in a new situation in which he or she doesn't know anybody will likely try to meet people and be outgoing. In sport psychology, the traits that have been studied include anxiety, aggressiveness, self-motivation, confidence, and achievement motivation.

An alternative to the trait approach is to focus on the situation. In the situational, or **state,** approach, behavior is expected to change from one situation to the next, and traits are given a subsidiary role in the explanation and prediction of behavior. *Psychological states are viewed as a function of the particular situation or environment in which a person is placed; thus, when the situation changes, so does the psychological state.* In essence, psychological states are transitory, potentially changing rapidly in a short time as the situation changes.

For example, assume you are a reserve member of a basketball team and usually just sit on the bench. Your team is playing in a championship game, but because you are not expected to play much, your state anxiety right before the game is low. However, your coach walks up to you and tells you that one of the starters is sick and won't be able to play and that you're going to be starting. In a few seconds, your state anxiety has greatly elevated, because you are now anxious about how well you will perform in this important game. The situation of starting on the team has caused a dramatic shift in your state anxiety level; this has little to do with your trait anxiety.

Describe the difference between the trait and state approaches to the study of personality and behavior.

Although sport psychologists make the distinction between traits and states, you need to consider both to attempt to understand and predict behavior in sport and exercise settings. The idea that traits and states are codeterminants of behavior is known as the interactionalist approach; it is the approach most widely endorsed today. This approach to personality and behavior contends that one's personality, needs, interests, and goals (i.e., traits), as well as the particular constraints of the situation (e.g., win–loss record, attractiveness of the facility, and crowd support), interact to determine behavior. Thus, from a measurement perspective, it is imperative that both traits and states be considered when we are attempting to understand and predict behavior in sport and exercise settings.

A study conducted by Sorrentino and Sheppard (1978) nicely demonstrated the usefulness of considering both trait and state variables in an interactionalist approach. Their study tested male and female swimmers by having them perform a 200 yd (182.8 m) freestyle time trial, once swimming individually and once as part of a relay team. The situational factor assessed was whether each swimmer had a faster individual split time when he or she swam alone or as part of a relay team. In addition, the personality characteristic of affiliation motivation was assessed. This personality trait represents whether the swimmers were more approval oriented—viewing competing with others as positive—or were more rejection threatened—feeling threatened because they might let their teammates down in a relay situation. As the researchers predicted, the approval-oriented swimmers demonstrated faster times in relays than in individual events. In contrast, the rejection-threatened swimmers swam faster in individual events than in relays. Thus, the swimmers' race times involved an interaction between their personalities (affiliation motivation) and the situation (individual versus relay).

GENERAL VERSUS SPORT-SPECIFIC MEASURES

Until recently, almost all the trait and state measures of personality and other psychological attributes used in sport psychology came from general psychological inventories. In essence, these inventories measured general or global personality traits and states, with no specific reference to sport or physical activity. Examples of such inventories are the State–Trait Anxiety Inventory (Spielberger, Gorsuch, and Lushene 1970), the Test of Attentional and Interpersonal Style (Nideffer 1976), the Profile of Mood States (McNair, Lorr, and Droppleman 1971, also see Morgan 1980), the Self-Motivation Inventory (Dishman and Ickes 1981), the Eysenck Personality Inventory (Eysenck and Eysenck 1968), and the Locus of Control (Rotter 1966).

Psychologists have found that situation-specific measures provide a more accurate and reliable predictor of behavior in specific situations. For example, Sarason (1975) observed that some students were doing poorly on tests, not because they weren't smart or hadn't studied, but simply because they became too anxious and froze up. These students were not particularly anxious in other situations, but taking exams made them extremely anxious. Sarason labeled such people test-anxious and devised a situation-specific test called test anxiety that measures how anxious a person feels before taking exams. This situation-specific test was a better predictor of immediate pretest anxiety than a general test of trait anxiety. Clearly, we can better predict behavior when we have more knowledge of the specific situation and how people tend to respond to it.

Along these lines, sport psychologists have recently begun to develop sport-specific tests to provide more reliable and valid measures of personality traits and states in sport, exercise, and physical activity contexts. For example, your coach might not be very concerned about how anxious you are before giving a speech or before taking a test, but he or she would surely be interested in how anxious you are before competition (especially if excess anxiety is detrimental to your performance). A sport-specific test of anxiety would provide a more reliable and valid assessment of an athlete's precompetitive anxiety than a general anxiety test. Some examples of psychological inventories developed specifically for use in sport and physical activity settings include the Sport Anxiety Scale (Smith, Smoll, and Shutz 1990), the Task and Ego Orientation Sport Questionnaire (Duda 1989), the Sport Motivation Scale (Briere, Vallerand, Blais, and Pelleteir 1995), the Physical Self-Perception Profile (Fox and Corbin 1989), the Sport Imagery Questionnaire (Hall, Mack, Paivio, and Hausenblas 1998), the Competitive State Anxiety Inventory–2 (Martens, Vealey, and Burton 1990), the Group Environment Questionnaire (Widmeyer, Brawley, and Carron 1985), and the Trait–State Confidence Inventory (Vealey 1986). Some sport psychologists have even gone a step further and developed tests for a specific sport rather than for sports in general. These include the Tennis Test of Attentional and Interpersonal Style (Van Schoyck and Grasha 1981), the Anxiety Assessment for Wrestlers (Gould, Horn, and Spreeman 1984), and Group Cohesion for Basketball (Yukelson, Weinberg, and Jackson 1984).

Mastery Item 14.3

Why are sport-specific tests more desirable than general psychological tests in sport and exercise settings?

CAUTIONS IN USING PSYCHOLOGICAL TESTS

Psychological inventories are crucial to sport psychologists from theoretical and applied perspectives. Such tests help to evaluate the accuracy of different psychological theories and provide practitioners with a tool for applying theory to their practice. We focus on the use of psychological tests in applied settings because it is here that abuses of test results and misconceptions in analysis are more likely.

Who is qualified to administer psychological tests to athletes? The American Psychological Association (APA) recommends that test givers have the following knowledge:

1. An understanding of testing principles and the concept of measurement error. A test giver needs a clear understanding of statistical concepts such as correlation, measures of central tendency (mean, median, and mode), variance, and standard deviation.

2. The ability and knowledge to evaluate a test's validity for the purposes (decisions) for which it is employed. A qualified test administrator will recognize that test results are not absolute or irrefutable and that there are potential sources of measurement error. He or she will do everything possible to eliminate or minimize such errors. For example, testers must be aware of the potential influences of situational factors as well as interpersonal factors that may alter the way test scores are interpreted. In addition, cultural, social, educational, and ethnic factors can all have a large impact on an athlete's test results. A test not only has to be reliable and valid but also needs to be validated for the particular sample and situation in which it is being conducted. For example, you might choose a test that was developed using adults and administer it to athletes aged 13 to 15 years. However, the wording of the test might be such that the younger athletes do not fully understand the questions; thus, the results are not relevant. Similarly, a test might have been developed on a predominantly white population, and your athletes happen to be mostly African American and Hispanic. This cultural difference might lead to problems in interpreting the results with this different population.

3. Self-awareness of one's own qualifications and limitations. Unfortunately, in sport psychology there have been cases in which individuals were not aware of their own limitations and were thus using tests and interpreting results in a manner that was

unethical and in fact potentially damaging to the athletes. For example, many psychological inventories are designed to measure psychopathology or abnormality. To interpret the test results, a test giver needs special training in psychological assessment and possibly in clinical psychology. Without this background, it would be unethical for an individual to use such tests with athletes. The best way to reduce measurement error is for the test giver to select reliable and valid tests.

Mastery Item 14.4

Why is it important to understand the concept of measurement error in relation to interpreting psychological tests? How is this error best reduced or controlled? Should psychological tests be used to select athletes for a team?

Some psychological tests are used inappropriately to determine whether an athlete should be drafted onto a team or to determine if an athlete has the "right" psychological profile for a certain position (such as a middle linebacker in American football). This practice was particularly rampant in the 1960s and 1970s but seems to have abated. These unethical uses of psychological tests can cause an athlete to be hastily eliminated from a team or not drafted simply because he or she does not appear to be mentally tough. In truth, however, it is difficult to predict athletic success from the results of these tests. An athlete's or a team's performance (often measured in terms of winning and losing) is a complicated issue affected by such factors as physical ability, experience, coach–player compatibility, ability of opponents, and teammate interactions. It would be naive to think that simply knowing something about an athlete's personality provides enough information to predict whether he or she will be successful.

What types of psychological tests should be given to athletes, and what conditions should be established for test administration and feedback? Psychological tests have been abused in sport settings both during their administration and when feedback is provided to the athletes. In several cases, athletes have been given psychological tests with no explanation as to why they were taking the test; furthermore, these athletes received no feedback about the results and interpretation of the tests. This, again, is unethical and violates the rights of the individuals taking the tests. *Before they actually take the tests, athletes should be told the purpose of the tests, what the tests measure, and how the tests are going to be used.* In most cases, the tests should be used to help coaches and athletes better understand the athletes' psychological strengths and weaknesses so that they can focus on building up the athletes' strengths and improving their weaknesses. In addition, athletes should be provided with specific feedback on the results of the testing process. *Administer feedback in a way that athletes can gain more insight into themselves and understand what the test indicates.* The results and feedback can then serve as a springboard to stimulate positive changes.

If athletes are not told the reason for the testing, they will typically become suspicious regarding what the test is going to be used for. In such cases it is not unusual for athletes to start thinking that the coach will use the test to select the starters or "weed out the undesirables." Given these circumstances, athletes will be more likely to attempt to exaggerate their strengths and minimize their weaknesses. This response style of "faking good" can distort the true results of the test and make its interpretation virtually useless. Thus, it is important that athletes be assured of confidentiality in whatever tests they take, because this increases the likelihood that they will answer truthfully. *Coaches should typically stay away from giving and interpreting psychological tests unless they have specific training in this area.* A sport psychology consultant who has formal training in psychological assessment and measurement is the best person to administer and interpret psychological tests.

Mastery Item 14.5

What are three key principles to follow in administering psychological tests to athletes and in providing them with feedback?

QUANTITATIVE VERSUS QUALITATIVE MEASUREMENT

Now that you know something about the field of sport psychology, including some measurement issues as well as cautions in using psychological tests, we discuss the two different approaches that sport psychologists use to gain a better understanding of the psychological factors involved in sport, exercise participation, and physical activity. These two general approaches are quantitative and qualitative methodologies.

Quantitative research, numerical in nature and the more traditional of the two approaches, involves experimental and correlational designs that typically use precise measurements, rigid control of variables (often in a laboratory setting), and statistical analyses. There are usually objectively measured independent and dependent variables, and psychological states and traits are assessed by reliable and valid psychological inventories. In quantitative research, the researcher tries to stay out of the data-gathering process by using laboratory measurements, questionnaires, and other objective instruments. The quantitative data are statistically analyzed with computations performed by computers.

Although **qualitative** research, which is textual in nature, is often depicted as the antithesis of the more traditional quantitative methods, it should be seen as a complementary method. Qualitative research methods generally include field observations, ethnography, and interviews seeking to understand the meaning of the experience to the participants in a particular setting and how the components mesh together as a whole. To this end, qualitative research focuses on the "essence of the phenomenon" and relies heavily on people's perceptions of their world. Hence, the objectives are primarily description, understanding, and meaning. Qualitative data are rich, because they provide depth and detail; they allow people to be understood in their own terms and in their natural settings (Patton 1980). The researcher does not manipulate variables through experimental treatment but rather is interested more in the process than in the product. Through observations and interviews, relationships and theories are allowed to emerge from the data rather than being imposed on them; thus induction is emphasized. This is in contrast to quantitative research, in which deduction is primary. Finally, in qualitative research, the researcher interacts with the subjects, and the sensitivity and perception of the researcher play crucial roles in procuring and processing the observations and responses.

A recent study by Holt and Sparkes (2001) provides an excellent example of the use of ethnography to study group cohesion in a team over the course of a season. The researcher (one of the authors) spent a season with a soccer team as a player and coach and collected data via participant observation, formal and informal interviews, a field diary, and a reflexive journal. Thus, being embedded in a group for an entire season allowed a richness of data and understanding that could not occur through the use of quantitative data collection techniques alone.

Most of the assessment of psychological traits and states in sport psychology has taken place through using inventories and questionnaires that have been carefully developed to provide high reliability and validity—a quantitative approach. We will be taking a close look at a number of these scales, including their psychometric development and their use in research and applied settings. But it should be noted that qualitative studies have become more popular in recent years (e.g., Bloom, Stevens, and Wickwire 2003; Culver, Gilbert, and Trudel 2003; Sparkes 1998; Strean 1998; Stuart 2003). Therefore, let us first examine two of the most common ways in which psychological data are gathered qualitatively: interview and observation.

Quantitative Methods

As noted earlier, most psychological assessment in the field of sport and exercise psychology uses the traditional quantitative methodology of the questionnaire (see Duda 1998 for a summary of advances in different areas of sport and exercise psychology measurement). There are a number of different types of quantitative questionnaires to choose from; two of the most popular employ Likert scales and semantic differential scales. Each of these scales can help define the multidimensional nature of the construct being assessed.

Likert Scales

A Likert scale is a 5- or 9-point scale (1-5 points or 1-9 points) that assumes equal intervals between responses. In the example provided here, the difference between *strongly agree* and *agree* is considered equivalent to the difference between *disagree* and *strongly disagree*. This type of scale is used to assess the degree of agreement or disagreement with statements and is widely used in attitude inventories. An example of an item using a Likert scale is as follows:

> *All high school students should be required to take 2 years of physical education classes.*

Strongly agree	Agree	Undecided	Disagree	Strongly disagree
5	4	3	2	1

A principal advantage of scaled responses is that they permit a wider choice of expression than categorical responses, which are typically dichotomous—that is, offering such choices as *yes* and *no* or *true* and *false*. The five, seven, or more intervals also increase the reliability of the instrument. In addition, different response words can be used in scaled responses (table 14.1): *excellent, good, fair, poor,* and *very poor,* or *very important, important, somewhat important, not very important,* and *of no importance.*

Table 14.1 Examples of Scaled Responses

1 Never	2 Sometimes	3 Often	4 Frequently	5 Always
1 Strongly agree	2 Agree	3 No opinion	4 Disagree	5 Strongly disagree

1 Always	2	3	4	5	6	7 Never

1 Agree	2	3	4	5	6	7	8	9	10	11 Disagree

1 Not important at all	2	3	4	5 Extremely important

Semantic Differential Scales

Another popular measuring technique involves using a semantic differential scale, which asks individuals to respond to bipolar adjectives—pairs of adjectives with opposite meanings, such as *weak–strong, relaxed–tense, fast–slow,* and *good–bad*—with scales anchored at the extremes. Using bipolar adjectives, respondents are asked to choose a point on the continuum that best describes how they feel about a specific concept. Refer to table 14.2, on attitudes toward physical activity; notice that you can choose any of seven points that best reflects your feelings about the concept.

The process of developing semantic differential scales involves defining the concept you want to evaluate and then selecting specific bipolar adjective pairs that best describe respondents' feelings and attitudes about the concept. Research has indicated that the semantic differential technique measures three major factors. By far, the most frequently used factor is evaluation—the degree of goodness you attribute to the concept or object being measured. Potency involves the strength of the concept being rated, and the activity factor uses adjectives that describe action. The following list shows some examples

Table 14.2   Semantic Differential Scales for Measuring Attitudes Toward Physical Activity

	Physical activity							
Good	___	___	___	___	___	___	___	Bad
Pleasant	___	___	___	___	___	___	___	Unpleasant
Relaxed	___	___	___	___	___	___	___	Tense
Hot	___	___	___	___	___	___	___	Cold
Healthy	___	___	___	___	___	___	___	Unhealthy
Nice	___	___	___	___	___	___	___	Awful
Delicate	___	___	___	___	___	___	___	Rugged
Active	___	___	___	___	___	___	___	Passive

of bipolar adjectives that measure the different evaluation components; this is an example of a differential scale for measuring attitudes toward competitive sports for children.

Evaluation

- ► Pleasant–unpleasant
- ► Fair–unfair
- ► Honest–dishonest
- ► Good–bad
- ► Successful–unsuccessful
- ► Useful–useless

Potency

- ► Strong–weak
- ► Hard–soft
- ► Heavy–light
- ► Dominant–submissive
- ► Rugged–delicate
- ► Dirty–clean

Activity

- ► Steady–nervous
- ► Happy–sad
- ► Active–passive
- ► Dynamic–static
- ► Stationary–moving
- ► Fast–slow

Mastery Item 14.6

Develop two items that measure attitudes toward physical activity. One should be a Likert scale, and the other should use a semantic differential scale.

Qualitative Methods

Qualitative methods are becoming more and more popular in sport psychology research because they provide a richness of information that is usually untapped when traditional questionnaires are used.

Interviews

The interview is undoubtedly the most common source of data in qualitative research. Interviews range from a highly structured style, in which questions are determined before the interview actually starts, to open-ended interviews, which allow for free responses. The most popular mode of interview used in sport psychology research is the semistructured interview. Each subject responds to a general set of questions, but the test administrator uses different probes and follow-up questions depending on the nature of the subject's response. A good interviewer must first establish rapport with subjects to allow them to open up and describe their true thoughts and feelings. It is also important that the interviewer remain nonjudgmental regardless of the content of the subject's responses. Above all, the interviewer has to be a good listener.

Using a tape recorder is probably the most common method of recording interviews, because it preserves the entire interview for subsequent data analysis. Although a small percentage of subjects are initially uncomfortable being recorded, this uneasiness typically disappears quickly. Taking notes during an interview is another frequently used method to record data; sometimes it is used in conjunction with recording when the interviewer wants to highlight certain important points. One drawback with taking notes without recording is that it keeps the interviewer very busy, thus interfering with his or her thoughts and observations of the subject.

A good example of the use of interviews to collect qualitative data in sport psychology is the work of Gould, Dieffenbach, and Moffett (2002). To better understand the psychological characteristics of Olympic athletes and their development, the researchers incorporated qualitative methods (along with some quantitative psychological inventories) and an inductive analytic approach to research. They studied 10 former Olympic champions (winners of 32 medals), 10 Olympic coaches, and 10 parents, guardians, or significant others (one for each athlete). The interviews focused on the mental skills needed to become an elite athlete as well as how these mental skills were developed during the early, middle, and later years of the athletes' careers. The interviews, transcribed verbatim from recordings, were analyzed by content analysis, a method that organizes the interview into increasingly more complex themes and categories representing psychological characteristics and their development. Among the numerous findings it was revealed that these Olympic athletes were characterized by the ability to focus and block out distractions, competitiveness, the ability to set and achieve goals, mental toughness, and ability to control anxiety and confidence. In addition, coaches and parents were particularly important (although other individuals were also involved to a lesser or greater extent) in assisting the development of these athletes. Specifically, coaches, parents, and other individuals provided support and encouragement, created an achievement environment, modeled, emphasized high expectations, provided motivation, taught physical and psychological skills, and enhanced confidence through positive feedback. Such depth could not have been accomplished through the strict use of questionnaires and other psychological inventories.

Observation

Observation is second only to the interview as a means of qualitative data collection. Although most early studies relied on direct observation with note taking and coding of certain categories of behavior, the current trend is to videotape. Just as using a tape recorder to conduct an interview allows the researcher to review everything the subject said, video-

Subjects who know that they are being watched and recorded may change their behavior.

taping likewise captures the subject's behavior for future analysis. It is important in field observations that the observer be unobtrusive to the subjects under study. Subjects who know that they are being watched and recorded may change their behavior. Observers can seem less obtrusive by hanging around for several days before actually starting to record their observations. It is important that the novelty of the observer's presence wear off so that behavior can occur naturally.

A classic example of using observations in sport psychology is seen in the seminal work of Smith, Smoll, and Curtis (1979) on the relationship between coaches' behaviors and young athletes' reactions to these behaviors. The first part of the investigation identified what coaches actually do. Observers were trained to watch Little League coaches during practice and games and carefully note what the coaches did. The observers recorded these behaviors over several months. After compiling literally thousands of data points, the researchers attempted to collapse the behaviors into some common categories. The end result of this process categorized coaching behaviors into those initiated by the coach (spontaneous) versus those that were reactions to a player's behavior (reactive).

For example, a coach's yelling at a player who made an error was a reactive behavior. However, a coach's instruction to his players on how to slide was considered a spontaneous behavior. Within these categories of reactive and spontaneous behaviors were subcategories, such as positive reinforcement, negative reinforcement, general technical instruction, general encouragement, and mistake-contingent technical instruction. These subcategories of coaching behaviors resulted in the development of an instrument called the Coaching Behavior Assessment System, which allowed the researchers to conduct several studies investigating the relationship between specific coaching behaviors and players' evaluative reactions to these behaviors. For example, in one study, team members who played for coaches who gave predominantly positive (as opposed to negative) reinforcement liked their teammates more, wanted to continue playing next year, and saw their coaches as more knowledgeable and as better teachers than did team members who played for coaches who did not favor positive reinforcement. The cornerstone of their research methodology, however, was collecting qualitative data through the use of observations and using these data to develop the Coaching Behavior Assessment System.

SCALES USED IN SPORT AND EXERCISE PSYCHOLOGY

Thus far we have provided an overview of sport psychology, the uses and abuses of psychological testing in sport and exercise settings, and some basic information on different types of scaling procedures. In this section, we highlight the most often used and most popular psychological tests used in sport, exercise, and physical activity settings. We focus on inventories that have been systematically developed with high standards of reliability and validity and provide examples of how these tests have been used in research and applied settings. For excellent reviews of the scales used in sport psychology research, see Anshel (1987) and Ostrow (1996).

Competitive Anxiety

One of the most popular topics of study in sport psychology concerns the relationship between anxiety and performance. Athletes, coaches, and researchers generally agree that there is an optimal level of anxiety associated with high levels of performance. Finding that optimal level is not necessarily easy, but the first step is to be able to measure an athlete's anxiety level. As noted earlier, there is a distinction in the general psychology literature between trait anxiety and state anxiety. This distinction is used in developing sport-specific measures of *competitive trait anxiety* and *competitive state anxiety*.

Sport Competition Anxiety Test

One of the most widely used tests in sport psychology is the Sport Competition Anxiety Test (SCAT) developed by Martens (1977). SCAT was developed to provide a reliable and valid measure of competitive trait anxiety. Competitive trait anxiety is a construct that describes individual differences in perception of threat, state anxiety response to perceived threat, or both. *SCAT was developed to provide a measure of how anxious athletes generally feel before competition.* The fact that SCAT is a good predictor of precompetitive state anxiety is important from a practical perspective, because it is not always feasible to test athletes right before competition to assess how anxious they feel at that moment (i.e., their competitive state anxiety). SCAT was initially developed for use with children between the ages of 10 and 15 years, and an adult form of the inventory was constructed shortly thereafter.

Mastery Item 14.7

What is the difference between competitive trait anxiety and precompetitive state anxiety?

Reliability and Validity   The internal structure, reliability, and validity of SCAT have been determined independently for the child and the adult forms on the basis of responses from more than 2500 athletes. SCAT's reliability has been assessed by test–retest and has produced correlation coefficients ranging from .73 to .88, with a mean of .81. Past research findings produced coefficients of internal consistency ranging from .95 to .97 (very high values) for both the adult version and the children's version of SCAT. Evidence for the construct validity of SCAT was obtained by demonstrating significant relationships between SCAT and other personality constructs in accordance with theoretical predictions. For example, SCAT is moderately correlated with other general anxiety scales but is not correlated with other general personality scales. Finally, a number of experimental and field studies support the construct validity of SCAT as a valid measure of competitive trait anxiety by providing results in accordance with theoretical predictions. For example, subjects scoring high on SCAT exhibit higher levels of precompetitive state anxiety than do low-SCAT subjects. SCAT was also found to correlate more strongly with state anxiety in competitive situations than in noncompetitive situations.

Norms and Scoring   The adult form of SCAT and norms for high school and college athletes are provided in tables 14.3 and 14.4. For each item, one of three responses is possible: (a) hardly ever, (b) sometimes, and (c) often. Eight of the test items—2, 3, 5, 8, 9, 12, 14, 15—are scored with the following point values:

Hardly ever = 1

Sometimes = 2

Often = 3

Note that items 6 and 11 are scored in reverse:

Hardly ever = 3

Sometimes = 2

Often = 1

Items 1, 4, 7, 10, and 13 are not scored; they are included in the inventory as filler items, to direct attention to elements of competition other than anxiety.

Table 14.3   **Sport Competition Anxiety Test**

Illinois Competition Questionnaire

Directions: Below are some statements about how people feel when they compete in sports and games. Read each statement and decide if you HARDLY EVER or SOMETIMES or OFTEN feel this way when you compete in sports and games. If your choice is HARDLY EVER, blacken the square labeled A; if your choice is SOMETIMES, blacken the square labeled B; if your choice is OFTEN, blacken the square labeled C. There are no right or wrong answers. Do not spend too much time on one statement. Rmember to choose the word that describes how you *usually* feel when competing in sports and games.

	Hardly ever	Sometimes	Often
1. Competing against others is socially enjoyable.	A ❏	B ❏	C ❏
2. Before I compete I feel uneasy.	A ❏	B ❏	C ❏
3. Before I compete I worry about not performing well.	A ❏	B ❏	C ❏
4. I am a good sport when I compete.	A ❏	B ❏	C ❏
5. When I compete I worry about mistakes.	A ❏	B ❏	C ❏
6. Before I compete I am calm.	A ❏	B ❏	C ❏
7. Setting a goal is important when competing.	A ❏	B ❏	C ❏
8. Before I compete I get a queasy feeling in my stomach.	A ❏	B ❏	C ❏
9. Just before competing I notice my heart beats faster than usual.	A ❏	B ❏	C ❏
10. I like to compete in games that demand considerable physical energy.	A ❏	B ❏	C ❏
11. Before I compete I feel relaxed.	A ❏	B ❏	C ❏
12. Before I compete I am nervous.	A ❏	B ❏	C ❏
13. Team sports are more exciting than individual sports.	A ❏	B ❏	C ❏
14. I get nervous wanting to start the game.	A ❏	B ❏	C ❏
15. Before I compete I usually get uptight.	A ❏	B ❏	C ❏

Reprinted from Martens (1997).

Table 14.4 **Sport Competition Anxiety Test: Norms for Male and Female High School and College Athletes**

Raw score	High school (percentiles)		College (percentiles)	
	Males	Females	Males	Females
30	98	98	97	99
29	96	95	93	98
28	93	89	93	96
27	88	83	88	93
26	78	77	82	90
25	65	70	77	86
24	53	60	70	82
23	43	51	63	77
22	33	43	56	69
21	24	34	51	62
20	17	26	43	53
19	12	19	34	43
18	9	14	27	34
17	7	11	21	25
16	6	9	16	18
15	5	6	8	14
14	5	5	4	10
13	4	4	2	6
12	2	2	1	3
11	1	1	0	1
10	0	0	0	0

Reprinted from Martens (1977).

Research and Practical Examples Let's look at a study in sport psychology that has used SCAT. Researchers have been interested in studying the differences in perceptions of threat between individuals who are high and low in competitive trait anxiety. Using male youth soccer players as subjects, Passer (1983) found that players high in competitive trait anxiety expected to play less well in the upcoming season and worried more frequently about making mistakes, not playing well, and losing than did players low in competitive trait anxiety. In addition, players high in competitive trait anxiety expected to experience greater emotional upset, shame, and criticism from parents and coaches after failing than did players low in competitive trait anxiety.

These findings demonstrate that athletes who are high and low in competitive trait anxiety significantly differ in their perceptions and reactions to threat and that there are important implications for coaches and parents. Specifically, because young athletes who have high competitive trait anxiety are more sensitive to criticism, failure, and making mistakes, it is important that coaches and parents not overreact when these young athletes do not perform well. Positive reinforcement, encouragement, and support are crucial in helping these young athletes deal with their mistakes and stay involved in sports.

Sport Anxiety Scale

Subsequent to the development of the SCAT, anxiety research determined that the construct was multidimensional with a cognitive and somatic component. To accommodate this advance, the Sport Anxiety Scale (SAS) was developed to provide a multidimensional assessment of competitive trait anxiety in keeping with contemporary theory. The three subscales that were obtained through confirmatory factor analysis included two factors relating to cognitive anxiety, named concentration disruption and worry, as well as a factor termed somatic anxiety.

Reliability and Validity   Test–retest reliability on 77 football players (18 days) was reported at .77 for the full scale and then was followed up with another 7-day recall (using 64 college athletes) that produced reliabilities of more than .85 on each subscale. Internal consistency reliability using Cronbach's alpha (performed on 382 high school athletes) found coefficients of .92 for somatic anxiety, .86 for worry, and .81 for concentration disruption with a coefficient of .93 for the total scale. These high coefficients were replicated in a second study of 490 high school athletes.

In terms of validity, Smith and colleagues (1990) assessed convergent and discriminant validity by correlating the SAS with other scales. Specifically, concurrent validity was indicated by the .81 correlation with the SCAT, with the highest correlation between the Somatic subscale and SCAT (because SCAT has primarily somatic items). In addition, as predicted, the SAS was more strongly correlated with SCAT (a sport-specific measure) than with the Trait Anxiety Inventory (a general measure of anxiety), once again indicating the usefulness of situation-specific scales.

Discriminant validity was demonstrated by the low to moderate correlations between the SAS and the Crowne–Marlowe Social Desirability Scale. This magnitude of correlation is similar to other anxiety scales and indicates that the tendency to present oneself in a positive light is negatively related to SAS scores.

Norms and Scoring   Means for the SAS (using approximately 850 high school athletes and 125 college athletes) were approximately 19 (*SD* = 5.80) for the Somatic Anxiety subscale (scores can range from 9 to 36), 15 (*SD* = 4.45) for the Worry subscale (scores range from 7 to 28), and 8 (*SD* = 2.55) for the Concentration Disruption subscale (scores range from 5 to 20).

There are 21 items in all, with items 1, 4, 8, 11, 12, 15, 17, 19, and 21 making up the Somatic Anxiety subscale; items 3, 5, 9, 10, 13, 16, and 18 comprising the Worry subscale; and items 2, 6, 7, 14, and 20 making up the Concentration Disruption subscale. All items are scored on a 1 *(not at all),* 2 *(somewhat),* 3 *(moderately so),* and 4 *(very much so)* scale with no reverse scoring.

Research and Practical Examples   Let's look at a couple of applications of the SAS. For example, in a study by Smith, Smoll, and Barnett (1995), an intervention designed to train coaches to reduce the stressfulness of the athletic environment by de-emphasizing winning and providing high levels of social support resulted in a significant decrease in the total score on the SAS. Thus, the SAS appears to be sensitive to interventions designed to decrease anxiety. In another study (Patterson, Smith, Everett, and Ptacek, 1998), SAS scores were related to the occurrence of injury under high levels of life stress. Specifically, those scoring high on all SAS subscales were more likely to incur injury when under high levels of life stress. These findings suggest that somatic anxiety, worry, and concentration disruption may all be capable of increasing the risk of anxiety in highly stressed individuals.

Download the data from the text World Wide Web site for Mastery Item 14.8. The scores presented are state anxiety scores from two groups, inexperienced and experienced skydivers. The State Anxiety Scale was administered as subjects boarded the airplane before a jump. The scores presented are from a newly created State Anxiety Scale. If the scale reflects construct validity, as presented in chapter 6, the inexperienced and experienced skydivers should differ on state anxiety. Because there are two groups, you can use an independent groups t test (as you learned in chapter 5) to see if the groups differ significantly. Use SPSS to confirm that the groups differ significantly (i.e., $t = 6.28$, $p < .001$). These results provide evidence of the construct validity of the newly created State Anxiety Scale.

Describe a sport situation in your life in which the use of SAS would have been helpful.

Competitive State Anxiety Inventory–2

The Competitive State Anxiety Inventory–2 (CSAI–2) was developed as a sport-specific inventory of state anxiety and as a revision of the earlier Competitive State Anxiety Inventory (Martens, Vealey, and Burton 1990). In fact, more than 40 published studies have used the CSAI–2 to investigate the relationship between state anxiety and performance (Craft, Magyar, Becker, and Feltz 2003). *The CSAI–2 measures precompetitive state anxiety, which is how anxious an athlete feels at a given moment in time—in this case, right before competition.* The CSAI–2 has three subscales: Somatic Anxiety, Cognitive Anxiety, and Confidence. As noted earlier regarding the SAS, somatic anxiety refers to the physiological component of anxiety and cognitive anxiety to the worry component. These subscales reflect the multidimensional nature of anxiety.

Reliability and Validity Test–retest reliability is inappropriate for state scales because, by definition, scores can change from moment to moment. Thus, the major source of reliability comes from examining the internal consistency of the scale—that is, the degree to which items in the same subscale are homogeneous. Alpha reliability coefficients (see chapter 6) have ranged from .79 to .90, demonstrating a high degree of internal consistency.

The concurrent validity of the CSAI–2 was examined by investigating the relationship between each of its subscales and eight selected state and trait personality inventories. Results strongly support the concurrent validity of the CSAI–2, because the correlations are highly congruent with hypothesized relationships among the CSAI–2 subscales and other personality constructs. For example, the CSAI–2 Cognitive subscale is highly correlated with the Worry–Emotionality Inventory (Morris, Davis, and Hutchins 1981). The worry subscale, and the CSAI–2 Somatic subscale is highly correlated with the Worry–Emotionality Inventory Emotionality subscale. The construct validity of the CSAI–2 was determined by a series of systematic studies demonstrating the relationships between the three subscales and other constructs (e.g., performance, situational variables, and individual differences) as predicted by theory.

Norms and Scoring The CSAI–2 and its norms for high school and college athletes are provided in tables 14.5 and 14.6. Scoring for the CSAI–2 is accomplished by computing a separate total for each of the three subscales, with scores ranging from a low of 9 to a high of 36. The higher the score, the greater the cognitive or somatic anxiety or self-confidence.

Table 14.5 Competitive State Anxiety Inventory-2

Illinois Self-Evaluation Questionnaire

Directions: A number of statements that athletes have used to describe their feelings before competition are given below. Read each statement and then circle the appropriate number to the right of the statement to indicate *how you feel right now*—at this moment. There are no right or wrong answers. Do not spend too much time on one statement, but choose the statement which describes your feelings *right now.*

	Not at all	Somewhat	Moderately so	Very much so
1. I am concerned about this competition.	1	2	3	4
2. I feel nervous.	1	2	3	4
3. I feel at ease.	1	2	3	4
4. I have self-doubts.	1	2	3	4
5. I feel jittery.	1	2	3	4
6. I feel comfortable.	1	2	3	4
7. I am concerned that I may not do as well in this competition as I could.	1	2	3	4
8. My body feels tense.	1	2	3	4
9. I feel self-confident.	1	2	3	4
10. I'm concerned about losing.	1	2	3	4
11. I feel tense in my stomach.	1	2	3	4
12. I feel secure.	1	2	3	4
13. I am concerned about choking under pressure.	1	2	3	4
14. My body feels relaxed.	1	2	3	4
15. I'm confident that I can meet the challenge.	1	2	3	4
16. I'm concerned about performing poorly.	1	2	3	4
17. My heart is racing.	1	2	3	4
18. I'm confident about performing well.	1	2	3	4
19. I'm concerned about reaching my goal.	1	2	3	4
20. I feel my stomach sinking.	1	2	3	4
21. I feel mentally relaxed.	1	2	3	4
22. I'm concerned that others will be disappointed with my performance.	1	2	3	4
23. My hands are clammy.	1	2	3	4
24. I'm confident because I mentally picture myself reaching my goal.	1	2	3	4
25. I'm concerned I won't be able to concentrate.	1	2	3	4
26. My body feels tight.	1	2	3	4
27. I'm confident of coming through under pressure.	1	2	3	4

Reprinted from Martens, Vealey, and Burton (1990).

Table 14.6 **Competitive State Anxiety Inventory-2: Norms for Male and Female High School and College Athletes**

Raw score	Male percentiles			Female percentiles		
	Cognitive	Somatic	Self-confidence	Cognitive	Somatic	Self-confidence
36	99	99	99	99	99	99
35	99	99	96	98	99	98
34	99	99	94	96	99	97
33	99	98	91	94	99	96
32	99	98	87	92	98	94
31	98	97	83	89	98	92
30	98	96	79	87	97	89
29	96	95	76	84	94	86
28	95	94	71	80	91	83
27	93	93	66	76	89	78
26	89	92	60	73	86	73
25	86	89	52	70	83	67
24	83	85	46	65	79	61
23	80	82	39	60	73	55
22	75	79	34	55	67	48
21	68	75	28	49	61	41
20	61	71	22	44	57	35
19	55	63	17	39	51	28
18	48	55	12	33	46	21
17	40	49	7	26	41	14
16	34	42	3	20	35	11
15	28	35	2	16	30	8
14	23	27	1	11	24	5
13	18	21	0	8	18	3
12	12	16	0	5	14	2
11	7	10	0	3	10	1
10	4	5	0	1	6	0
9	1	1	0	0	1	0

Reprinted from Martens, Vealey, and Burton (1990).

▶ The Cognitive Anxiety subscale is scored by adding the responses to items 1, 4, 7, 10, 13, 16, 19, 22, and 25.

▶ The Somatic State Anxiety subscale is scored by adding the responses to items 2, 5, 8, 11, 14, 17, 20, 23, and 26 (scoring for item 14 must be reversed, i.e., 4–3–2–1).

▶ The State Self-Confidence subscale is scored by adding up the responses to items 3, 6, 9, 12, 15, 18, 21, 24, and 27.

Research and Practical Examples   A study by Burton (1988) on the relationship between precompetitive state anxiety and performance of collegiate swimmers provides a good illustration of the use of the CSAI–2. Swimmers completed the CSAI–2 right before competing in three separate meets during the season. Results revealed different relationships between each of the subscales and performance in accordance with theoretical predictions. Specifically, cognitive anxiety was negatively related to performance, confidence was positively related to performance, and somatic anxiety exhibited a curvilinear relationship to performance in the shape of an inverted U (i.e., there was an optimal level of somatic anxiety, with low and high levels producing decrements in performance).

This would be useful information for a coach or athlete in trying to get the athlete emotionally ready for competition. Specifically, it would seem important to reduce worry and fear before competition while building confidence to as high a level as possible. Furthermore, it appears that getting emotionally psyched up and physiologically activated is good up to a point, but that too much arousal decreases performance. Finally, results revealed that athletes reacted differently in terms of the anxiety–performance relationship; thus, group pep talks do not have as much value or sensitivity to each athlete's individual needs (i.e., zone of optimal performance).

One final point relating to anxiety measurement appears appropriate. Over the past 10 years, researchers have been measuring the direction of anxiety as well as the previously discussed intensity. That is, how much anxiety you have would represent the intensity level, but is this anxiety facilitating or debilitating? This represents the direction of anxiety. Thus, having a high level of anxiety is not necessarily detrimental to performance because it might depend more on how people interpret their levels of anxiety. Thus, measures of direction should accompany measures of intensity.

Attitudes

Much of the early research in physical education on the affective domain focused on attitudes and their measurement. Attitudes are feelings about things—physical objects, types of people, particular persons, social institutions, and government policies (Nunnally 1978). *A large number of attitude scales have been developed in physical education; unfortunately, many of them have not been constructed in a scientific manner, and often only limited information is provided about reliability and validity.* One problem with establishing concurrent validity for an attitude scale, for example, is that research has indicated low correlations between attitude and behavior. Thus, you may have a favorable attitude toward **physical fitness** yet not exercise regularly.

Another limitation of attitude measurements (as with other self-report inventories) is the problem of *social desirability*—that is, people often want to look good and thus distort their responses accordingly. For instance, you might be asked to indicate your level of agreement with this statement: *If the teacher walked out of a room during an exam, it would be all right to cheat if you saw other students cheating.* You might respond by strongly disagreeing with the statement merely because this is the socially desirable response, regardless of your true feelings. Let's briefly discuss a few attitude scales in physical education that have been carefully constructed to avoid these problems.

Mastery Item 14.10

What are two limitations of attitude measurement? What can be done to deal with these problems?

Attitudes Toward Physical Activity

Kenyon's (1968b) Attitudes Toward Physical Activity (ATPA) Inventory was constructed to measure six dimensions of active and passive involvement in physical activity. This scale was one of the first in physical education to demonstrate that attitude must be considered to be multidimensional. That is, there may be several types of attitudes toward physical activity, so instead of obtaining one composite score, as with most previous attitude instruments, the researcher splits the total score into several scores to validly measure each dimension. The six dimensions are as follows:

- ▶ Physical activity as a social experience
- ▶ Physical activity for health and fitness
- ▶ Physical activity as the pursuit of vertigo (thrill or excitement)
- ▶ Physical activity as an aesthetic experience
- ▶ Physical activity as a catharsis
- ▶ Physical activity as an ascetic experience

Reliability and Validity The ATPA was one of the first attitude scales in physical education to demonstrate both good reliability and validity. Whereas most previous attitude scales were atheoretical in nature, the ATPA is based on a theoretical model proposing that attitude toward physical activity is relatively stable and that positive attitudes are manifested by active participation or by watching others perform (passive involvement). As a result, content validity was established by expert opinion. Construct validity was established through factor analysis and the group difference method. For example, a group of people such as athletes on a sport team exhibited more positive attitudes toward the different dimensions of physical activity than a group of artists not participating in athletics. Internal consistency reliability estimates for each subscale were consistently high, ranging from .70 to .87. In addition, test–retest reliability coefficients ranged from .78 to .91 across the six subscales.

Items from the Health and Fitness subscale are displayed in table 14.7. Responses range from very *strongly agree* to *very strongly disagree*. The ATPA scale for men consists of 59 items, and the parallel scale for women consists of 54 items.

Table 14.7 Health and Fitnesss Scale Items From the Attitudes Toward Physical Activity Scale

VSA	SA	A	U	D	SD	VSD	Of all physical activities, those whose purpose is primarily to develop physical fitness would not be my first choice.[a]
VSA	SA	A	U	D	SD	VSD	I would usually choose strenuous physical activity over light physical activity if given the choice.
VSA	SA	A	U	D	SD	VSD	A large part of our daily lives must be committed to vigorous activity.
VSA	SA	A	U	D	SD	VSD	Being strong and highly fit is not the most important thing in my life.[a]
VSA	SA	A	U	D	SD	VSD	The time spent doing daily calisthenics could probably be used more profitably in other ways.[a]

Note: VSA = very strongly agree; SA = strongly agree; A = agree; U = undecided; D = disagree; SD = strongly disagree; VSD = very strongly disagree; [a] = reversed scored.

Research and Practical Examples Zaichkowsky (1975) studied the differences in attitudes toward physical activity after participation in a college-level lifetime sports curriculum (e.g., swimming, badminton, or bowling) versus a foundations curriculum that consisted primarily of jogging. The ATPA was completed by males and females before and after participation in one of these programs. Results indicated that participants in the lifetime sports curriculum had more positive attitudes toward health and fitness than did participants in the foundations curriculum. This was particularly true for the women in the sample. In addition, women tended to view their participation more as an aesthetic experience (focus on beauty and artistry of physical activity), whereas men tended to view their participation in physical activity as an ascetic experience (emphasizing hard training and dedication) and a pursuit of vertigo (i.e., thrill, excitement).

These results indicate that it is important for human performance professionals to understand that males and females have different attitudes toward physical activity and to structure their presentations accordingly. In addition, it also seems that physical activity programs that foster lifetime skills will have a greater impact on developing positive attitudes toward health and fitness. This is of great importance if we want to promote healthy living and fitness throughout life.

Children's Attitudes Toward Physical Activity

Sometimes scales need to be modified to be more appropriate for different populations. Because this was the case with the ATPA Inventory, Simon and Smoll (1974) developed the Children's Attitudes Toward Physical Activity (CATPA) Inventory. The researchers changed the wording of the dimension descriptions to be more appropriate for young children. They used a semantic differential scale with each of the six dimensions and evaluated each dimension on the basis of eight pairs of bipolar adjectives. The CATPA Inventory has shown high internal consistency, with reliabilities ranging from .80 to .89, and test–retest reliabilities (with a 6-week time interval) of approximately .60 (Simon and Smoll 1974). Table 14.8 provides an example of using the semantic differential in the domain of physical activity as a social experience.

Table 14.8 Items for the Social Dimension of the Children's Attitudes Toward Physical Activity

What does the following idea mean to you?
PHYSICAL ACTIVITY AS A SOCIAL EXPERIENCE
Physical activities that give you a chance to meet new people and be with your friends.
Always think about the idea above.

1. Good	___ ___ ___ ___ ___ ___ ___	Bad
2. Of no use	___ ___ ___ ___ ___ ___ ___	Useful
3. Not pleasant	___ ___ ___ ___ ___ ___ ___	Pleasant
4. Bitter	___ ___ ___ ___ ___ ___ ___	Sweet
5. Nice	___ ___ ___ ___ ___ ___ ___	Awful
6. Happy	___ ___ ___ ___ ___ ___ ___	Sad
7. Dirty	___ ___ ___ ___ ___ ___ ___	Clean
8. Steady	___ ___ ___ ___ ___ ___ ___	Nervous

Reprinted from Simon and Smoll (1974).

Mastery Item 14.11

Describe a study in which you would use the ATPA or CATPA to measure attitudes toward physical activity.

Physical Self-Perception Scale

The notion of the physical self and its relationship to feelings of self-esteem and self-concept have long been a focus of researchers. For example, Sonstroem (1974) developed the Physical Estimation and Attraction Scale (PEAS) to measure the motivational properties of physical self-esteem (estimation) and interest in vigorous physical activity (attraction). The PEAS is based on the theory that attitude toward physical activity is modifiable by participation in physical activity. Sonstroem (1978) incorporated the PEAS into a model explaining the psychological benefits of physical activity and motivation to participate in physical activity.

More recently, the multidimensional nature of the physical self has been established, and the Physical Self-Perception Profile (PSPP) has been one of the leading instruments developed to assess the physical self. The four specific subscales of the physical self (30 items in all) include Body Attractiveness, Sport Competence, Physical Strength, Physical Conditioning and a generalized subscale (Physical Self-worth). The items from the PSPP are presented in a structured-alternative format to reduce socially desirable responses. For example, a sample item would be, "some people believe that they are very strong and have well-developed muscles compared with most people" but "others believe that they are not so strong and their muscles are not very well developed." For each sentence, the respondent checks either "really true of me" or "sort of true of me."

Finally, the PSPP is also accompanied by the Perceived Importance Profile, which includes two-item subscales designed to assess the relative centrality to the self of each subdomain content.

Reliability and Validity   Reliability and validity research has been conducted with a variety of populations including male and female college students, middle-aged adults, and overweight adults. Test–retest reliability has been established over 16- and 23-day periods with a range of reliability coefficients from .74 to .89. Alpha reliabilities have consistently scored in the range of .80 to .95. Predictive validity was established by showing the links between the PSPP and choice of activity involvement. Convergent validity was established by evidence of logical and strong links between various exercise behaviors and PSPP subscales. In addition, the PSPP was shown not to be susceptible to social desirability because this type of scale might produce socially desirable responses.

Research and Practical Examples   The PSPP has been used a great deal since its development to investigate relationships between how people feel about their physical self and numerous physical activity outcome measures as well as mental health measures. Specifically, scores on the PSPP have been correlated with depression, health complaints, positive and negative affect, feelings of competence, feelings of self-esteem, amount of physical activity, and body image. The critical point is that how people feel about themselves can have an important influence on their mental–emotional states as well as their levels of physical activity. Thus, making activities more enjoyable so that there is a higher level of participation will help build individuals' sense of physical self. This, in turn, would be positively related to improving self-esteem and a variety of other mental health behaviors.

Trait and State Sport Confidence Inventories

Vealy (1986) developed the Trait Sport Confidence Inventory (TSCI) and the State Sport Confidence Inventory (SSCI) to measure the benefit or certainty that people have about their ability to be successful in sport. The TSCI indicates the belief an athlete usually has, and the SSCI measures the degree of certainty an athlete has at a particular moment in time about his or her ability to be successful in sport. The TSCI and SSCI scales are similar to SCAT and the CSAI–2 in that they measure trait and state anxiety, respectively. The two confidence scales were developed based on an interactionalist paradigm in which the individual difference construct of trait self-confidence interacts with the objective sport situation to produce state self-confidence. That is, individual differences in trait self-confidence are predicted to influence how athletes perceive factors within an objective sport situation and predispose them to respond to sport situations with certain levels of state sport confidence.

Reliability and Validity   Vealey (1986) established reliability and validity in a series of studies. High school and college athletes took both the SSCI and TSCI inventories twice, either 1 day, 1 week, or 1 month apart. Results revealed test–retest reliability ranging from

.63 for the 1-month retest to .89 for the 1-day retest. The internal consistency was high, with alpha coefficients of .93 and .95 for the TSCI and SSCI, respectively. Construct validity was established by analyzing the relationships between the SSCI and TSCI with other personality constructs; all correlations were significant in the predicted direction. For example, both the TSCI and SSCI were positively correlated with other measures of perceived physical ability and self-esteem but negatively correlated with somatic and cognitive state anxiety. Evidence supporting the relationship between the SSCI and TSCI and other constructs in the theoretical model established construct validity. For instance, trait self-confidence was found to be a good predictor of precompetitive and postcompetitive state self-confidence. Table 14.9 illustrates sample items from the TSCI. Responses to all items are simply added together to get a total score.

Table 14.9 Sample Items From the Trait Sport Confidence Inventory

When you compete, how confident do you *generally feel*? (Circle number.)

	Low			Medium			High		
1. Compare your confidence in *your ability to execute the skills necessary to be successful* to the most confident athlete you know.	1	2	3	4	5	6	7	8	9
2. Compare your confidence in *your ability to make critical decisions during competition* to the most confident athlete you know.	1	2	3	4	5	6	7	8	9
3. Compare your confidence in *your ability to perform under pressure* to the most confident athlete you know.	1	2	3	4	5	6	7	8	9
4. Compare your confidence in *your ability to execute successful strategy* to the most confident athlete you know.	1	2	3	4	5	6	7	8	9
5. Compare your confidence in *your ability to concentrate well enough to be successful* to the most confident athlete you know.	1	2	3	4	5	6	7	8	9

Reprinted from Vealey (1986).

Research and Practical Examples As part of the validation of the TSCI and SSCI, Vealey (1986) had elite gymnasts complete the TSCI 24 hr before a national competition and the SSCI approximately 1 hr before the competition. In addition, the gymnasts completed the Competitive Orientation Inventory, which measures whether they are more performance oriented (focusing on improving their own performance) or more outcome oriented (focusing on winning the competition). Gymnasts who were high in trait confidence and also performance oriented exhibited the highest state confidence just before competition. This illustrates the importance of the interaction between an athlete's self-confidence and his or her competitive orientation. *In essence, it appears that athletes who are generally confident and also focus on doing their best instead of winning will be most confident when the time comes to perform.* Because a great deal of research in sport psychology has shown that higher levels of confidence are related to higher levels of performance, it is imperative that coaches and parents foster an orientation that focuses on self-improvement rather than merely on winning.

Mastery Item 14.12

Describe a study that would use both TSCI and SSCI to measure a person's confidence in sport situations.

Group Environment Questionnaire

Although the focus of measurement in sport and exercise psychology is on individual state and traits, some researchers have focused their efforts on measuring certain attributes of groups and teams. The one attribute of teams that has received the most attention is group cohesion. Researchers made a number of earlier attempts to measure this concept, but it wasn't until the seminal work of Carron, Widmeyer, and Brawley (1985) on the development of the Group Environment Questionnaire (GEQ) that a reliable and valid measure was developed. The conceptual model that is the basis of the GEQ is that group cohesion is a multidimensional construct comprising both task (achieving the group's objectives—common goals) and social (developing social relationships) aspects. In addition, the group dynamics literature distinguishes between groups and individuals in terms of group integration (an individual's perception of the closeness and similarity within the group) and individual attraction to the group (an individual's perceptions about his or her attraction to the group).

If we combine these aspects, the four major constructs measured by the GEQ are the following:

▶ Group integration–task—e.g., *Our team is united in trying to reach its goals for performance.*

▶ Group integration–social—e.g., *Members of our team do not stick together outside of practices and games.*

▶ Interpersonal attraction to the group–task—e.g., *I do not like the style of play on this team.*

▶ Interpersonal attraction to the group–social—e.g., *Some of my best friends are on this team.*

The GEQ has a total of 18 items scored on a 1 to 9 Likert scale ranging from *strongly disagree* to *strongly agree*. There are four items each for the Group Integration–Social and Individual Attraction to Group–Task subscales and five items each for the Individual Attraction to Group–Social and Group Integration–Task subscales. Table 14.10 provides sample items.

Reliability and Validity Reliability and validity of the GEQ have been established in a series of systematic studies by the authors. Construct validity of the GEQ was demonstrated through its close relationship with theory. For example, differences in attributional patterns have been found in low versus high task-cohesive athletes. Construct validity was established by correlating the GEQ to other similar or dissimilar instruments. For example, as expected, Brawley, Carron, and Widmeyer (1987) found that the GEQ was moderately correlated (correlations ranging from .40 to .55) with the Sport Cohesiveness Questionnaire (Martens, Landers, and Loy 1971), which is a similar instrument. Conversely, the GEQ exhibited low correlations (correlations ranging from .03 to .28) with various scales of the Bass Orientation Inventory (Bass 1962), which is a dissimilar instrument measuring motivational orientations. The predictive validity of the GEQ has been demonstrated in a number of studies that have found that the GEQ was able to predict such variables as exercise adherence, resistance to disruption, performance, leadership behavior, and team building (Carron, Brawley, and Widmeyer 1998). Internal consistency reliability was established using more than 250 athletes across 26 different sports with alpha reliability estimates ranging from .64 to .76 on the various GEQ subscales (Carron, Widmeyer, and Brawley 1985).

Research and Practical Examples The GEQ has been widely used to assess cohesion in sport teams and exercise groups, relating cohesion to such variables as adherence to exer-

Table 14.10   **Sample Items From the Group Environment Questionnaire**

	Strongly disagree						Strongly agree		
1. This team is one of my most important social groups.	1	2	3	4	5	6	7	8	9
2. Our team members rarely party together.	1	2	3	4	5	6	7	8	9
3. Some of my best friends are on this team.	1	2	3	4	5	6	7	8	9
4. I'm unhappy with my team's level of desire to win.	1	2	3	4	5	6	7	8	9
5. Our team is united in trying to reach its performance goals.	1	2	3	4	5	6	7	8	9
6. Members of our team do not stick together outside of practice and games.	1	2	3	4	5	6	7	8	9
7. Our team would like to spend time together during the off-season.	1	2	3	4	5	6	7	8	9
8. If members of our team have problems in practice, everyone wants to help them so we can get back together.	1	2	3	4	5	6	7	8	9

Adapted from Wildmeyer, Brawley, and Carron (1985).

cise, group size, leadership, team building, and role involvement. For example, in a series of studies by Spink and Carron (1992, 1994), group cohesion was measured in exercise class participants approximately midway through the class. Results revealed that participants with higher levels of perceived group cohesion (as measured by the GEQ) had higher attendance rates, lower dropout rates, less lateness, and less absenteeism than participants who perceived their exercise group to be lower in cohesion. More specifically, as indicated by the subscales of the GEQ, participants who scored highest on individual attraction to the group–task subscales and individual attraction to the group–social subscales had the best attendance and lowest number of tardies.

From a practical point of view, this has tremendous implications for exercise leaders and fitness directors. Specifically, statistics reveal that a large percentage of American adults either do not exercise at all or do not exercise regularly enough to gain health benefits. In addition, of those individuals starting an exercise program, 50% drop out within 6 months. But the previously described results indicate that if exercise leaders can develop a feeling of group cohesion within their exercise groups (especially feelings of individual attraction toward the group), then increased attendance and adherence rates will likely result. This will begin to especially help those people who have traditionally not been able to maintain a regular exercise program and start them on the road to enhanced fitness and increased overall health.

GENERAL PSYCHOLOGICAL SCALES USED IN SPORT AND EXERCISE

Although the trend in sport psychology has been to develop sport-specific versions of psychological inventories, several general psychological inventories have been used regularly in physical activity and competitive sport settings. These inventories have added a great deal to the sport psychology literature, in terms of both performance enhancement and psychological well-being.

Self-Motivation Inventory

The Self-Motivation Inventory (SMI) was designed to measure a person's self-motivation to persist and was originally developed to be used in exercise adherence studies (Dishman and Ickes 1981). Because approximately 50% of all people who start exercise programs drop out in the first 6 months, it is important to know what types of people might be more likely to adhere to or drop out of a formal exercise program.

Reliability and Validity

Internal consistency reliability was established with a sample of 400 undergraduate men and women; the items were found to be highly reliable ($r = .91$). Test–retest reliability (with a 1-month time interval) was also found to be high ($r = .92$), thus demonstrating the stability of the construct of **self-motivation,** the desire to take action. Construct validity was found to be high: The SMI has consistently discriminated among people who will and will not adhere to exercise programs in athletic and adult fitness environments. In addition, correlations between self-motivation and social desirability, achievement motive, locus of control, and ego strength provided discriminant and convergent evidence for the construct validity of the SMI.

The SMI consists of 40 items in a Likert format that ask you to describe how characteristic the statement is of you. Responses can range from 1 = *extremely uncharacteristic of me* to 5 = *extremely characteristic of me*. There are 21 positively keyed items and 19 negatively keyed items to reduce response bias. Table 14.11 provides sample items from the SMI.

Table 14.11   **Sample Items From the Self-Motivation Inventory**

1. I'm not very good at committing myself to doing things.

2. Whenever I get bored with projects I start, I drop them to do something else.

3. I can persevere at stressful tasks, even when they are physically tiring or painful.

4. If something gets to be too much of an effort to do, I'm likely to just forget it.

5. I'm really concerned about developing and maintaining self-discipline.

6. I'm good at keeping promises, especially the ones I make to myself.

7. I don't work any harder than I have to.

8. I seldom work to my full capacity.

Note: Responses are scored on a scale of 1 to 5

    1. Extremely uncharacteristic of me

    2. Somewhat uncharacteristic of me

    3. Neither characteristic nor uncharacteristic of me

    4. Somewhat characteristic of me

    5. Extremely characteristic of me

Reprinted from Dishman and Ickes (1981).

Research and Practical Examples

As noted, the SMI was originally devised as a potential dispositional determinant of adherence to exercise programs. Along these lines, Dishman and Ickes (1981) administered the SMI along with a series of other psychological and physiological assessments to a group of people before the start of an exercise program. Rates of adherence to a regularly scheduled exercise program were assessed over a 20-week period. *It was found that only self-moti-*

vation and percent body fat were able to predict those people who would adhere to or drop out of the exercise program. In fact, self-motivation and percent body fat (when taken together) were able to accurately classify participants into actual adherence and dropout groups in approximately 80% of cases.

This result has important implications for you as a health and fitness leader. Specifically, if a person is low in self-motivation, it is more likely that he or she will drop out of the program. Knowing this, you must provide such individuals with extra reinforcement and encouragement.

Profile of Mood States

McNair, Lorr, and Droppleman (1971) initially developed the Profile of Mood States (POMS) to provide a measure of **mood** states—that is, one's emotional state of mind, feeling, inclination, or disposition. The scale has six different subscales, each representing a different mood: Vigor, Confusion, Anxiety, Tension, Anger, and Fatigue. The POMS can be used with instructions that ask subjects to state how they feel right now or have felt for the last week or the last month. Thus, it can be used as both a state measure and a trait measure.

The use of the POMS in sport and exercise settings has been so extensive that it prompted a special issue in the *Journal of Applied Sport Psychology* (Terry 2000). The POMS has been the major measure of mood, providing a link between physical activity and mental health based on some of the groundbreaking research of William Morgan (Morgan 1980). Although the measure is not without controversy (e.g., Rowley, Landers, Kyllo, and Etnier 1995) in its prediction of performance for successful and less successful athletes, it still remains as one of the most consistent measures of mood in sport and exercise settings.

Reliability and Validity

McNair, Lorr, and Droppleman (1971) found that the reliability and validity of the POMS do not vary for the three different time frames. Although such a mood scale would be expected to change over time, test–retest reliability coefficients for the six subscales were found to range from .65 for vigor to .74 for depression. Internal consistency reliability was shown to be consistently high within each subscale, with reliability estimates of approximately .90. Construct validity was established by relating the six POMS subscales to other personality measures; results were consistently in the predicted directions for the different subscales. The scale consists of 65 items scored on a 5-point Likert scale from *not at all* to *extremely*. Table 14.12 presents items from the Vigor subscale.

Table 14.12   **Vigor Subscale From the Profile of Mood States**

Below is a list of words that describe feelings people have. Answer *how you feel right now.*

	Not at all	A little	Moderately	Quite a bit	Extremely
1. Lively	0	1	2	3	4
2. Active	0	1	2	3	4
3. Energetic	0	1	2	3	4
4. Alert	0	1	2	3	4
5. Full of pep	0	1	2	3	4
6. Carefree	0	1	2	3	4
7. Vigorous	0	1	2	3	4

Reprinted from McNair, Lorr, and Droppleman (1971).

Research and Practical Examples

The POMS has been widely used to study the moods of elite athletes over the course of a season. In a series of studies investigating mood states of elite wrestlers, distance runners, swimmers, and rowers (Morgan and Johnson 1978; Morgan and Pollock 1977), athletes completed the POMS at different times during a competitive season. *One consistent finding was that successful and less successful elite athletes differed in their mood profiles.* The more successful athletes were high on vigor (a positive attribute) but low on all the other scales (negative attributes), whereas the less successful athletes were higher on all the negative mood states and lower on the one positive state of vigor. More recently, in a thorough review (Prapavessis 2000) it was found that individual differences really needed to be considered when one is investigating the relationship between mood (as measured by the POMS) and athletic performance. Specifically, the amount of divergence for each athlete from his or her "optimal" mood state was related to performance. In essence, the more an athlete's mood deviated from his or her optimal state (whether positive or negative), the poorer the performance. Thus, from a coaching perspective, it is important to individualize athletes' mood states in relationship to what produces the best performance. This might involve several assessments of mood states and performance to determine what types and levels of mood states are associated with peak performance.

Test of Attentional and Interpersonal Style

Nideffer (1976) developed the Test of Attentional and Interpersonal Style (TAIS) to measure a person's attentional and interpersonal characteristics, the appropriateness of one's attentional focus, and the ability or inability to shift from one attentional focus to another. The basis of the TAIS is a theory that seeks to predict behavior based on the interaction between interpersonal and attentional processes and physiological arousal. Attention is hypothesized to vary along two dimensions: width (broad to narrow) and direction (internal to external). It is postulated that both attentional and interpersonal characteristics have state and trait components and that these scales can be examined by manipulating arousal. The test is composed of six attentional subscales—two subscales reflecting behavioral and cognitive control—and nine interpersonal subscales. In sport psychology, most researchers and practitioners have focused on the attentional subscales.

Reliability and Validity

The reliability and validity of the TAIS have been demonstrated in a variety of studies using different samples, although Nideffer (1976) initially developed the test using college students. Test–retest reliability based on a 2-week interval ranged from .60 to .93 with a median of .83. Construct validity was examined by correlating TAIS scale scores with scores on other psychological instruments. The overall pattern of results indicated that the TAIS subscales were correlated with conceptually similar scales but were not correlated with scales measuring different constructs. Correlating the TAIS subscales with future behaviors demonstrated predictive validity. For example, swimmers scoring high on the subscale measuring the tendency to make errors of underinclusion were rated by their coaches as choking under pressure, falling apart when they make early performance errors, and becoming worried about one thing and being unable to think about anything else. Poor attentional control and the tendency to make errors of underinclusion were found to be associated with performance deficiencies. Table 14.13 describes the six attentional subscales. In recent years (see Abernethy, Summers, and Ford 1998 for a review) some psychometric properties of the TAIS have been called into question, although the TAIS has contributed much to the understanding of attentional processes in sport.

Research and Practical Examples

The TAIS has been used with a number of different sports and athletes, providing researchers with a great deal of information on the relationship between attentional processes and performance. In a study conducted by Martin (1983), high school basketball players com-

Table 14.13   Attentional Subscales of the Test of Attentional and Interpersonal Style

Scale	Description
Broad-external	High scores indicate an ability to effectively integrate many external stimuli simultaneously.
External-overload	High scores indicate a tendency to become confused and overloaded with external stimuli.
Broad-internal	High scores indicate an ability to effectively integrate several ideas at one time.
Internal-overload	High scores indicate a tendency to become overloaded by internal stimuli.
Narrow focus	High scores indicate an ability to effectively narrow attention when it is appropriate.
Reduced focus	High scores indicate chronically narrowed attention.

Reprinted from Nideffer (1976).

pleted the Narrow Focus and External-Overload subscales of the TAIS before the season. The players who scored high on the Narrow Focus subscale had far superior free throw percentages compared with the players who scored high on external overload. Such a finding can be useful for coaches in that it indicates potentially good or poor free throw shooters. A player's high score on external overload would indicate to a coach that the player needs help to cope more effectively with the many potential distractions when shooting free throws. Attentional control training designed to focus attention on the relevant cues (the rim) while eliminating irrelevant cues (crowd noises and gestures) could be used at the beginning of the season to help the player develop better attentional skills while shooting free throws.

Stages of Change for Exercise and Physical Activity

Marcus, Selby, Niaura, and Rossi (1992) developed the Stages of Change Scale for Exercise and Physical Activity to measure the specific stage of exercise that individuals might be in at a particular moment in time; stages fall somewhere between states and traits. Specifically, although stages may last for considerable lengths of time, they are open to change. This is the nature of most high-risk behaviors—stable over time yet open to change. The stages of change or transtheoretical model was developed as a framework to describe the different phases involved in the acquisition and maintenance of a behavior (Prochaska and DiClemente 1983; Velicer and Prochaska 1997). Specifically, Marcus et al. (1994) suggested that individuals who engage in a new lifestyle behavior (such as physical activity, smoking cessation, condom use, seat-belt wearing) progress in an orderly manner through the following stages:

- **Precontemplation**—no intention to change behavior
- **Contemplation**—intention to change behavior
- **Preparation**—preparing for action
- **Action**—involved in behavior change
- **Maintenance**—sustained behavior change

Individuals are thought to progress through these stages as they adopt lifestyle behaviors at varying rates; some may move right through each stage, others might get "stuck" at certain stages, and others may relapse to earlier stages.

Reliability and Validity

Marcus and colleagues (1992) conducted three studies to develop and refine the validity and reliability of the Stages of Change Instrument (SCI). The original SCI was developed by modifying an existing instrument that had previously been developed for smoking cessation. A test–retest reliability estimate of .78 was obtained for the new scale. In addition, concurrent validity was demonstrated by showing that the SCI was significantly related to a 7-day physical activity recall questionnaire (Marcus and Simkin 1993).

The SCI has been used to classify individuals into different stages of exercise change and thereby to develop specific interventions to help people in these different stages. A 5-point Likert scale is used to rate each item, ranging from 1 *(strongly agree)* to 5 *(strongly disagree)*. Individuals are placed into the stage corresponding to the item that they endorsed most strongly (i.e., *agree* or *strongly agree*). Any individual not endorsing any items with *agree* or *strongly agree* is not placed into a stage. Table 14.14 presents the specific stages and corresponding scale items.

Table 14.14   **Stages of Change in Exercise and Physical Activity**

Stage	Item
Precontemplation	I currently do not exercise and I do not intend to start exercising in the next 6 months.
Contemplation	I currently do not exercise , but I am thinking about starting to exercise in the next 6 months.
Preparation	I currently exercise some, but not regularly.
Action	I currently exercise regularly, but I have only begun doing so within the last 6 months.
Maintenance	I currently exercise regularly, and have done so for longer than 6 months.

Note: Regular exercise = 3 or more times per week for 20 minutes or more at each time.

Reprinted from Marcus, Selby, Niaura, and Rossi (1992).

Mastery Item 14.13

You are a fitness leader in a corporate wellness program. You have been assigned to increase employee participation in the physical activity program. How could you use the Stages of Change Instrument to increase participation?

Research and Practical Examples

As noted previously, one of the main benefits of the Stages of Change model is that it helps practitioners individualize behavioral interventions to increase exercise by identifying exactly what stage an individual happens to be in at a given time. Using this approach, Marcus and colleagues (1994) tested 610 adults aged 18 to 82 and, using the SCI, classified them into one of five stages of change identified by the model. The researchers devised 6-week interventions using written resource materials and specific exercise opportunities targeted at an individual's specific stage of readiness to adopt or continue an exercise program. Results revealed that 65% of participants in the contemplation stage became active, and 61% of people in the preparation stage became more active. From a practical point of view, practitioners may be able to achieve greater compliance by targeting interventions based on an individual's specific stage of exercise change. In essence, practitioners can devise specific exercise programs and educational materials that would be particularly motivating and relevant for people in certain stages of exercise change. Thus, different processes can be used to promote transitions from one stage to the next. This, in turn, should enhance both exercise participation and maintenance.

Measurement and Evaluation Challenge

After reading this chapter, Coach Keller is more aware of some of the different types of psychological inventories at his disposal. He knows about the guidelines for using psychological tests with athletes, and he knows that he can get the best prediction if he uses tests that are sport specific. Armed with this information, Bill decides on the following approach.

1. He will hire a qualified sport psychologist to coordinate the psychological evaluations and to interpret all psychological inventories.

2. He and the test administrator will tell his athletes specifically what the inventories are being used for and will provide feedback about their results.

3. He will use a few trait inventories before the season starts, such as the Sport Competition Anxiety Test (SCAT) and the Trait Sport Confidence Inventory (TSCI), to get a better understanding of his athletes' general psychological profiles.

4. He will assess his athletes just before competition using state measures such as the Competitive State Anxiety Inventory–2 (CSAI–2) and the State Sport Confidence Inventory (SSCI) to determine how they are feeling just before games.

5. With this information, Coach Keller, in conjunction with the sport psychologist, will devise a mental training program to help his athletes practice and develop their mental skills.

By combining this psychological testing and training with his work and practice on the physical aspects of training, Coach Keller hopes that his football team will be maximally ready to perform at their optimal level.

SUMMARY

The field of sport and exercise psychology has been expanding rapidly in the last 10 to 20 years, with the two major areas being performance enhancement and mental health. Part of this expansion has involved developing and refining the measurement of psychological traits and states. This has highlighted issues involving the use and abuse of psychological inventories in sport and exercise settings. The American Psychological Association provides guidelines for the use of psychological testing to ensure that athletes are treated in an ethical manner and that test administration and feedback are conducted responsibly.

Although much early work in sport psychology used standardized psychological scales to assess personality and other psychological constructs, recently more qualitative methods have been used, including in-depth interviews and observation. Along with the increased emphasis on qualitative methods, more accurate and reliable prediction of behaviors in sport and exercise settings has resulted from the development of sport-specific psychological inventories. (Although there are still many scales in use without established reliability and validity, more and more sport-specific psychological inventories are being carefully developed from a psychometric standpoint.) Some of the more noteworthy scales are the Sport Competition Anxiety Test (SCAT), the Competitive State Anxiety Inventory–2 (CSAI–2), and the Trait Sport Confidence and State Sport Confidence Inventories. Attitude inventories such as the Physical Estimation and Attraction Scale (PEAS) and the Attitudes Toward Physical Activity (ATPA) scale have demonstrated that attitudes are multidimensional in nature and that attitude scales can be developed with construct validity. In addition, the Stages of Change questionnaire has recently been developed to identify an individual's specific stage of exercise behavior so that interventions can be targeted for that particular stage. Finally, although most researchers have focused on measuring personality attributes of individuals, the Group Environment Questionnaire (GEQ) focuses on the measurement of cohesiveness in sport and exercise groups.

Although the trend is to develop sport-specific tests, several general psychological inventories have also been used extensively in sport and exercise settings; these have added to our understanding of sport-related behavior. Such scales include the Self-Motivation Inventory (SMI), the Test of Attentional and Interpersonal Style (TAIS), and the Profile of Mood States (POMS).

It is important to note that most instruments consist of several scales. This is because psychological measures need to reflect the multidimensionality of personality, perception, and other psychological factors. Personality and its associated constructs are multifactorial in nature. Sub scales help researchers get at these various factors.

Glossary

ability—A general, innate psychomotor trait.

absolute endurance—A measure of repetitive performance against a fixed resistance (e.g., number of repetitions of a 100 lb bench press).

absolute rating—Evaluating performance on a fixed scale.

absolute risk—The risk (proportion, percentage, rate) of mortality or morbidity in a population that is exposed or not exposed to a risk factor.

accuracy-based skills test—A skills test that assesses one's ability to serve or project an object (e.g., ball, bird) into a prescribed area or for distance and accuracy.

achievement test—A test designed to measure the extent to which an examinee comprehends some body of knowledge.

adapted physical education—Physical education adjusted to accommodate children with physical or mental limitations.

aerobic power—The body's ability to supply oxygen to the working muscles during physical activity.

affective domain—Involves attitudes and perceptions.

alpha level—The probability of falsely rejecting a null hypothesis (i.e., claiming that there is a relationship between the variables when, in fact, there is no such relationship). Also, the probability of making a type I error.

alternative assessment—An assessment technique that is different from traditional, standardized testing. Also called *authentic assessment.*

analytic assessment—Assessment typically conducted for diagnostic purposes, which focuses on the separate elements of a performance or product.

analytic scoring—A method of scoring responses to essay questions that involves identifying specific facts, points, or ideas in the answer and awarding credit for each one located.

anxiety—Distress and tension caused by apprehension.

assessment—The process of collecting information and judging what it means.

athletic fitness—Physical fitness related to sport performance (e.g., 50 yd dash, speed, agility).

attributable risk—The risk of mortality and morbidity directly related to a risk factor. It can be thought of as the reduction in risk related to removing a risk factor.

authentic assessment—Assessment designed to take place in a real-life setting that provides authenticity and contextualized meaning.

behavioral objectives—Goals with specific measurable steps for their achievement.

body composition—The physical makeup of the body, including weight, lean weight, and percent fat.

cardiovascular endurance—The body's ability to extract and use oxygen in a manner that permits continuous exercise, physical work, or physical activities (e.g., jogging).

central tendency—Statistics that are located near the center of a set of scores.

central tendency error—The type of rating scale error that results from the hesitancy of raters to assign extreme ratings.

checklist—A typically dichotomous rating of a trait.

coefficient of determination—A measure of variation in common between two variables. Interpreted as a percent, it is the square of the correlation (r^2).

cognitive domain—Involves knowledge and mental achievement.

composite score—A total score developed from the scores of a set of separate tests or performances.

concurrent validity—The relationship between a test (a surrogate measure) and a criterion when the two measures are taken relatively close together in time. It is based on the Pearson product-moment (PPM) correlation coefficient between the test and criterion.

construct-related validity—The highest form of validity; it combines both logical and statistical evidence of validity through the gathering of a variety of statistical information that, when viewed collectively, adds evidence for the existence of the theoretical construct being measured.

content-related validity—Evidence of truthfulness based on logical decision making and interpretation. Also called *face validity* or *logical validity*.

contingency table—A table used for cross-referencing two nominal variables.

**correlation coefficient *(r)*—An index of the linear relationship between two tests, indicating the magnitude, or amount of relationship, and the direction of the relationship.

correlation—A measure of the relationship between two variables *(r)*. Must be between –1.00 and 1.00.

criterion-referenced standard—A specific, predetermined level of achievement.

criterion-referenced test—A test with specific, predetermined performance or health standards.

criterion-related validity—Evidence that a test has a statistical relationship with the trait being measured; also called *statistical validity* and *correlational validity*.

curvilinear relationship—An association among the variables that can be depicted by a curve.

cutoff scores—Scores that establish identifiable groups or levels of performance—typically used with criterion-referenced testing.

dependent variable—The variable used as a criterion or that you are trying to predict *(Y)*.

depression—A mental condition of gloom, sadness, or dejection.

descriptive statistics—Mathematics used to organize, summarize, and describe data.

direct relationship—A positive relationship between two variables such that higher scores on one variable are associated with higher scores on the other variable. Similarly, lower scores on one variable are associated with lower scores on the other variable.

distance or power performance test—A skills test that assesses one's ability to project an object for maximum displacement or force.

distractor—An incorrect response to a multiple-choice test item.

epidemiology—The study of the incidence, distribution, and frequency of diseases (e.g., the study of the effects of inactivity on coronary heart disease).

error score—An unobservable but theoretically existent score that contributes to an inaccurate estimate of individual differences.

evaluation—A dynamic decision-making process that places a value judgment on the quality of what has been measured (e.g., a test score or physical performance).

Excel—A spreadsheet product of Microsoft that results in rows of subjects and columns of variables.

exhibition—A public presentation or performance in which an individual displays his or her knowledge and skills.

extrinsic ambiguity—The characteristic of a test item that appears ambiguous to a respondent who does not understand the concept that the item is testing.

fairness—A characteristic of an assessment, meaning the absence of bias, that allows all participants an equal opportunity to perform to the best of their capability.

flexibility—The range of motion of a joint or group of joints.

formative evaluation—An evaluation conducted during (as opposed to at the end of) an instruction or training program.

frequency distribution—A list of the observed scores and their frequency of occurrence.

functional capacity—The ability to perform the normal activities of daily living.

general motor ability (GMA)—The overall proficiency to perform a variety of psychomotor tasks.

global rating—Evaluating overall performance rather than assessing individual components of the action.

global scoring—A method of assessing an answer to an essay question that involves converting into a score the general impression obtained from reading the answer.

halo effect—The tendency to elevate a person's score because of positive bias. May also work in reverse (reduce a score because of negative bias).

health-related physical fitness—The physical fitness parameters associated with health (e.g., cardiovascular endurance, body composition, muscular fitness).

histogram—A graph using vertical bars to present the frequency distribution of the observed scores.

holistic assessment—Analysis based on the overall quality of a performance or product.

hypothesis—A statement of relationship between at least two variables.

incidence—The number, proportion, rate, or percentage of new cases of mortality and morbidity. Incidence could be calculated in a randomized clinical trial or a prospective, longitudinal cohort study.

independent variable—The variable related to a dependent variable often used as a predictor (X).

index of difficulty—A mathematical expression used in item analysis to estimate the percentage of examinees answering a test item correctly.

index of discrimination—A mathematical expression used in item analysis to estimate how well a test item discriminates among examinees who have been categorized by some criterion.

index of reliability—The theoretical correlation between observed score and true score; the square root of the reliability coefficient.

indirect relationship—A negative relationship between two variables such that higher scores on one variable are associated with lower scores on the other variable. Similarly, lower scores on one variable are associated with higher scores on the other variable.

inferential statistics—Statistics used to test a hypothesis within a small group (sample) to infer to a larger group (population).

Internet—A network of computers for high-speed transmission of information.

intrinsic ambiguity—The characteristic of a test item that is truly ambiguous, even to the respondent who understands the concept that the item is testing.

item analysis—A prescribed process used to examine the quality (e.g., the difficulty and the discrimination) of individual items on a written test.

kappa (K)—A measure of agreement or association between categorical variables that is adjusted for chance.

keyed response—The correct response to a test item.

kurtosis—An indication of the shape of a distribution that specifies how flat or peaked the distribution is.

linear relationship—An association between two variables that can be depicted by a straight line.

marginal—The sum of observations across a specific row or column of a contingency table.

mastery test—A test designed to measure whether an examinee has achieved enough knowledge to satisfy a prescribed standard or criterion. Typically used with criterion-referenced testing.

maximal exercise test—A fitness test that requires the subject to exercise until reaching exhaustion (e.g., treadmill stress test).

maximal oxygen consumption ($\dot{V}O_2$max)—The criterion measure of aerobic capacity.

measurement—The act of assessing (e.g., assessing a knowledge or psychomotor test score or one's attitude toward physical activity).

mentally disabled—Having mental or psychological limitations (e.g., autism).

microcomputer—A small but powerful computer, used by one person at a time.

mood—An emotional state of mind, feeling, inclination, or disposition.

motor educability—The ability to learn motor skills.

motor fitness—See *athletic fitness.*

multiple correlation—The relationship between one outcome (dependent) variable and multiple predictor (independent) variables.

muscular endurance—The physical ability to perform work.

muscular strength—The force that can be generated by the musculature that is contracting.

negative correlation—A correlation in which high scores on one variable are associated with low scores on another variable and low scores are associated with high scores; also called *indirect* or *inverse correlation.*

Net D—An index of discrimination for written test items that indicates the proportion of good discriminations remaining after neutral and bad discriminations are removed.

normal distribution—A bell-shaped, symmetric probability distribution.

norm-referenced standard—A level of achievement relative to a clearly defined subgroup, such as all women or women your age.

objectivity—The degree of interrater reliability; the ability of two or more raters to equivalently score a test.

observed score—One's score on a test. The observed score is the sum of an individual's true score and error score.

odds ratio—An estimate of relative risk used in prevalence studies.

older adults—Individuals 65 years of age and older.

parameter—A statistic in the population of interest (e.g., the population mean).

Pearson product-moment correlation coefficient (PPM) *(r)*—See *correlation coefficient.*

perceived exertion—The mental perception of the intensity of physical work.

percentile—The percentage of observations occurring at or below a given score.

performance assessment—Testing method that requires the participant to create a product or performance that demonstrates his or her knowledge or skills.

performance criteria—Standards for judging a performance or product.

performance ratio—Dividing a performance score by another measure to better compare performance between subjects (e.g., weight, speed).

personality—The totality of a person's distinctive psychological traits.

Phi coefficient—The Pearson product-moment correlation between two dichotomously scored variables, each of which can take on the value of 0 or 1.

physical activity—The act of bodily movement that requires the contraction of muscles and the expenditure of energy.

physical fitness—A set of attributes that people have or achieve that relates to the ability to perform physical activity.

physically disabled—Having physical or organic limitations (e.g., cerebral palsy).

population—The target group of individuals or observations for which research findings are to be inferred.

portfolio—A systematic, purposeful, and meaningful collection of one's work that has been assembled over time.

positive correlation—A correlation in which high scores on one variable are associated with high scores on another variable and low scores are associated with low scores; also called a *direct correlation*.

power—The amount of work performed in a fixed amount of time.

prediction—The ability to estimate the value of one variable from one or more other variables.

predictive validity—The relationship between a test (a surrogate measure) and a criterion when the criterion is measured in the future. It is based on the PPM correlation coefficient between the test and the criterion.

prevalence—The number, proportion, rate, or percentage of total cases of mortality and morbidity. Prevalence would be calculated in a cross-sectional study.

proportion of agreement *(P)*—Percentage of agreement on two measures.

psychomotor domain—Involves physiological and physical performance.

qualitative—Perceptive measurement, typically textual in nature.

quantitative—Precise measurement, typically numerical in nature.

range—A measure of variability obtained by subtracting the lowest score from the highest score.

relationship—The statistical association between two or more variables.

relative endurance—A measurement of repetitive performance related to maximum strength.

relative risk—The risk of mortality (death) or morbidity (disease) associated with one group compared with another (e.g., smokers vs. nonsmokers). Also, the ratio of risks between the exposed or unexposed populations. This statistic is calculated with incidence measures.

relative scale—Evaluating performance relative to others in a specific group.

relative scoring—A method of scoring responses to essay questions that involves reading all the answers to one question and arranging the papers in order according to degree of adequacy.

relevance—The degree to which a test pertains to the objectives of the measurement.

reliability—The degree to which repeated measurements of the same trait are reproducible under the same conditions; consistency.

repetitive-performance test—A skills test that involves continuous performance of an activity for a specified period of time (e.g., volleying).

residual volume—The air left in the lungs after maximal forced expiration of air.

sample—A subgroup of the population in which scientific research is conducted.

scatterplot—A graphic representation of the relationship or correlation between two variables.

scientific method—A method of inquiry that requires the development of a hypothesis and the subsequent test of its plausibility.

scoring rubric—Rating scale that defines performance criteria for an alternative assessment; consists of a fixed scale and a list of characteristics describing performance for each point on the scale.

self-motivation—The desire to take action.

significance—A probability decision of rejecting the null hypothesis when it is true (alpha). See *type I error*.

simple linear prediction—Using one variable *(X)* to predict the criterion *(Y)*.

skewness—An indication of the shape of a distribution that specifies a lack of symmetry.

skill—A learned trait based on the abilities a person has.

specific determiner—A word or a phrase in a written test item that provides an unintended clue to the correct answer.

specificity—Refers to motor abilities or skills that are unique to individual psychomotor tasks.

SPSS—Computer software developed to conduct statistical analyses.

stability reliability—Consistency of measures across time.

stakes—The magnitude, or consequences, of the decisions made from assessment information.

standard deviation—A linear measure of variability that takes into account every score in the distribution; the square root of the variance.

standard error (SE)—The type of rating scale error that results from differences in the standards of evaluation applied by raters of the same performance.

standard error of estimate (SEE)—An indication of the amount of error when predicting *Y* from *X*; the standard deviation of the errors of prediction. Also called *standard error (SE)* or *standard error of prediction (SEP)*. A validity statistic.

standard error of measurement (SEM)—A value reflecting the degree to which a person's observed score fluctuates as a result of errors of measurement; it is interpreted in the same way as a standard deviation. A reliability statistic.

standard score—A score resulting from the conversion of observed values into a score with a given mean and a standard deviation.

state—A psychological attribute that is related to situational changes.

statistic—A numerical value calculated in the sample to estimate a population parameter (e.g., a sample mean).

subjective rating—A value that an instructor places on a skill or performance based on personal observation.

submaximal exercise test—A fitness test that requires the subject to put forth less than maximal effort (e.g., using cycle ergometry to assess heart rate).

summative evaluation—A final, comprehensive evaluation conducted near the end of an instruction or training program.

surrogate measure—A test used to estimate the criterion (e.g., skinfold measurement is a surrogate measure of the criterion percent body fat that is obtained by underwater weighing).

table of specifications—A test blueprint that indicates the proportion of test items dealing with each combination of content objective and educational objective.

target weight—The body weight required to achieve a target percent body fat.

taxonomy—A classification system based on common characteristics.

test—A measuring instrument (e.g., written test, performance test, or a wide variety of other instruments) used to make a particular measurement.

torque—A force producing rotation about an axis.

total body movement test—A test that assesses the speed at which a performer completes a task that involves movement of the whole body in a restricted area (e.g., basketball defensive shuffle test).

trait—A relatively stable, common, consistent psychological attribute.

trials-to-criterion testing—The performance of a skill until a standard of achievement is reached.

true score—An unobservable but theoretically existent score that contributes to one's observed test score; it contributes to an accurate estimate of individual differences.

type I error—A false rejection of the null hypothesis. Deciding there is a relationship between variables when there actually is no relationship. Expressed as a probability (α).

type II error—A false rejection of the alternative or research hypothesis. Failing to discern a relationship between variables when a relationship actually exists. Expressed as a probability (β).

validity—The degree of truthfulness in a test.

variability—The spread, or dispersion, of scores in a set of data, the result of the fact that not all scores are exactly the same.

variance *(s²)*—A measure of variability; a measure of the spread of a set of scores based on the squared deviation of each score from the mean.

waist–hip girth ratio—The waist circumference divided by the hip circumference. This measure provides an estimate of body fat distribution, which is a risk factor for cardiovascular disease.

work—The result of the physical effort that is performed; the product of the amount of force applied and the distance over which it is applied.

World Wide Web—A specific application of the Internet that provides locations of Web sites, information, and commercial capabilities.

youth fitness battery—A group of fitness tests to provide an overall assessment of physical fitness (e.g., FITNESSGRAM).

zero correlation *(r = 0)*—An indication that there is no linear relationship between two variables.

References

Abernethy, B., J. Summers, and S. Ford. 1998. Issues in the measurement of attention. In *Advances in sport and exercise psychology measurement,* ed. J. Duda. Morgantown, WV: Fitness Information Technology.

Ainsworth, B., W. Haskell, A. Leon, D. Jacobs, H. Montoye, J. Sallis, and R. Paffenbarger. 1993. Compendium of physical activities: Classification of energy costs of human physical activities. *Medicine and Science in Sports and Exercise* 25:71-80.

Ainsworth, B.E., W.L. Haskell, M.C. Whitt, M.L. Irwin, A.M. Swartz, S.J. Strath, W.L. O'Brien, D.R. Bassett, Jr., K.H. Schmitz, P.O. Emplaincourt, D.R. Jacobs, Jr., and A.S. Leon. 2000. Compendium of physical activities: An update of activity codes and MET intensities. *Medicine and Science in Sports and Exercise* 32(Suppl.):498-504.

Ainsworth, B.E., and C.E. Matthews. 2001. Descriptive research in physical activity epidemiology. In *Research methods in physical activity,* 4th ed., ed. J.R. Thomas and J.K. Nelson, 291-308. Champaign: Human Kinetics.

American Alliance for Health, Physical Education, Recreation, and Dance. 1980. *Health-related physical fitness test manual.* Reston, VA: AAHPERD.

———. 1985. *Norms for college students: Health-related physical fitness test.* Reston, VA: AAHPERD.

———. 1988. *Physical best.* Reston, VA: AAHPERD.

American College of Sports Medicine (ACSM). 2005. *ACSM's health-related physical fitness assessment manual.* Philadelphia: Lippincott, Williams & Wilkins.

———. 1988. *Resource manual for guidelines for exercise testing and prescription.* Philadelphia: Lea & Febiger.

———. 2001. *ACSM's resource manual for guidelines for exercise testing and prescription.* 4th ed. Philadelphia: Lea & Febiger.

———. 1991. *Guidelines for exercise testing and prescription.* 4th ed. Philadelphia: Lea & Febiger.

———. 1995. *ACSM's guidelines for exercise testing and prescription.* 5th ed. Philadelphia: Lea & Febiger.

———. 2000. *ACSM's guidelines for exercise testing and prescription.* 6th ed. Philadelphia: Lea & Febiger.

American Heart Association. 1994. *Heart and stroke facts.* Dallas: American Heart Association.

American Psychological Association. 1999. *Standards for educational and psychological testing.* Washington, DC: American Psychological Association.

American Red Cross. 1992. *American Red Cross water safety instructor's manual.* St. Louis: Mosby Lifeline.

Anshel, M. 1987. Psychological inventories used in sport psychology research. *Sport Psychologist* 1:331-349.

Åstrand, P., and I. Rhyming. 1954. A nomogram for calculation of aerobic capacity (physical fitness) for pulse rate during submaximal work. *Journal of Applied Physiology* 7:218-221.

Barlow, D.A. 1970. Relation between power and selected variables in the vertical jump. In *Selected topics on biomechanics,* ed. J.M. Cooper, 233-241. Chicago: Athletic Institute.

Barrow, H.M. 1954. Test of motor ability for college men. *Research Quarterly* 25:253-260.

Bartlett, J., L. Smith, K. Davis, and J. Peel. 1991. Development of a valid volleyball skills test battery. *Journal of Physical Education and Dance* 62(2):19-21.

Bass, B.M. 1962. *The orientation inventory.* Palo Alto, CA: Consulting Psychologists Press.

Bass, R.I. 1939. An analysis of the components of tests of semicircular canal function and static and dynamic balance. *Research Quarterly* 2:33-52.

Battinelli, T. 1984. From motor ability to motor learning, the generality–specificity connection. *Physical Educator* 41(3):108-113.

Baumgartner, T.A., A.S. Jackson, M.T. Mahar, and D.A. Rowe. 2003. *Measurement for evaluation in physical education and exercise science.* 7th ed. Dubuque, IA: McGraw-Hill.

Berger, R.A. 1966. Relationship of chinning strength to total dynamic strength. *Research Quarterly* 37:431-432.

Blair, S. 1992. Are American children and youth fit? The need for better data. *Research Quarterly for Exercise and Sport* 63:120-123.

Blair, S., W. Kannel, H. Kohl, and N. Goodyear. 1989. Surrogate measures of physical activity and physical fitness: Evidence for sedentary traits of resting tachycardia, obesity, and low vital capacity. *American Journal of Epidemiology* 129:1145-1156.

Blair, S., H. Kohl, R. Paffenbarger, D. Clark, K. Cooper, and L. Gibbons. 1989. Physical fitness and all-cause mortality: A prospective study of healthy men and women. *Journal of the American Medical Association* 262:2395-2401.

Blair, S.N., J.B. Kampert, H.W. Kohl, III, C.E. Barlow, C.A. Macera, R.S. Paffenbarger, Jr., and L.W. Gibbons. 1996. Influence of cardiorespiratory fitness and other precursors on cardiovascular disease and all-cause mortality in men and women. *Journal of the American Medical Association* 276:205-210.

Bloom, B.S., ed. 1956. *Taxonomy of educational objectives: Cognitive domain.* New York: McKay.

Bloom, G., D. Stevens, and T. Wickwire. 2003. Expert coaches' perceptions of team building. *Journal of Applied Sport Psychology* 15:129-143.

Booth, M.L., A. Okely, T. Chey, and A. Bauman. 2002. The reliability and validity of the Adolescent Physical Activity Recall Questionnaire. *Medicine and Science in Sports and Exercise* 34:1986-1995.

Borg, G. 1962. *Physical performance and perceived exertion.* Lund, Sweden: Gleerup.

———. 1998. *Borg's perceived exertion and pain scales.* Champaign, IL: Human Kinetics.

Brace, D.K. 1927. *Measuring motor ability.* New York: Barnes.

Brawley, L.R., A.V. Carron, and W.N. Widmeyer. 1987. Assessing the cohesion of teams: Validity of the Group Environment Questionnaire. *Journal of Sport Psychology* 9:275-294.

Briere, N., R. Vallerand, M. Blais, M., and L. Pelleteir. 1995. Development and validation of the French form of the Sport Motivation Scale. *International Journal of Sport Psychology* 26:465-489.

Brown, L.E., ed. 2000. *Isokinetics in human performance.* Champaign, IL: Human Kinetics.

Bungum, T.J., D.L. Peaslee, A.W. Jackson, and M.A. Perez. 2000. Exercise during pregnancy and type of delivery in nulliparae. *Journal of Obstetric, Gynecologic, and Neonatal Nursing* 29:258-264.

Burton, D. 1988. Do anxious swimmers swim slower? Reexamining the elusive anxiety-performance relationship. *Journal of Sport and Exercise Psychology* 10:45-61.

Carron, A.V., L.R. Brawley, and W.N. Widmeyer. 1998. The measurement of cohesiveness in sport groups. In *Advances in sport and exercise psychology measurement,* ed. J. Duda, 213-226. Champaign, IL: Human Kinetics.

Carron, A.V., W.N. Widmeyer, and L.R. Brawley. 1985. The development of an instrument to assess cohesion in sport teams: The Group Environment Questionnaire. *Journal of Sport Psychology* 7:244-266.

Caspersen, C. 1989. Physical activity epidemiology: Concepts, methods, and applications to exercise science. In *Exercise and sport science reviews,* ed. K. Pandolph, 423-473. Baltimore: Williams & Wilkins.

Clarke, H.H., and H.A. Bonesteel. 1935. Equalizing the ability of intramural teams at a small high school. *Research Quarterly Supplement* 6(March):193-196.

Clarke, H.H., and R. Munroe. 1970. *Test manual: Oregon cable tension strength test batteries for boys and girls from fourth grade through college.* Eugene, OR: Microcard Publications in Health, Physical Education, and Recreation.

Coleman, R., S. Wilkie, L. Viscio, S. O'Hanley, J. Porcari, G. Kline, B. Keller, S. Hsieh, P. Freedson, and J. Rippe. 1987. Validation of 1-mile walk test for estimating $\dot{V}O_2$max in 20-29 year olds. *Medicine and Science in Sports and Exercise* 19(Suppl. 2):S29.

Considine, W.J. 1970. A validity analysis of selected leg power tests utilizing a force platform. In *Selected topics on biomechanics,* ed. J.M. Cooper, 243-250. Chicago: Athletic Institute.

Cooper Institute for Aerobics Research. 1987. *FITNESSGRAM.* Dallas: Cooper Institute for Aerobics Research.

———. 1992. *FITNESSGRAM.* Dallas: Cooper Institute for Aerobics Research.

———. 1999. *FITNESSGRAM test administration manual.* 2nd ed. Champaign, IL: Human Kinetics.

———. 2004. *FITNESSGRAM/ACTIVITYGRAM test administration manual.* 3rd ed. Champaign, IL: Human Kinetics.

Cooper, K. 1968. A means for assessing maximal oxygen intake. *Journal of the American Medical Association* 203: 201-204.

Corbin, C., and R. Lindsey. 1999. Amateur Athletic Union Physical Fitness Program; FITNESSGRAM, *Fitness for Life,* 4th edition, AAHPERD Physical Best: Cooper Institute for Aerobic Research.

Cox, J. 1997. *Your opinion, please! How to build the best questionnaires in the field of education.* Thousand Oaks, CA: Sage.

Craft, L., M. Magyar, B. Becker, and D. Feltz. 2003. The relationship between the Competitive State Anxiety Inventory-2 and sport performance: A meta-analysis. *Journal of Sport and Exercise Psychology* 25:44-65.

Culver, D., W. Gilbert, and P. Trudel. 2003. A decade of qualitative research in sport psychology journals—1990-1999. *Sport Psychologist* 17:1-15.

Cunningham, G.K. 1986. *Educational and psychological measurement.* New York: Macmillan.

Cureton, K.J., and G.L. Warren. 1990. Criterion-referenced standards for youth health-related fitness tests: A tutorial. *Research Quarterly for Exercise and Sport* 61:7-19.

Dennison, B., J.H. Straus, D. Mellits, and E. Charney. 1988. Childhood physical fitness tests: Predictor of adult physical activity levels? *Pediatrics* 82:324-330.

Disch, C.F., and J.G. Disch. 1979. Predictive analysis of a battery of anthropometric and motor performance tests for classifications of male volleyball players. *Volleyball Technical Journal* 4:93-98.

Disch, J.G. 1978. The construction and analysis of a test battery related to volleyball playing capacity in females. Report No. ED 148815. Washington, DC: ERIC Clearinghouse in Teacher Education.

———. 1979. A factor analysis of selected tests for speed of body movement. *Journal of Human Movement Studies* 5:141-151.

Dishman, R.K., and W. Ickes. 1981. Self-motivation and adherence to therapeutic exercise. *Journal of Behavioral Medicine* 4:421-436.

Docherty, D., ed. 1996. *Measurement in pediatric exercise science.* Champaign, IL: Human Kinetics.

Doyle, J., and G. Parfitt. 1996. Performance profiling and predictive validity. *Journal of Applied Sports Psychology* 8:160-170.

Duda, J. 1989. Relationship between task and ego orientation and the perceived purpose of sport among high school athletes. *Journal of Sport and Exercise Psychology* 11:318-335.

Ebel, R. 1965. *Measuring educational achievement.* Englewood Cliffs, NJ: Prentice-Hall.

Ekelund, L., W. Haskell, J. Johnson, F. Whaley, M. Criqui, and D. Sheps. 1988. Physical fitness as a predictor of cardiovascular mortality in asymptomatic North American men. *New England Journal of Medicine* 319:1379-1384.

Ellenbrand, D.A. 1973. Gymnastics skills tests for college women. Unpublished master's thesis, Indiana University, Bloomington.

Engelman, M.E., and J.R. Morrow, Jr. 1991. Reliability and skinfold correlates for traditional and modified pull-ups in children grades 3-5. *Research Quarterly for Exercise and Sport* 62:88-91.

Eysenck, H.J., and S.B.G. Eysenck. 1968. *Eysenck Personality Inventory manual.* London: University of London Press.

Ezzell, G., J. Smith, and A. Jackson. 1991. One-mile run results in youth: A comparison of natural criterion-referenced standards. *Medicine and Science in Sports and Exercise* 23:S30.

Fiatarone, M.A., E.F. O'Neill, N.D. Ryan, K.M. Clements, G.R. Solares, M.E. Nelson, S.B. Roberts, J.J. Kehayias, L.A. Lipsitz, and W.J. Evans. 1994. Exercise training and nutritional supplementation for physical frailty in very elderly people. *New England Journal of Medicine* 330:1769-1775.

Fleishman, E.A. 1964. *The structure and measurement of physical fitness.* Englewood Cliffs, NJ: Prentice-Hall.

Fleishman, E.A., and M.K. Quaintance. 1984. *Taxonomies of human performance.* New York: Academic Press.

Fox, K., and C. Corbin. 1989. The Physical Self-Perception Profile: Development and preliminary validation. *Journal of Sport and Exercise Psychology* 11:408-430.

Getchell, L.H., D. Kirkendall, and G. Robbins. 1977. Prediction of maximal oxygen uptake in young adult women joggers. *Research Quarterly* 48:61-67.

Glaser, R., and D.J. Klaus. 1962. Proficiency measurement: Assessing human performance. In *Psychological principles in systems development,* ed. R.M. Gagne, 419-474. New York: Holt, Rinehart & Winston.

Glass, G.V., and K.D. Hopkins. 1996. *Statistical methods in education and psychology.* 3rd ed. Englewood Cliffs, NJ: Prentice-Hall.

Golding, L., C. Myers, and W. Sinning. 1989. *Y's way to physical fitness.* Champaign, IL: Human Kinetics.

Gould, D., K. Dieffenbach, and A. Moffatt. 2002. Psychological characteristics and their development in Olympic champions. *Journal of Applied Sport Psychology* 14: 172-204.

Gould, D., R. Horn, and J. Spreeman. 1984. Competitive anxiety in junior elite wrestlers. *Journal of Sport Psychology* 5:58-71. (Available from Dr. Daniel Gould, Department of Exercise and Sport Science, University of North Carolina, Greensboro, NC 27412)

Graves, J., M. Pollock, D. Carpenter, S. Leggett, A. Jones, M. MacMillan, and M. Fulton. 1990. Quantitative assessment of full range-of-motion isometric lumbar extension strength. *Spine* 15:289-294.

Green, K.N., W.B. East, and L.D. Hensley. 1987. A golf skills test battery for college males and females. *Research Quarterly for Exercise and Sport* 58:72-76.

Gronlund, N.E. 1993. *Assessment of student achievement.* 6th ed. Boston: Allyn & Bacon.

Grove, J.R. 2001. Practical screening tests for talent identification in baseball. *Applied Research in Coaching and Athletics Annual* 16:63-77.

Grunbaum J.A., L. Kann, S. Kinchen, J. Ross, J. Hawkins, R. Lowry, W.A. Harris, T. McManus, D. Chyen, and J. Collins. 2004. Youth risk behavior surveillance—United States, 2003. *MMWR Surveill Summ (MMWR. Surveillance summaries: Morbidity and mortality weekly report. Surveillance summaries / CDC.)* May 21, 53(2): 1-96.

Hagberg, J.M., J.E. Graves, M. Limacher, D.R. Woods, S.H. Leggett, C. Cononie, J. Gruber, and M.L. Pollock. 1989. Cardiovascular responses of 70-79 year old men and women to exercise training. *Journal of Applied Physiology* 66:2589-2594.

Hall, G., R. Hetzler, D. Perrin, and A. Weltman. 1992. Relationship of timed sit-up tests to isokinetic abdominal strength. *Research Quarterly for Exercise and Sport* 63:80-84.

Hall, C., D. Mack, A. Paivio, and H.A. Hausenblas. 1998. Imagery use by athletes: Development of the Sport Imagery Questionnaire. *International Journal of Sport Psychology* 29:73-89.

Harris, M.L. 1969. A factor analytic study of flexibility. *Research Quarterly* 40:62-70.

Harrow, A.J. 1972. *A taxonomy of the psychomotor domain.* New York: McKay.

Henry, F.M. 1956. Coordination and motor learning. In *59th Proceedings of the Annual College Physical Education Association,* 68-75. Washington, DC.

———. 1958. Specificity vs. generality in learning motor skills. In *61st Annual Proceedings of the College Physical Education Association,* 126-128. Washington, DC.

Hensley, L.D., ed. 1989. *Tennis for boys and girls: Skills test manual.* Reston, VA: AAHPERD.

Hensley, L.D., and W.B. East. 1989. Testing and grading in the psychomotor domain. In *Measurement concepts in physical education and exercise science,* ed. M.J. Safrit and T.M. Wood, 297-321. Champaign, IL: Human Kinetics.

Hensley, L.D., W.B. East, and J.L. Stillwell. 1979. A racquetball skills test. *Research Quarterly* 50:114-118.

Herman, J.L., P.R. Aschbacher, and L. Winters. 1992. *A practical guide to alternative assessment.* Alexandria, VA: Association for Supervision and Curriculum Development.

Hopkins, D.R., J. Schick, and J.J. Plack. 1984. *Basketball for boys and girls: Skills test manual.* Reston, VA: AAHPERD.

Holt, N., and A. Sparkes. 2001. An ethnographic study of cohesiveness in a college soccer team over a season. *Sport Psychologist* 15:237-259.

Howley, E.T., and B.D. Franks. 2003. *Health fitness instructor's handbook.* 4th ed. Champaign, IL: Human Kinetics.

Jackson, A.S., S.N. Blair, M.T. Mahar, L.T. Wier, R.M. Ross, and J.E. Stuteville. 1990. Prediction of functional aerobic capacity without exercise testing. *Medicine and Science in Sports and Exercise* 22:863-870.

Jackson, A.S., and R.J. Frankiewicz. 1975. Factorial expressions of muscular strength. *Research Quarterly* 46:206-217.

Jackson, A.S., and M. Pollock. 1978. Generalized equations for predicting body density of men. *British Journal of Nutrition* 40:497-504.

Jackson, A.S., M. Pollock, and A. Ward. 1980. Generalized equations for predicting body density of women. *Medicine and Science in Sports and Exercise* 12:175-182.

Jackson, A.W., and A. Baker. 1986. The relationship of the sit and reach test to criterion measures of hamstring and back flexibility in young females. *Research Quarterly for Exercise and Sport* 57:183-186.

Jackson, A.W., A.S. Jackson, and J. Bell. 1980. A comparison of alpha and the intraclass reliability coefficients. *Research Quarterly for Exercise and Sport* 51:568-571.

Jackson, A.W., and N. Langford. 1989. The criterion-related validity of the sit and reach test: Replication and extension of previous findings. *Research Quarterly for Exercise and Sport* 60:384-387.

Jackson, A.W., J.R. Morrow, Jr., P.A. Brill, H.W. Kohl, III, N.F. Gordon, and S.N. Blair. 1998. Relations of sit-up and sit-and-reach tests to low back pain in adults. *Journal of Orthopaedic and Sports Physical Therapy* 27(1):22-26.

Jackson, A.W., J. Solomon, and M. Stusek. 1992. One-mile walk test: Reliability, validity, and norms for young adults [Abstract]. *Research Quarterly for Exercise and Sport* 63:A52.

Jackson, A.W., M. Watkins, and R. Patton. 1980. A factor analysis of twelve selected maximal isotonic strength performances on the universal gym. *Medicine and Science in Sports and Exercise* 12:274-277.

Jacobs, D., B. Ainsworth, T. Hartman, and A. Leon. 1993. A simultaneous evaluation of 10 commonly used physical activity questionnaires. *Medicine and Science in Sports and Exercise* 25:81-91.

Jensen, C., and C. Hirst. 1980. *Measurement in physical education and athletics.* New York: Macmillan.

Johnson, B., and J. Nelson. 1979. *Practical measurements for evaluation in physical education.* 3rd ed. Minneapolis: Burgess.

Kelly, G.A. 1955. *The psychology of personal constructs.* New York: Norton.

Kenyon, G.S. 1968a. A conceptual model for characterizing physical activity. *Research Quarterly* 39:96-105.

———. 1968b. Six scales for assessing attitude toward physical activity. *Research Quarterly* 39:566-574.

Kirby, R.F., ed. 1991. *Kirby's guide to fitness and motor performance tests.* Cape Girardeau, MO: Ben Oak.

Kirk, M.F. 1997. Using portfolios to enhance student learning and assessment. *Journal of Physical Education, Recreation, and Dance* 68(7):29-33.

Kline, G., J. Porcari, R. Hintermeister, P. Freedson, A. Ward, R. McCarron, J. Ross, and J. Rippe. 1987. Estimation of $\dot{V}O_2$max from a one-mile track walk, gender, age, and body weight. *Medicine and Science in Sports and Exercise* 19:253-259.

Krathwohl, D.R., B.S. Bloom, and B.A. Masia. 1964. *Taxonomy of educational objectives: Handbook II: The affective domain.* New York: McKay.

Larson, L.A. 1941. A factor analysis of motor ability variables and tests, with test for college men. *Research Quarterly* 12:499-517.

Last, J. 1992. *Dictionary of epidemiology.* 2nd ed. New York: Oxford University.

Leone, M., G. Lariviere, and A.S. Comtois. 2002. Discriminant analysis of anthropometric and biomotor variables among elite adolescent female athletes in four sports. *Journal of Sport Sciences* 20:443-449.

Linn, R., E. Baker, and S. Dunbar. 1991. Complex, performance-based assessment: Expectations and validation criteria. *Educational Researcher* 20(8):15-21.

Lohman, T. 1989. Assessment of body composition in children. *Pediatric Exercise Science* 1:19-30.

Looney, M., and S. Plowman. 1990. Passing rates of American children and youth on the FITNESSGRAM criterion-referenced physical fitness standards. *Research Quarterly for Exercise and Sport* 61:215-223.

Lund, J.L. 2000. *Creating rubrics for physical education.* Reston, VA: National Association for Sport and Physical Education.

Lund, J.L., and M.F. Kirk. 2002. *Performance-based assessment for middle and high school physical education.* Champaign, IL: Human Kinetics.

MacDougall, J.D., and H.A. Wenger. 1991. The purpose of physiological testing. In *Physiological testing of the high performance athlete,* 2nd ed., ed. J.D. MacDougall, H.A. Wenger, and H.J. Green, 1-5. Champaign, IL: Human Kinetics.

Mager, N.F. 1962. *Preparing instructional objectives.* Palo Alto, CA: Feason.

Mahar, M.T., D.A. Rowe, C.R. Parker, F.J. Mahar, D.M. Dawson, and J.E. Holt. 1997. Criterion-referenced and norm-referenced agreement between the mile run/walk and PACER. *Measurement in Physical Education and Exercise Science* 1(4):245-258.

Marcus, B.H., S.W. Banspach, R.C. Lefebvre, J.S. Rossi, R.A. Carleton, and D.B. Abrams. 1994. Using the stages of change model to increase the adoption of physical activity among community participants. *American Journal of Health Promotion* 6:424-429.

Marcus, B.H., V.C. Selby, R.S. Niaura, and J.S. Rossi. 1992. Self-efficacy and the stages of exercise behavior change. *Research Quarterly for Exercise and Sport* 63:60-66.

Marcus, B.H., and L. Simkin. 1993. The stages of exercise behavior. *Journal of Sports Medicine and Physical Fitness* 33:83-88.

Martens, R. 1977. *Sport competition anxiety test.* Champaign, IL: Human Kinetics.

Martens, R., D.M. Landers, and J. Loy. 1971. *Sport cohesiveness questionnaire.* Champaign: University of Illinois, Department of Physical Education.

Martens, R., R. Vealey, and D. Burton. 1990. *Competitive anxiety in sport.* Champaign, IL: Human Kinetics.

Martin, R.H. 1983. Effectiveness of attentional focus and basketball free-throw percentage: An attempt at prediction. Unpublished master's thesis, California State University, Fullerton.

Marzano, R.J., D. Pickering, and J. McTighe. 1993. *Assessing student outcomes: Performance assessment using the dimensions of learning model.* Alexandria, VA: Association for Curriculum and Development.

Matanin, M., and D. Tannehill. 1994. Assessment and grading in physical education. *Journal of Teaching in Physical Education* 13:395-405.

Mayhew, T., and J. Rothstein. 1985. Measurement of muscle performance with instruments. In *Measurement in physical therapy,* ed. J. Rothstein, 57-102. New York: Churchill Livingstone.

McCloy, C.H. 1932. *The measurement of athletic power.* New York: Barnes.

McKenzie, T.L., S.J. Marshall, J.F. Sallis, and T.L. Conway. 2000. Leisure-time physical activity in school environments: An observational study using SOPLAY. *Preventive Medicine* 30:70-77.

McKenzie, T.L., J.F. Sallis, and P.R. Nader. 1991. SOFIT: System for Observing Fitness Instruction Time. *Journal of Teaching in Physical Education* 11:195-205.

McKenzie, T.L., J.F. Sallis, T.L. Patterson, J.P. Elder, C.C. Berry, J.W. Rupp, C.J. Atkins, M.J. Buono, and P.R. Nader. 1991. BEACHES: An observational system for assessing children's eating and physical activity behaviors and associated events. *Journal of Applied Behavior Analysis* 24:141-151.

McNair, D.M., M. Lorr, and L.F. Droppleman. 1971. *EDITS manual for POMS.* San Diego: Educational and Industrial Testing Service.

Melograno, V.J. 1998. *Professional and student portfolios for physical education.* Champaign, IL: Human Kinetics.

Messick, S. 1995. Standards of validity and the validity of standards in performance assessment. *Educational Measurement: Issues and Practice* 14(1):5-8.

Miller, M.D., and S.M. Legg. 1993. Alternative assessment in a high-stakes environment. *Educational Measurement: Issues and Practice* 12(3):9-15.

Miller, P. 1985. Assessment of joint motion. In *Measurement in physical therapy,* ed. J. Rothstein, 103-136. New York: Churchill Livingstone.

Mintah, J.K. 2003. Authentic assessment in physical education: Prevalence of use and perceived impact on students' self-concept, motivation, and skill achievement. *Measurement in Physical Education and Exercise Science* 7(3):161-174.

Montoye, H.J., H.C. Kemper, W.H.M. Saris, and R.A. Washburn. 1996. *Measuring physical activity and energy expenditure.* Champaign, IL: Human Kinetics.

Morgan, W.P. 1980. Test of champions: The iceberg profile. *Psychology Today* (July):92-93, 97-99, 102, 108.

Morgan, W.P., and R.W. Johnson. 1978. Psychological characteristics of successful and unsuccessful oarsmen. *International Journal of Sport Psychology* 11:38-49.

Morgan, W.P., and M.L. Pollock. 1977. Psychologic characterization of the elite distance runner. *Annals of the New York Academy of Science* 301:382-403.

Morris, L., D. Davis, and C. Hutchins. 1981. Cognitive and emotional components of anxiety: Literature review and revised worry-emotionality scale. *Journal of Education Psychology* 73:541-555.

Morrow, J.R., Jr., and H.B. Falls. 2003. Physical fitness standards for children. Ed. G. Welk, J.R. Morrow, Jr., and H.B. Falls, 45-54. Available: http://www.cooperinst.org/shopping/PDF%20format/FITNESSGRAM%20Reference%20Guide.pdf [accessed June 10, 2004]

Morrow, J.R., Jr., T. Fridye, and S. Monaghen. 1986. Generalizability of the AAHPERD health-related skinfold test. *Research Quarterly for Exercise and Sport* 57:187-195.

Morrow, J.R., Jr., A. Jackson, P. Bradley, and H. Hartung. 1986. Accuracy of measured and predicted residual lung volume on body density measurement. *Medicine and Science in Sports and Exercise* 18:647-652.

Murray, T.D., J.L. Walker, A.S. Jackson, J.R. Morrow, Jr., J.A. Eldridge, and D.L. Rainey. 1993. Validation of a 20-minute steady-state jog as an estimate of peak oxygen uptake in adolescents. *Research Quarterly for Exercise and Sport* 64:75-82.

National Association for Sport and Physical Education. (NASPE). 1995. *Moving into the future: National standards for physical education.* Reston, VA: NASPE.

National Center for Health Statistics. 2002. *Health, United States, 2002 with chartbook on trends in the health of Americans.* Hyattsville, MD: National Center for Health Statistics.

Neilson, N.P., and F.W. Cozens. 1934. *Achievement scales in physical education activities for boys and girls in elementary and junior high schools.* New York: Barnes.

Nelson, J.K., S.H. Yoon, and K.R. Nelson. 1991. A field test for upper body strength and endurance. *Research Quarterly for Exercise and Sport* 62:436-441.

Nideffer, R.M. 1976. Test of attentional and interpersonal style. *Journal of Personality and Social Psychology* 34:394-404.

Nieman, D.C. 1995. *Fitness and sports medicine: A health-related approach.* 3rd ed. Mountain View, CA: Mayfield.

Nitko, A.J. 1984. Defining "criterion-referenced test." In *A guide to criterion-referenced test construction,* ed. R.A. Berk, 8-28. Baltimore: Johns Hopkins University Press.

Ostrow, A. 1996. *Directory of psychological tests in the sport and exercise sciences.* 2nd ed. Morgantown, WV: Fitness Information Technology.

Passer, M.W. 1983. Fear of failure, fear of evaluation, perceived competence, and self-esteem in competitive-trait-anxious children. *Journal of Sport Psychology* 5:172-188.

Pate, R. 1988. The evolving definition of physical fitness. *Quest* 40:174-179.

Pate, R., J. Ross, C. Dotson, and G. Gilbert. 1985. The new norms: A comparison with the 1980 AAHPERD norms. *Journal of Physical Education, Recreation, and Dance* 56(1):70-72.

Patterson, E.L., R.E. Smith, J.J., Everett, and J.T. Ptacek. 1998. Psychosocial factors as predictors of ballet injuries: Interactive effects of life stress and social support. *Journal of Sport Behavior* 21:101-112.

Patton, M.Q. 1980. *Qualitative evaluation methods.* Beverly Hills, CA: Sage.

Perusse, A., C. Tremblay, C. Leblanc, and C. Bouchard. 1989. Genetic and environmental influences on level of habitual physical activity and exercise participation. *American Journal of Epidemiology* 129:1012-1022.

Plowman, S.A. 1992a. Criterion referenced standards for neuromuscular physical fitness tests. *Pediatric Exercise Science* 4:10-19.

Pollock, M., R. Bohannon, K. Cooper, J. Ayres, A. Ward, S. White, and A. Linnerud. 1976. A comparative analysis of four protocols for maximal treadmill stress testing. *American Heart Journal* 92:39-46.

Prapavessis, H. 2000. The POMS and sports performance: A review. *Journal of Applied Sport Psychology* 12:34-48.

President's Council on Physical Fitness and Sports (PCPFS). 1999. *The Presidential Physical Fitness Award program.* Washington, DC: PCPFS.

Prochaska, J.O., and C.C. DiClemente. 1983. Stages and processes of self-change in smoking: Toward an integrative model of change. *Journal of Consulting and Clinical Psychology* 51:390-395.

Reiff, G., W. Dixon, D. Jacoby, G. Ye, C. Spain, and P. Hunsicker. 1985. *The President's Council on Physical Fitness and Sports 1985 National School Population Fitness Survey.* Washington, DC: President's Council on Physical Fitness and Sports.

Rikli, R.E. 1991. *Softball for boys and girls: Skills test manual.* Reston, VA: AAHPERD.

Rikli, R.E., and C.J. Jones. 1999a. Development and validation of a functional fitness test for community-residing older adults. *Journal of Aging and Physical Activity* 7:129-161.

———. 1999b. Functional fitness normative scores for community-residing older adults, ages 60-94. *Journal of Aging and Physical Activity* 7:162-181.

Rikli, R.E, C. Petray, and T.A. Baumgartner. 1992. The reliability of distance run tests for children in grades K-4. *Research Quarterly for Exercise and Sport* 63:270-276.

Robertson, L., and H. Magnusdottir. 1987. Evaluation of criteria associated with abdominal fitness testing. *Research Quarterly for Exercise and Sport* 58:355-359.

Ross, J., C. Dotson, G. Gilbert, and S. Katz. 1985. New standards for fitness measurement. *Journal of Physical Education, Recreation and Dance* 56(1):62-66.

Ross, J., R. Pate, L. Delby, R. Gold, and M. Svilar. 1987. New health-related fitness norms. *Journal of Physical Education, Recreation and Dance* 58(9):66-70.

Rotter, J.B. 1966. Generalized expectancies for internal versus external control of reinforcement. *Psychological Monographs 80* (No. 609).

Rowland, T. 1990. *Exercise and children's health.* Champaign, IL: Human Kinetics.

Rowley, A., D. Landers, B. Kyllo, and J. Etnier. 1995. Does the iceberg profile discriminate between successful, and less successful athletes? A meta-analysis. *Journal of Sport and Exercise Psychology* 17:185-199.

Safrit, M. 1986. *Introduction to measurement in physical education and exercise science.* St. Louis: Mosby.

Safrit, M.J., T.A. Baumgartner, A.S. Jackson, and C.L. Stamm. 1980. Issues in setting motor performance standards. *Quest* 32:152-162.

Safrit, M., L. Hooper, S. Ehlert, M. Costa, and P. Patterson. 1988. The validity generalization of distance run tests. *Canadian Journal of Sport Sciences* 13:188-196.

Safrit, M., and M. Looney. 1992. Should the punishment fit the crime? A measurement dilemma. *Research Quarterly for Exercise and Sport* 63:124-127.

Safrit, M., and C. Pemberton. 1995. *Complete guide to youth fitness testing.* Champaign, IL: Human Kinetics.

Sallis, J.F., M.J. Buono, J.J. Roby, F.G. Micale, and J.A. Nelson. 1993. Seven-day recall and other physical activity self-reports in children and adolescents. *Medicine and Science in Sports and Exercise* 25:99-108.

Sarason, I.G. 1975. Test anxiety and the self-disclosing coping model. *Journal of Consulting and Clinical Psychology* 43:148-153.

Sargent, D.A. 1921. The physical test of man. *American Physical Education Review* 26(April):188-194.

Schick, J., and N.G. Berg. 1983. Indoor golf skill test for junior high boys. *Research Quarterly for Exercise and Sport* 54:75-78.

Seaman, J., and K. DePauw. 1989. *The new adapted physical education: A developmental approach.* Mountain View, CA: Mayfield.

Shephard, R. 1990. *Fitness in special populations.* Champaign, IL: Human Kinetics.

Shifflett, B., and B.J. Shuman. 1982. A criterion-referenced test for archery. *Research Quarterly for Exercise and Sport* 53:330-335.

Siedentop, D. 1996. Physical education and education reform: The case for sport education. In *Student learning in physical education: Applying research to enhance instruction,* ed. S.J. Silverman and C.D. Ennis, 247-267. Champaign, IL: Human Kinetics.

Simon, J.A., and F.L. Smoll. 1974. An instrument for assessing children's attitudes toward physical activity. *Research Quarterly* 45:407-415.

Singer, R.N. 1968. *Motor learning and human performance.* New York: Macmillan.

Siri, W. 1956. Gross composition of the body. In *Advances in biological and medical physics,* ed. J. Lawrence, 239-280. New York: Academic Press.

Smith, R. E., F. L. Smoll, and N. P. Barnett. 1995. Reduction of children's sport performance anxiety through social support and stress-reduction training for coaches. *Journal of Applied Developmental Psychology* 16:125-142.

Smith, R., F.L. Smoll, and B. Curtis. 1979. Coach effectiveness training: A cognitive behavioral approach to enhancing relationship skills in youth and sport coaches. *Journal of Sport Psychology* 1:59-75.

Smith, R.E., F.L. Smoll, and R.W. Schutz. 1990. Measurement and correlates of sport-specific cognitive and somatic trait anxiety: The Sport Anxiety Scale. *Anxiety-Research* 2:263-280.

Sonstroem, R.J. 1974. Attitude testing examining certain psychological correlates of physical activity. *Research Quarterly* 45:35, 103.

———. 1978. Physical estimation and attraction scales: Rationale and research. *Medicine and Science in Sports* 10:97-102.

Sorrentino, R.M., and B.H. Sheppard. 1978. Efforts of affiliation-related motives on swimmers in individual versus group competition: A field experiment. *Journal of Personality and Social Psychology* 36:704-714.

Sparkes, A. 1998. Validity in qualitative inquiry and the problem of criteria: Implications for sport psychology. *Sport Psychologist* 12:333-345.

Spielberger, C.D., R.L. Gorsuch, and R.F. Lushene. 1970. *Manual for the State–Trait Anxiety Inventory.* Palo Alto, CA: Consulting Psychologists Press.

Spink, K.S., and A.V. Carron. 1992. Group cohesion and adherence in exercise classes. *Journal of Sport and Exercise Psychology* 14:78-86.

———. 1994. Group cohesion effects in exercise classes. *Small Group Research* 25:26-42.

SPSS for Windows Version 9.0.1. 1999. Chicago: SPSS.

Stiggins, R. 1987. Design and development of performance assessment. *Educational Measurement: Issues and Practice* 6(3):33-42.

Stone, D.B., W.R. Armstrong, D.M. Macrina, and J.W. Pankau. 1996. *Introduction to epidemiology.* Madison, WI: Brown and Benchmark.

Strand, B., and R. Wilson. 1993. *Assessing sports skills.* Champaign, IL: Human Kinetics.

Strean, W. 1998. Possibilities for qualitative research in sport psychology. *Sport Psychologist* 12:333-345.

Stuart, M. 2003. Sources of subjective task value in sport: An examination of adolescents with high or low value for sport. *Journal of Applied Sport Psychology* 15:239-255.

Suzuki, N., and S. Endo. 1983. A quantitative study of trunk muscle strength and fatigability in the low-back pain syndrome. *Spine* 8:69-74.

Telama, R., X. Yang, L. Laasko, and J. Viikari. 1997. Physical activity in childhood and adolescence as predictor of physical activity in young adulthood. *American Journal of Preventive Medicine* 13:317-323.

Terry, P. 2000. Perspectives on mood in sport and exercise. *Journal of Applied Sport Psychology* 12:1-4.

Thissen-Milder, M., and J.L. Mayhew. 1991. Selection and classification of high school volleyball players from performance tests. *Journal of Sports Medicine and Physical Fitness* 31:380-384.

Thomas, J.R., and J.K. Nelson. 2001. *Research methods in physical activity.* 4th ed. Champaign, IL: Human Kinetics.

U.S. Congress, Office of Technology Assessment. 1992. *Testing in American schools: Asking the right questions* (OTA-SET-519; February). Washington, DC: U.S. Government Printing Office.

U.S. Department of Health and Human Services (USDHHS). 1996. *Physical activity and health: A report of the Surgeon General.* Atlanta: USDHHS, Centers for Disease Control and Prevention, National Center for Chronic Disease Prevention and Health Promotion.

Van Schoyck, R.S., and A.F. Grasha. 1981. Attentional style variations and athletic ability: The advantage of a sport-specific test. *Journal of Sport Psychology* 2:149-165.

Vealey, R.S. 1986. Conceptualization of sport-confidence and competitive orientation: Preliminary investigation and instrument development. *Journal of Sport Psychology* 8:221-246.

Velicer, W.F., and J.O. Prochaska. 1997. Introduction: The transtheoretical model. *American Journal of Health Promotion* 12:6-7.

Verducci, F.M. 1980. *Measurement concepts in physical education.* St. Louis: Mosby.

Vincent, W.J. 1999. *Statistics in kinesiology.* 2nd ed. Champaign, IL: Human Kinetics.

Welk, G.J., ed. 2002. *Physical activity assessments for health-related research.* Champaign. IL: Human Kinetics.

Weston, A.T., R. Petosa, and R.R. Pate. 1997. Validation of an instrument for measurement of physical activity in youth. *Medicine and Science in Sports and Exercise* 29:138-143.

Widmeyer, W.N., L.R. Brawley, and A.V. Carron. 1985. *The measurement of cohesion in sport teams: The group environment questionnaire.* (Available from Sports Dynamics, 11 Ravenglass Crescent, London, ON, Canada N6G 3X7)

Winnick, J.P., and F.X. Short. 1999. *The Brockport physical fitness test manual.* Champaign, IL: Human Kinetics.

Wood, T.M. 1996. Evaluation and testing: The road less traveled. In *Student learning in physical education: Applying research to enhance instruction,* ed. S.J. Silverman and C.D. Ennis, 199-219. Champaign, IL: Human Kinetics.

———. 2003. Assessment in physical education: The future is now. In *Student learning in physical education: Applying research to enhance instruction,* 2nd ed., ed. S.J. Silverman and C.D. Ennis, 187-203. Champaign, IL: Human Kinetics.

Yukelson, D., R. Weinberg, and A. Jackson. 1984. A multidimensional group cohesion instrument for intercollegiate basketball teams. *Journal of Sport Psychology* 6:103-117.

Zaichkowsky, L. 1975. Attitudinal difference in two types of physical education programs. *Research Quarterly* 46: 364-370.

Index

Note: The italicized *f* or *t* following a page number denotes a figure or table on that page. The italicized *ff* or *tt* following a page number denotes multiple figures or tables on that page. Boldfaced locators indicate the glossary page that term can be found on.

About the Authors

James R. Morrow, Jr, PhD, is a regents professor in the department of kinesiology, health promotion, and recreation at the University of North Texas at Denton. Dr. Morrow regularly teaches courses in measurement and evaluation in human performance. He has authored more than 100 articles and chapters on exercise physiology, measurement, and computer use, and he has conducted significant research using the techniques presented in the text.

In addition to teaching, Dr. Morrow is president of the American Academy of Kinesiology and Physical Education, as well as a fellow of the American College of Sports Medicine (ACSM) and a research fellow of the American Alliance for Health, Physical Education, Recreation and Dance (AAHPERD). Dr. Morrow has chaired the AAHPERD Measurement and Evaluation Council and is a recipient of that council's Honor Award. He has produced four fitness-testing software packages, including the AAHPERD Health-Related Physical Fitness test, and was editor in chief of *Research Quarterly for Exercise and Sport* from 1989 to 1993. He currently serves as the co-editor of the *Journal of Physical Activity and Health.* He enjoys playing golf, reading, and traveling.

Allen W. Jackson, EdD, is a regents professor at the University of North Texas, where he has taught kinesiology with research, statistics, and computer applications since 1979. He has published extensively in measurement and evaluation, including more than 100 articles, and has presented more than 200 scientific papers. He is a reviewer for *Measurement in Physical Education and Exercise Science* and served as associate editor and statistical consultant for *Medicine and Science in Sport and Exercise.* He has also served as section editor for *Research Quarterly for Exercise and Sport.*

Dr. Jackson earned his EdD in 1978 at the University of Houston. He is an ACSM fellow, AAKPE fellow, and a member of AAHPERD. Dr. Jackson's favorite leisure activities include jogging, weightlifting, and walking.

James G. Disch, PED, is an associate professor in the kinesiology department at Rice University. From 1986 to 1991 he served as master of Richardson College at Rice. Dr. Disch has authored numerous articles, chapters, manuals, and texts in the areas of applied measurement, prediction in sport, and applied sport science. A member of AAHPERD since 1974, he has served as chair, secretary, and advisory board member of the measurement and evaluation council of AAHPERD. He is also a reviewer for *Research Quarterly for Exercise and Sport* and *Medicine and Science in Sport and Exercise.*

Dr. Disch has coordinated several workshops and symposia on measurement and evaluation and, along with Dr. Morrow, cochaired the Third National Measurement and Evaluation Symposium in Houston in 1980. Dr. Disch was a major contributor to the development of AAHPERD Health-Related Fitness norms in 1980 and has worked as a consultant and advisor for Olympic and professional teams. He is currently on the Educational Advisory Committee of USA Volleyball. In 1999 he was named recipient of the National Measurement and Evaluation Council Honor Award. Dr. Disch earned his PED in biomechanics and measurement from Indiana University in 1973. He directs several youth sport clinics and competes in men's senior baseball.

Dale P. Mood, PhD, is a professor and former associate dean of arts and sciences at the University of Colorado at Boulder. Dr. Mood has taught measurement and evaluation, statistics, and research methods courses since 1970 and has published extensively in the field, including 47 articles and 5 books. He has served as a consultant to five NFL football teams and chair of the Measurement and Evaluation Council of AAHPERD, and he is a former president of AAALF. He is a reviewer for *Medicine and Science in Sport and Exercise, Measurement in Physical Education and Exercise Science,* and *The Research Quarterly for Exercise and Sport.* In his leisure time, Dr. Mood enjoys reading, managing a private swimming and tennis facility, and participating in a variety of physical activities.

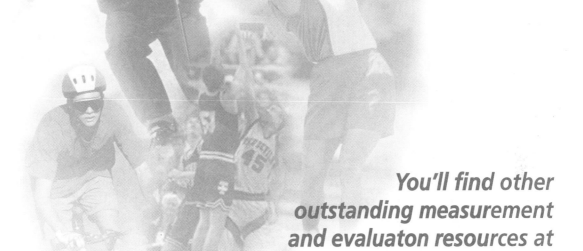